Moral Entanglements

Moral Entanglements

CONSERVING BIRDS IN BRITAIN AND GERMANY

Stefan Bargheer

The University of Chicago Press CHICAGO & LONDON

The University of Chicago Press, Chicago 60637
The University of Chicago Press, Ltd., London

Published 2018
Printed in the United States of America

27 26 25 24 23 22 21 20 19 18   1 2 3 4 5

ISBN-13: 978-0-226-37663-9 (cloth)
ISBN-13: 978-0-226-54382-6 (paper)
ISBN-13: 978-0-226-54396-3 (e-book)
DOI: https://doi.org/10.7208/chicago/9780226543963.001.0001

Library of Congress Cataloging-in-Publication Data

Names: Bargheer, Stefan, author.
Title: Moral entanglements : conserving birds in Britain and Germany / Stefan Bargheer.
Description: Chicago ; London : The University of Chicago Press, 2018. | Includes bibliographical references and index.
Identifiers: LCCN 2017039380 | ISBN 9780226376639 (cloth : alk. paper) | ISBN 9780226543826 (pbk. : alk. paper) | ISBN 9780226543963 (e-book)
Subjects: LCSH: Birds—Conservation—Moral and ethical aspects—England. | Birds—Conservation—Moral and ethical aspects. | Ethics—England. | Ethics—Germany.
Classification: LCC QL676.5.B255 2018 | DDC 333.95/8—dc23
LC record available at https://lccn.loc.gov/2017039380

♾ This paper meets the requirements of ANSI/NISO Z39.48-1992 (Permanence of Paper).

For Marlene

CONTENTS

ABBREVIATIONS

BfV	Bund für Vogelschutz
BOU	British Ornithologists' Union
BTO	British Trust for Ornithology
DBV	Deutsche Bund für Vogelschutz
DDA	Dachverband Deutscher Avifaunisten
DO-G	Deutsche Ornithologen-Gesellschaft
IBA	Important Bird Area
ICBP	International Council for Bird Preservation
IUCN	International Union for the Conservation of Nature
NABU	Naturschutzbund Deutschland
RSPB	Royal Society for the Protection of Birds
SCI	Sites of Community Importance
SPA	Special Protection Area
SPEC	Species of European Conservation Concern
WWF	World Wildlife Fund, later World Wide Fund for Nature
YOC	Young Ornithologists' Club

ACKNOWLEDGMENTS

This book was in the making for a long time. If my mentors at the University of Chicago had not talked me out of some of the initial excesses in case selection, I would in all likelihood still be writing it. I thank in particular Andrew Abbott for his unwavering moral support—not to mention his sociological advice—for this project from start to finish. His seminar on historical research methods was the beginning of this project. I have also greatly benefited from working with Andreas Glaeser, who provided crucial feedback throughout the research process. Elisabeth Clemens has assisted with a special field exam that informed the outline of this project. Working with Hans Joas has proven insightful, and his intellectual stimulation found its way into this book. The Max Planck Institute for the History of Science in Berlin was an ideal place for working on the manuscript. I benefited greatly from presenting at the institute's colloquium, and Lorraine Daston's comments on an early version of the entire manuscript were invaluable.

More than anything else, however, I have to thank the people whose work I studied and whom I interviewed for this project. It is a truism that research cannot be done without the help of others; yet more than this, at times I felt that my main purpose was to channel the voices of others out to a larger audience. I found the people I interviewed to produce perfectly clear and plausible narrative accounts of and reflections on how they act, why it matters to them, and what kind of impact it has. Stephen Moss, Norbert Schäffer, and Martin Flade made particular efforts to communicate their experiences with bird watching, field ornithology, and nature conservation to me. It was not only the knowledge of these and many other people that I met but also their enthusiasm for birds that made the research for this project personally gratifying. While conducting this research, I was asked more than once by fellow sociologists "Are you a bird watcher yourself?" I am not, but this is probably my own loss, since the people I encountered made a convincing case that having a preoccupation one is passionate about can be genuinely enriching. While not passionate about birds myself, I did, however, become very fond of watching bird watchers.

In addition to the people I interviewed, I am particularly indebted to the staff at the libraries and archives of the organizations I studied. At the Naturschutzbund Deutschland (NABU), Helge May was instrumental in giving me access to his collection of documents on the history of the organization. I also thank the environmental historian Anna-Katharina Wöbse for countless discussions on the history of the NABU and its founder, Lina Hähnle. Hans-Werner Frohn at the Stiftung Naturschutzgeschichte was particularly helpful in providing information on the early history of bird conservation. At the Royal Society for the Protection of Birds (RSPB), I would like to thank Ian Dawson for assistance with locating sources and for sharing his file of archival materials with me. I am further indebted to the librarians at BirdLife International and the Bundesamt für Naturschutz.

Discussing this research at conferences, workshops, and seminars has greatly helped me in formulating my ideas. I thank the students in my courses taught at the University of Chicago, the University of Göttingen, NYU Berlin, and UCLA for their thoughts on morality and nature conservation. For much the same reasons, I owe gratitude to former fellow students at the University of Chicago. I want to thank in particular the members of a writing group, first of all Sida Liu, Maria Medvedeva, Daniel Menchik, Josh Pacewicz, and Xiaoli Tian. I received substantial assistance from Jessica Feldman, Melissa Kew, Zohar Lechtman, Gregory Liegel, Etienne Ollion, and Paki Reid-Brossard. While all errors of fact do, of course, remain exclusively my own, I am more than happy to share the responsibility for all shortcomings in sociological perspective with the people in my graduate department. If faculty and students at this place were not so insistent that Chicago sociology is a worthwhile approach, I might well have overlooked it.

At UCLA, Gail Kligman, Hannah Landecker, Aaron Panofsky, and Stefan Timmermans gave helpful advice on the manuscript and the publishing process. I thank Gabriel Abend and Nicholas Wilson for reading versions of the revised manuscript. I am equally grateful to Doug Mitchell and Kyle Wagner at the University of Chicago Press and to the reviewers, in particular Marion Fourcade, for making the transition from manuscript to book possible. The initial field research for this project was supported by a research grant from the Nicholson Center for British Studies at the University of Chicago and the Andrew W. Mellon Early Career Fellowship Program provided by the American Council of Learned Societies. Despite this assistance, the project would not have been possible without the support of my parents, Karl and Hildegard, as well as that of several credit card companies. I have meanwhile been able to repay my debts to the latter, but I am not certain that I will ever be able to do the same for all of the aforementioned people.

PROLOGUE

A Bird's Eye View

Heinrich Gätke caused a stir in the world of ornithology in 1891 when he first published *Heligoland as an Ornithological Observatory*. It included a list of the breeding birds and irregular visitors to this island in the North Sea, as collected by Gätke and reported in the leading ornithological journals of the time (Cordeaux 1875; Hartlaub 1894; Seebohm 1877, 1892). The author had been the prime recorder of birdlife on Heligoland for fifty years, and by the time his observations were published, the number of species on his list was, in fact, astonishing: no less than three hundred and ninety-six species were reported on an island less than a square mile in size. Initially, the wealth of birdlife came as a surprise to Gätke. When he started comparing his findings with the established collections on the mainland it slowly dawned on him "what never expected riches of noteworthy items come together at this place, and how infinitely this little rock overshadows the proudest Empires in this respect" (Gätke 1891, preface; 1895).

The reason for these remarkable numbers was not hard to find. Heligoland's location in the middle of the North Sea is a prime spot along the route of birds migrating from northern Europe and Russia to the Mediterranean and sub-Saharan Africa. The island, located thirty miles from the German coast and 270 miles from the British Isles, is one of the few high-seas islands in the North Sea, making it a natural resting place and point of orientation for migrating birds. Gätke, a painter and natural historian, had come to the island as a secretary to the British government, under whose jurisdiction the island stood until 1890, when it became part of Germany. Gätke assembled a museum collection in his spare time and began comparing the birds procured on Heligoland to those in standard handbooks on European birdlife. When these publications proved inadequate for birds that came from as far as North America and Siberia, he sought the expert advice of ornithologists on the European continent and the British Isles.

Gätke and his bird observatory (*Vogelwarte*) in Heligoland gained a reputation in the process. The most venerable ornithologists at the time came to visit his collection and reported enthusiastically about the riches in their writings (Blasius 1890; Naumann 1846; Rohweder 1905; Schulz 1947; Stresemann 1967; Vauk 1977). Gätke and other ornithologists pointed out that the reason for the high number of species recorded on this little island was not only due to its location but also a result of the copious attention the birds were beginning to attract. Once the ornithological status of Heligoland became apparent, ornithologists such as Gätke were willing to pay high prices for newly acquired birds, and they became duly sought-after by the islanders, who had previously collected migrating birds mainly for food. With the arrival of Gätke, collecting rare migrants turned into a minor industry, and few migrating species that crossed the island escaped the guns, nets, and traps of the islanders. The number of noteworthy observations from Heligoland continued to grow and filled the pages of the major ornithological journals in Britain and Germany. Gätke's long-awaited book (published in German in 1891) was soon translated into English and included the addition of two new species discovered since the first edition. A second German edition, published posthumously in 1900, added yet another species.

One of Gätke's early ornithological surprises, brought to him by a local hunter, was a specimen of what was in all likelihood a collared dove, called *Türkentaube* in German and sometimes also identified as *Lachtaube*. The collared dove had first been named in 1834 by the Hungarian ornithologist Frivaldszky as *Columba risoria*, variation *decaocto*, now *Streptopelia decaocto*, based on a specimen from a region in Turkey, today part of Bulgaria. At first Gätke thought the bird—which at the time had a breeding range from Turkey to southern China and from India to Sri Lanka—must be an escaped cage bird, but eventually he came to the conclusion that it was a stray migrant that had been blown off course by unfavorable weather conditions. The record was reported in Gätke's publication and stood on its own for almost half a century. Within the first half of the twentieth century the species became an unprecedented example of species dispersion. The spread can be traced in hundreds of reports in popular and professional ornithological journals (Fisher 1953; Nowak 1965; Stresemann and Nowak 1958; Vauk 1957).

The first breeding pairs of collared doves outside their known range were recorded in Hungary in 1928, the Czech Republic in 1938, Austria and Poland in 1943, West Germany in 1945, East Germany in 1946, Italy in 1947, Denmark in 1948, the Netherlands and Switzerland in 1950, Sweden in 1951, France in 1952, Belgium, Great Britain, and Norway in 1955, Luxembourg in 1956, and Ireland in 1959. Thus followed an extensive proliferation, which, by the end of the twentieth century, stretched north beyond the Arctic Circle in Norway, east to the Ural Mountains in Russia, and southwest to the Canary Islands and northern Africa from Morocco to Egypt. Its eastern spread had also reached most of central and northern China and Japan. The collared dove was introduced into the Bahamas in the 1970s and spread from there to Florida by 1982. Its stronghold in North America is the Gulf Coast, and it is found as far south as Veracruz, as far west as the Carolinas, and as far north as Alaska, the Great Lakes, and Nova Scotia. The collared dove has transformed from a true rarity on Heligoland at the time of Gätke to one of the most common birds all across the globe (Bezzel 1985; Roselaar 1985).

The data on the spread of the collared dove has been compiled by passionate bird collectors such as Gätke. Over the period of the species' global spread, the initial collection of such records with guns, traps, and nets has given way to their collection with binoculars, cameras, and notebooks. These records are submitted to local bird-watching societies, their authenticity is judged by national rarity committees, and the records are then compiled by international ornithological organizations. Birds account for 10,027 species out of about 1.7 million total animal and plant species identified so far, while the estimate for all species living on earth ranges from 3 to 30 million (R. May 1988, 1990). The rate of new discoveries indicates that only in the case of birds, mammals, and amphibians can there be relative certainty that almost all living species have already been discovered and named. In the case of insects, on the other hand, most species are still expected to be unknown. This uneven distribution in attention paid to different taxonomic groups is not restricted to new discoveries, but also reflected in the amount of data available on population size. In the 2010 Red List of Threatened Species compiled by the International Union for the Conservation of Nature, birds and mammals are the only taxonomic groups for which population totals and trends have been assessed for all named species. Fungi rank last in this regard, with only one out of 31,496 species assessed in 2010 (IUCN 2010).

Over the time of such data's existence, very few bird species have seen an increase in range and population size comparable to that of the collared dove. In 2010 there were 1,240 bird species considered to be endangered to various degrees, and 132 bird species have become extinct since the year 1600. More than a hundred of these are considered to have been lost in the last two centuries alone (Lawton and May 1995; Luther 1986). Throughout the same period, bird-conservation organizations have been founded across the globe, legal measures for their protection have been established, and nature reserves have been organized to conserve the more endangered species. Almost half of Heligoland, for instance, is currently legally designated as a bird reserve and is administered by a local bird-conservation organization (Jordsand 2007; Meise 1957).

The work of national bird-conservation organizations is coordinated by BirdLife International, which represents more than 2.5 million people and 111 national organizations in its global partnership for conservation (BirdLife International 2010). Although focused exclusively on birds, the organization rivals Friends of the Earth International's (2011) more than 2 million members, Greenpeace International's (2011) 2.8 million members, and the World Wide Fund for Nature's (2011) more than 5 million global supporters (McTaggart and Hunter 1978; Groth 2003; Powell 2004; Russell 1996). BirdLife has a partner in virtually every country that has a voluntary organization involved in the conservation of birds. Concern for bird conservation, however, is distributed unevenly: The three largest partner organizations as reported in the early 2000s—the Royal Society for the Protection of Birds (RSPB) in Great Britain with 1,050,000 members; the Audubon Society in the United States with 550,000 members; and the Naturschutzbund (NABU) in Germany with 420,000 members—account for four fifths of the global membership (BirdLife International 2010). Only one additional organization in the global partnership has a membership of more than one hundred thousand, the Dutch Vogelbescherming Nederland with 125,000 members. Only three other partner organizations, those in Belgium, Israel, and Switzerland, count more than fifty thousand members. The smallest partner "organization" is to be found in Bahrain, where Dr. Saeed A. Mohammed is the BirdLife affiliate. Organized bird conservation in Britain stands out no matter which measure is chosen—total membership or membership in proportion to the population. Although Luxembourg's Lëtzebuerger Natur- a Vulleschutzliga (LNVL) ranks higher in proportional membership, it counts no more than 14,000 members and therefore does not exert the same influence in the international arena as the RSPB, which accounts for more than 40 percent of the global membership of BirdLife International and is considered the lead partner of the organization.

FORMULATING THE QUESTION

The way in which birds are valued differs over time and between countries. Of the three major partner organizations of BirdLife International the contrast in the valuation of birds is most pronounced between the British organization, founded in 1889 (Clarke 2004; Samstag 1988), and the German organization, founded in 1899 (Hanemann and Simon 1987; H. May 2003). This study analyzes the emergence and transformation of bird conservation in these two countries from the late eighteenth to the early twenty-first century. During the late eighteenth and early nineteenth centuries there was little noteworthy concern for the conservation of birds, in sharp contrast with the high concern visible during the late twentieth and early twenty-first century. Using life-history data derived from written narratives and oral histories, I follow the development of bird conservation from the point in time at which the greatest declines in birdlife took place to the current efforts in large-scale biodiversity conservation and environmental policy. In doing this I address three interrelated questions: first, why are birds the most popular aspect of nature, second, what accounts for the differences in the value attributed to birds in the two countries, and, finally, how do we explain the timing of the emergence of organized bird conservation and its transformation over time?

In analyzing these questions I aim to contribute to the sociology of morality. I advance a pragmatist theory of valuation that shows how the value attributed to birds derives from their relational position within a set of practices and institutions as social forms. This theory contrasts with one of the most famous approaches in the social sciences to the question of how meaning is attributed to objects in nature, which provides a useful heuristic for highlighting the central argument of the pragmatist approach. Claude Lévi-Strauss coined the now classic dictum that "animals are good to think with" (Lévi-Strauss [1962] 1963; Lévi-Strauss [1962] 1966). Following this dictum, the meaning attributed to animals and nature is considered to be a mirror image or projection of the meaning given to the relations between human beings. Nature, in short, is a social phenomenon. The central assumption of this school of thought is that, as far as the emergence or origin of meaning is concerned, society is a primary and nature a secondary or derivative category. Nature—and the way human beings act in it and experience it—does not produce meaning in its own right but merely serves as an empty screen onto which meanings originally attributed to social relations are projected.

The most famous and clear-cut example of this line of thought is Clifford Geertz's work on Balinese cockfights. According to Geertz, the Balinese see a reflection of themselves and their social order in the fighting cocks. Like a written text, the cockfight has the benefit of rendering ordinary, everyday experiences comprehensible by presenting them in terms of acts and objects that have their practical consequences removed and are reduced to a level of sheer symbolism, and thus their meaning can be more powerfully articulated and more accurately perceived. For Balinese men, fighting cocks depict how social life is perceived to be; that is, they provide what Geertz calls a vocabulary of sentiment or a metasocial commentary. In this view, only people's relations to their fellow human beings have true substance and meaning; the relations of human beings to nature are merely a reflection or enactment of these more genuine social relations. Or, as Geertz puts it, the "cockfight is 'really real' only to the cocks" (Geertz 1973, 443).

Based on the example of the status of wild birds in Great Britain and Germany, it can be shown that the process by which meaning is attributed to objects of nature is far from exhausted by such a one-dimensional explanation. Nature is not merely an empty screen on which social relations are projected, but the experience of nature is itself a source of the meanings and valuations attributed to it. These meanings and valuations differ with the social forms of practices and institutions that guide the experience of nature. Work and play were the most crucial social forms in the development of conservation in Britain and Germany. In Britain, birds derived their meaning in the context of the practices and institutions of the game of bird watching. Here birds are the most popular and best-protected taxonomic group of wildlife due to their particularly suitable status as "toys" on the "playground" of nature. In Germany, by contrast, birds were part of an economic set of practices and institutions in the world of work. They were initially protected as "labor birds" for their usefulness in agriculture and forestry. The study looks at the interaction between conservationists in these two countries that resulted in a gradual adaptation to each other over time and ultimately informed the recontextualization of bird conservation within the legal framework of the European Union.

By tracing this development, this study contributes to the sociology of morality. Debates in this field have changed substantially since the demise of structural functionalism that dominated the topic in the mid-twentieth century. Recent debates have addressed the questions of how morality relates to nonmoral valuations, how widely it is shared, how stable it is over time, and last but not least how it relates to bodily practices, material objects, and geographical space. I elaborate and extend on the pragmatist theory of valuation of John Dewey in order to contribute to these debates. The aim is to make this pragmatist tradition relevant to the recent rejuvenation of research on the sociology of morality. This approach resonates with many of the theoretical considerations and empirical case studies produced by scholars associated with the Chicago school of sociology, whose work was temporarily neglected due to the dominance of structural functionalism. The central argument of this study is that moral valuations are embedded in practices and institutions. I show that changes in moral valuations follow from the transformation of practices and the way these practices become institutionalized, not from innovations in moral discourse, with discourse understood as a text. Such a model of culture as a text informed Geertz's account of Balinese cockfighting and was heavily indebted to the structural functionalism of his mentor Talcott Parsons.

A pragmatist approach is well suited to provide answers to the three questions listed above. Regarding the first question, a pragmatist account of the practice of play (as opposed to work) explains why wild birds are currently the most popular and best protected taxonomic group of wildlife. The key to an understanding of the appeal of birds is its embeddedness in the practices and institutions of bird watching as a game. What constitutes bird watching as a game is not the mere incidence of witnessing the presence of a bird, however far or remote, but the coordinated effort to spot, identify, and record wild birds on a regular basis. As such, bird watching is a form of fieldnote collecting and record keeping. Being part of a game, it is not the symbolic status of birds within a wider system of social relations but their peculiar pattern of diversity and distribution that gives meaning to them, as exemplified by Gätke's bird collection at Heligoland. Birds, in other words, are so popular not because they are good to think with but because they are good to play with.

What makes bird watching such a good game has to do with birds' material, that is, their biological characteristics. It is the fact that birds can fly, are distributed unevenly in space, and are available in fairly high yet manageable numbers. In these characteristics birds differ from other species. In both Britain and Germany, beetles are available in thousands of species, and mammals only in a few dozen. Collecting species that belong to these taxonomic groups can be either overburdening or boring. Birds, by contrast, offer just the right amount of challenge. In addition, birds, unlike for instance trees, do not stay put. Their ability to fly makes their exact whereabouts unpredictable and adds to the excitement of spotting them. Bird migration, often coupled with the effects of severe weather events, can furthermore produce complete surprise sightings, such as the collared dove at Heligoland during Gätke's days. This is not to deny that a rich symbolic meaning can be attached to birds: if one wants to understand why a pub is called the black swan, symbolic meaning is the way to go. If, however, one wants to account for why millions of people watch birds, it is not first and foremost birds' symbolic meaning, but their ambiguous status of being "not too few and not too many" and being "neither here nor there" that provides an explanation. It is this ambiguity of birds, their twofold in-between status, that fuels the game of bird watching.

The argument holds true for the past as much as for the present. Bird collectors continue to be among the most frequent visitors to Heligoland. Today, these collectors are bird watchers who collect sight records with binoculars and notebooks, rather than museum ornithologists who collect specimens with guns. A busy period is fall migration, when many rarities—that is, species that do not breed in the country—can be seen. The influx of bird watchers during migration season is in fact so high that the number of rarity records reported from other parts of Germany drops during this time. Field ornithologists dub the phenomenon the Heligoland effect. So many records are collected at Heligoland that the island established its own branch of the German Rarities Commission, the Heligoland Avifaunistic Commission (Helgoländer Avifaunistische Kommission). The commission establishes what counts as a legitimate ornithological sight record and what does not. Watching birds is not about the experience of nature as such. What makes it so interesting is that these field-ornithological observations, which collectors tick off their lists, constitute game scores that they can compare to records collected by others.

Conservation organizations use this game to turn people into bird watchers. The NABU, for instance, organizes field outings at Heligoland as part of the World Birdwatch event, organized annually by BirdLife International. The aim of the event is to attract people to the game of bird watching, with the expectation that those who become bird watchers will likely join bird conservation organizations. These organizations thus do not try to gain new members by teaching people abstract moral lessons about the environment but by engaging people in practices that will produce emotional attachments to birds.

The attachment to birds produced in this way derives from their status as collectibles in the game of bird watching. Collecting as a game is different from the simple amassing of objects. In a collection, objects are valued according to diversity and rarity. All collections have in common that they aim for a full or complete set of collectibles. Within this full set special emphasis is put on those items that are difficult to obtain. Bird collectors such as Gätke did accordingly try to obtain as many different species as possible and displayed particular pride in those that were rare.

Differences in practices make it possible to answer the second question, what accounts for the differences in value attributed to birds in the two countries? When birds are not part of a collecting game, their value usually derives from other practices. At the time when Gätke was collecting specimens for his museum, the local population at Heligoland used birds largely as food. Rarity did not make a species exempt from use. Among the birds used as food by the locals was the only colony of common guillemots breeding in the country.

While this species was treated with much care to ensure sustainable use, residents initially harvested migrating birds free from any restriction. No regulation limited how many birds one could shoot, trap, or lime for the simple reason that local practices did not affect the availability of this food supply. No special emphasis was put on rare species until Gätke and other bird collectors began to offer high prices for specimens in their role as collectibles.

The initial focus on the economic utility of birds as food was not peculiar to the local population at Heligoland. It was widespread throughout the country and much of continental Europe at the time. In turn-of-the-century Germany birds were part of an economic set of practices and institutions in the world of work (as opposed to play). Ornithological handbooks frequently reported on the practical or economic utility of birds, including their taste and the dishes they were commonly used for. Not diversity and rarity but utility and abundance mattered most, and those species that were particularly suitable as food were accordingly the most cherished. These two forms of valuation of birds—that is, diversity and rarity in the world of play, and utility and abundance in the world of work—persisted over time and account for major differences in bird conservation between Britain and Germany.

Despite this major difference in the form of valuation of birds, both Gätke and the local population at Heligoland contributed to the destruction and decline of birdlife. Whether collected or utilized, birds had first of all to be captured and killed. This was to change over time: the very same people who initially killed birds and contributed to their extinction were also the first to protect them. This seeming contradiction becomes understandable if one looks at the motivation for conservation not as something that springs from internalized moral ideals or abstract ethical principles but as something that is embedded in concrete practices and their larger institutional settings.

This insight makes it possible to answer the third question, how to explain the timing of the emergence of organized bird conservation and its transformation over time. The way bird conservation develops over time differs depending on the peculiarities of the practices and institutions in which it is embedded. Bird conservation in Germany developed in the context of economic practices and institutions in agriculture and forestry. Habitats for birds decreased in the wake of the large-scale industrial rationalization of both agriculture and forestry in the late eighteenth and early nineteenth centuries. With the decline of suitable habitats came the decline in species abundance, which in turn made the negative side effects of a diminished birdlife apparent: it became increasingly recognized that songbirds played a vital role in the destruction of so-called insect pests. The transformation of agriculture and forestry did in consequence also transform the economic status of birds: they were initially valued as units of consumption (i.e., as food), and subsequently as units of production (i.e., as pest control).

Organized bird conservation as promoted by the NABU grew out of this concern for birds as units of production. Practical bird conservation aimed for the discovery of ever more effective ways to increase the number of useful birds through the mass production and distribution of nest boxes and bird feeders. The organization heralded these tools as a solution to the loss of natural habitats. Bird conservation in Germany was thus part of an economic arrangement of practices and institutions. Nature was a factory that produced food, and the countryside was accordingly not treated as a playground, as in those instances in which the practice of play dominated, but as a shop floor. Birds were accordingly used as tools, not as toys.

Economic ornithology was almost completely absent in turn-of-the-century Britain. In this country, it was a transformation of the practice of collecting birds in the field that had the most decisive impact. The valuation of rare birds as collector's items displayed in private showcases and public museums gradually gave way to the concern for the conservation of bird species living in nature. The transformation was brought about by changes in technology, most importantly improvements in cameras and binoculars, that made it for the first time possible to collect reliable sight records and take pictures of moving objects such as birds. Collectors who went into the field with guns tried to get a hold of bird bodies, while collectors who now used binoculars and cameras were interested in collecting records of living birds. The mass production of improved binoculars and cameras toward the end of the nineteenth century thus literally made the difference between life and death for birdlife. The focus of this collecting practice remained on diversity and rarity. The RSPB established nature reserves as spaces akin to outdoor natural-history museums, selecting islands and coastal areas that, like the German island of Heligoland, displayed a particularly large diversity of species.

It is against this background that one should evaluate the developments that took place in the time from the 1960s to 1980s, the period frequently credited with the emergence of environmental activism. Viewed from a long-term perspective, the events during this time were actually nonevents. All major components of contemporary environmentalism were already fully institutionalized by the 1930s. The available membership numbers of the major organizations, albeit with certain gaps and variations in the data, illustrate the continuity of the development (see also appendix 1).

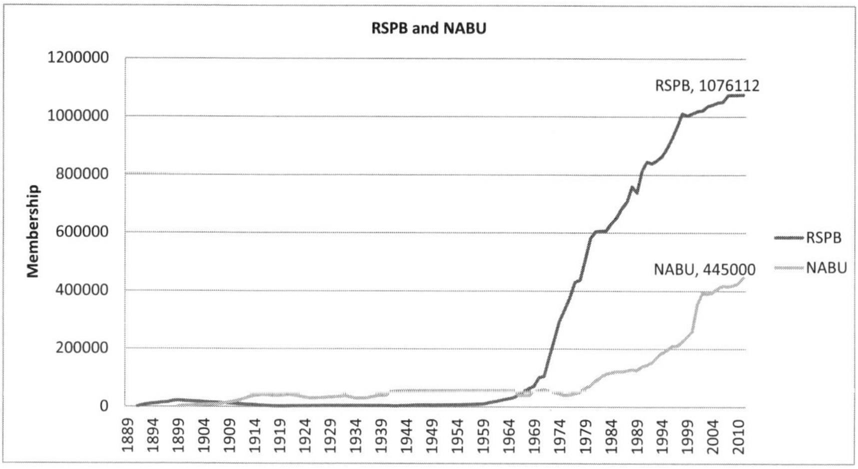

FIGURE 0.1 Membership RSPB and NABU, 1889–2010.

There are three noteworthy things about the membership graphs. First, they indicate the continuity of the two organizations over time. Second, they show a sharp rise in membership in the time from the 1960s to 1980s. Finally, one can see that while the RSPB has by far more members than the NABU in the second half of the twentieth century, the relation was the exact opposite for most of the first half of the century. These three aspects indicate that there is truth to the story that major changes took place in the second half of the twentieth century, yet these changes took place within an organizational framework that had already been established many decades before. All too often nature conservation of the past is compared with environmentalism of the present. Yet nature conservation is not a thing of the past—the organizations founded in the late nineteenth century have their high time at the present. Environmental organizations joined the picture, but they did not replace what had already been established. The membership numbers of nature conservation organizations rose throughout the same period in which new environmental organizations were created, such as Friends of the Earth International in 1971 and Greenpeace International in 1979.

In Britain this rise in membership stands out for being particularly steep, and there is the accompanying danger of reading this finality into the past—that is, of ignoring the fact that when both organizations were still relatively small the NABU was for a considerable period of time more successful than the RSPB. There was accordingly nothing in the first half of the twentieth century that indicated that the RSPB, rather than the NABU, would turn out to be the more successful organization in the long run.

The different degrees of success of these two organizations over time is not accidental but a central element of the development analyzed in this study. The difference is explained by the fact that bird conservation in Germany and bird conservation in Britain were not about the same thing. The valuation of wild birds was embedded in different practices and institutions and flourished or declined in conjunction with these practices and institutions. This also holds true if one compares bird and nature conservation to environmentalism. The various concerns for nature and the environment are worlds apart if viewed from the point of view of the practices and institutions in which they are embedded. There are no nature conservation parties to match Green political parties just as there are no pollution watchers that would give environmentalism a base in amateur science comparable to the role played by bird watching in conservation. Looking at this development from the point of view of the constitutive practices and institutions thus highlights the futility of any effort to attribute this development to a generalized set of environmental values.

What truly changed from the 1960s through the 1980s was not a set of abstract values or attitudes toward nature and the environment held by the population at large but the number of people who watched birds. The RSPB was not more successful during this period than the NABU because it was more progressive. The reason that the RSPB's membership exploded while the NABU's membership stayed stagnant and even declined was not a change in organizational direction but the fact that bird watching became increasingly popular. Economic ornithology (the use of birds for pest control), on the other hand, lost importance after World War II due to the large-scale use of pesticides such as DDT in agriculture, which, in addition, produced problems for birdlife that had not existed before. In response, the German organization began to copy the by then far more successful British example. Throughout this period bird conservation in Germany moved from the world of work to the world of play.

European environmental policies are an additional reason why bird conservation in the two countries looks more similar at present than in the past. The RSPB and other national bird-conservation organizations, the NABU included, were crucial in creating the EU Birds Directive in 1979, one of the most effective transnational environmental policies ever enacted. In 2010, 11.4 percent of the total land of the European Union was protected under this legislation, including about half of Heligoland. The international umbrella organization BirdLife International grew out of the almost dormant International Council for Bird Preservation (ICBP; initially founded in 1922) to provide the field-ornithological data needed for the implementation of the Birds Directive. The fact that nature conservation policies in most European countries look so similar at present does not reflect a common underlying social structure or an identical set of overarching cultural values; it is a consequence of the widening of the arena in which the peculiarities of a British concern for birds play out. The British model turned into the European model. By analyzing this development with a pragmatist theory of valuation, I am able to show that these changes were possible not because of any high-flying cultural values but for the very mundane reason that birds are good to play with.

CHAPTER OUTLINE

The analysis of this development in the chapters to follow proceeds in a chronological order, beginning with the late eighteenth century and ending with the early twenty-first. The chapters are not neatly divided into decades or periods. There is a substantial amount of temporal overlap between chapters. In accordance with the argument that the development under investigation is a gradual transformation that unfolds along multiple dimensions of practices and institutions, the story cannot be divided into sharply demarcated time periods. While one element in the relational process of temporal unfolding changes, others stay constant. Each chapter has accordingly a conceptual focus, highlighting an aspect that is central to the development, and looks at other dimensions in the light of this aspect. A temporal overlap between chapters is thus fully intended, each chapter going back into the period discussed before and looking ahead into the period to follow.

Chapter 1 outlines the contribution of this study to the sociology of morality. Research and theorizing in this field has in recent years moved away from the approach to morality that dominated the discipline in the heyday of structural functionalism. The study is designed to contribute to the central questions raised in recent debates in the field that followed the demise of structural functionalism. It is based on the pragmatist theory of valuation of John Dewey. The chapter outlines his theory and advances it with a theory of institutions. I argue that valuations are embedded in practices and institutions, and I show the implications of this argument for the sociology of morality.

Chapter 2 investigates the different practices and institutions that influence the valuation of birds. The focus is on the time from the late eighteenth to the mid-nineteenth century. The dominant forms of valuation during this time differed between the two countries. In Britain it was the logic of natural-history collecting in private and public museums that made for the valuation of birds. Natural-history museums did of course exist in Germany as well, yet in this country many ornithologists had gained their knowledge about birds as bird catchers, not as natural-history collectors. Wild birds were a major ingredient in the country's traditional diet and had a status as consumer goods rather than as collector's items. The logic of valuation entailed in this institutional arrangement put emphasis on the usefulness of birds, while any valuation along the lines of diversity and rarity was absent. The emerging concern for conservation that developed throughout the nineteenth century was part and parcel of these two forms of valuation.

Chapter 3 looks at the development of organized bird conservation in the late nineteenth and early twentieth century. The focus is on the factors that made for the foundation of the Society for the Protection of Birds (currently Royal Society for the Protection of Birds or RSPB) in Britain in 1889 and the Bund für Vogelschutz (later Deutsche Bund für Vogelschutz and currently Naturschutzbund or NABU) in Germany in 1899. Concern for conservation during this period did not develop from the often-assumed aim of the population in the urban industrialized centers to go back to nature but rather from a process of technology coming to the people living in the countryside. Concern for birds developed from the transformation of the experience of birds in the context of already-established practices and institutions. The experience of birds was transformed through the mass production of prismatic binoculars, cameras, bird tables, and nest boxes. In Germany bird conservation was advanced mainly for economic ends. Schemes for so-called rational bird conservation, as they were promoted around the turn of the century, aimed for the discovery of ever more effective ways to increase the number of birds useful to agriculture, forestry, and gardening through the mass production of bird tables and nest boxes. Organized bird conservation heralded this system of mass production as the solution to the loss of natural habitats. In Britain, on the other hand, the valuation of rare bird as collector's items displayed in the museum was gradually replaced by an interest in birds in their natural environment, accompanied by concern for their conservation. With the introduction of binoculars and cameras into the field in the second half of the nineteenth century, collectors' interest in birds transformed from collecting bird bodies to collecting field notes and photographic images of living birds. Yet one thing remained the same: concern focused on rare and endangered species as collector's items, not on the population numbers of useful species.

Chapter 4 analyzes how in the mid-twentieth century these different logics of valuation became tied to the emerging science of field ornithology. The

Royal Society for the Protection of Birds transformed into a club of bird watchers throughout this period and formed a tie to professional field ornithology. Field ornithology became institutionalized through the foundation in 1932 of the British Trust for Ornithology (BTO) as a volunteer body aiming to organize amateur bird watchers for the goals of science and conservation. Amateur bird watching and professional field ornithology in Britain was driven by a logic of collecting and focused on rare and endangered species. In Germany, by contrast, there developed no such close link between amateur bird watching and bird conservation that the work of the Bund für Vogelschutz could rely on. Instead, initially privately owned research stations for bird protection, which were later administered by the state, became the central institutional places for translating the concern for birds into action. Bird Protection Stations were experimental sites for the development of ever more refined measures and techniques to determine use and harm of various bird species to benefit practical bird conservation. Throughout this period there was no army of amateur bird watchers in Germany comparable to that in Britain. The German organizations aimed for the cooperation of farmers, fruit growers, and foresters, not predominantly for that of amateur bird watchers.

Chapter 5 analyzes the transformation of bird collecting in the time from the 1950s to 1970s. In Britain the Royal Society for the Protection of Birds turned to the creation of nature reserves as its major strategy for conservation. The organization targeted those places for acquisition that had the highest proportion of rare species or displayed a particularly high diversity in birdlife. Nature reserves were museums for living birds, and like museums they were used for the display of birds to a larger public. The aim was to attract people to the practice of bird watching to turn them into future bird lovers. The already-established institutional logic of conservation determined how the large-scale poisoning of birds through pesticides such as DDT was addressed. Research on the topic had begun in Britain already well before the American naturalist Rachel Carson published her bestselling *Silent Spring* in 1962. The most important consequence of this episode, other than the eventual ban of DDT, was the establishment of a population survey on common breeding birds by the British Trust for Ornithology. German conservationists, by contrast, were latecomers in reacting to the pesticides issue and largely followed the British and American examples. Also following the British example, endangered (i.e., rare) birds rather than common ones were for the first time put at center stage by organized bird conservation. The different outcomes of the DDT crisis in the two countries are why it makes more sense to speak of an environmental revolution in Germany than in Britain. This difference between historical continuity and discontinuity is further indicated by the relationship between the main bird-conservation organization in each country and their youth organizations: in Britain the relationship between the two was a cooperative one; in Germany it was marked by conflict between the generations.

Chapter 6 shifts the focus away from the goals of conservation and its transformation over time and instead looks at the rise of membership numbers of organizations from the 1960s through the early 2000s. One of the main strategies of organized bird conservation for recruiting new members is to turn people into bird watchers. Beginning with the 1970s and continuing until the present, the Royal Society for the Protection of Birds and later also the Deutsche Bund für Vogelschutz organized various popular events that make people "look" at birds—that is, move them as far as possible from just "seeing" birds as a mere by product of everyday experience to purposefully "watching" them as an organized activity. Larger conservation messages are tied to the popular bird species. Bird conservationists use flagship species (bird species known to a wider population) to communicate their conservation message. The way in which bird conservation relates to other topics of nature conservation and environmentalism differs between the two countries. The large membership of the Royal Society for the Protection of Birds is based exclusively on the appeal of birds; the organization utilizes this appeal to raise awareness for larger environmental issues such as global warming, which affects the habitats of birds and is considered a major potential cause of species extinction. The RSPB is engaged in environmental issues, yet it maintains birds as its focal concern. In Germany, on the other hand, the organizational strategy is almost the exact opposite. Bird conservation has become one issue among many others in the work of the organization throughout the second half of the twentieth century. In 1990, the Deutsche Bund für Vogelschutz (German League for Bird Protection) changed its name to Naturschutzbund Deutschland (League for Nature Protection Germany) to appeal to a larger audience.

Chapter 7 analyzes how, as a result of the establishment of a European legal framework for nature conservation, the differences between the two countries have diminished. A European Birds Directive passed in 1979 and was extended in 1992 by a Habitats Directive. The two directives together established a network of protected areas known as NATURA 2000. The data for selecting these areas is administered by BirdLife International and provided by thousands of bird watchers and field ornithologists across Europe. The field-ornithological data collected by British bird watchers since the 1960s provided the institutional model. Countries such as Germany that did not have the necessary field-ornithological monitoring schemes in place had to follow suit when the

European law was enacted. The fact that the conservation policies in the two countries and in the rest of Europe currently look so similar is due not to a shared underlying social structure or an overarching scheme of cultural values but to the fact that the arena in which the British peculiarities of the game of bird watching play out has widened. Following European legislation, bird watchers in Germany and other European member states have followed the British example. In some German regions, where not enough voluntary bird watchers are available, paid professionals fill in. Providing the required data is partially work in Germany, while it continues to be play in Britain.

The final chapter discusses the implication of this analysis for the study of morality within sociology. Three central insights are highlighted: first, morality is embedded in practices rather than in discourse, and different forms of moral valuation differ with the peculiarities of the practice in which they are embedded and the way these practices are institutionalized; second, these practices and institutions can be purposefully designed in order to produce specific valuations, while moral valuations that are not embedded in any practices and institutions are likely to remain mere discourse without material effects; and, last but not least, deeply ingrained moral valuations are the outcome of larger historical processes of transformation in practices and institutions, not their causes. I argue that taking these insights into account has far reaching implications for the way we study morality as sociologists.

CHAPTER 1

A Pragmatist Theory of Morality

The mid-twentieth century heralded morality as the defining element of sociology as a scientific discipline. Within structural-functionalist theorizing as advanced in particular by Talcott Parsons, morality, then conceptualized as norms and values, was used to solve the explanatory problem of social order (Blake and Davis 1964, 760–61). Leonard Broom and Philip Selznick, for example, in what is considered to be the most popular sociological textbook of the era (Manza, Sauder, and Wright 2010), defined the aim of sociology as discovering the basic structure of human society—that is, identifying the main forces that hold groups together or weaken them. Norms and values were presented as the central concepts in this endeavor: "All societies have rules or *norms* specifying appropriate and inappropriate behavior, and individuals are rewarded or punished as they conform to or deviate from the rules. . . . Norms are based on cultural *values*, which are justified by moral standards, reasoning, or aesthetic judgment. . . . It is impossible to imagine a normless society, because . . . the standards of conduct contained in the norms give order to social relations" (Broom and Selznick 1963, 68–69).

Much of the temporary neglect of morality as an object of sociological inquiry is tied to the demise of structural functionalism and the accompanying definitions of norms and values as provided in this textbook passage (Spates 1983). In recent years, however, morality has resurfaced as an object of sociological inquiry. Scholars currently working on morality have made an effort to distinguish their approaches to the topic from the structural-functionalist account of morality associated with the work of Parsons. Vaisey and Hitlin have analyzed some of the more recent departures from Parsonian theorizing and argued for the existence of a new sociology of morality (Hitlin and Vaisey 2013).

The new sociology of morality is not a coherent school of thought or paradigm, and it is not the work of any single scholar or group of scholars. Instead of providing one consistent set of theoretical answers, its common denominator is better described as opening the issues covered by the old sociology of morality for empirical research and reconceptualization. Five topics of inquiry that depart from the conceptualization of morality within structural functionalism stand out:

1. How does morality relate to nonmoral valuations?
2. How widely shared are moral values within a given group, and to what extent are different moral values either consistent with each other or contradictory?
3. How stable or persistent is morality over time, and what causes it to either reproduce or change?
4. How does morality relate to practice?
5. How does morality relate to material objects and space?

Structural functionalism settled all five of these questions by theoretical fiat; that is, their answers were independent of the empirical case under investigation. Parsons and other structural functionalists contributed to a form of theorizing according to which morality constitutes a phenomenon sui generis that (1) is sharply separated from nonmoral valuations, (2) is collectively shared and internally consistent, (3) contributes to the maintenance of the existing social order, (4) is the motivating force of social action, and (5) finds expression in material objects and the arrangement of space yet is not causally influenced by them. This rather rigid way to think about morality (and its settling of all central questions on the topic) was among the main reasons for the temporary disappearance of morality as a topic of sociological inquiry.

The new sociology of morality that surfaced in recent years has returned these questions to the theoretical agenda and reopened them for empirical inquiry. These interventions and debates are no longer solely carried out within in the domain of sociological theory, as under the predominance of structural functionalism, but have emerged in several subfields of the discipline, most salient among them economic sociology, cultural sociology, and the sociology of science, knowledge, and technology. Scholars working in these fields have shown (1) that moral valuations and economic valuations (i.e., moral values and economic value) can mutually reinforce each other

(Fourcade 2009a, 2009b; Fourcade and Healy 2007; Healy 2000, 2004, 2006; Krause 2014; Stark 2009; Zelizer 1979, 1985, 1994, 2005); (2) that morality is not just a source of social order and consensus but can likewise be a source of ambiguity, disagreement, and conflict (Boltanski and Thévenot 2006; Lainer-Vos 2013), and that it can create symbolic boundaries that reproduce social inequalities (Lamont 1992; Lamont and Molnár 2002; Lamont and Thévenot 2000); (3) that not only can the content and strength of moral valuations change over time but the very category of morality can transform (Abend 2011, 2014); (4) that morality can motivate behavior as much as it can be a tool or repertoire for action (Swidler 1986, 2001a, 2001b; Vaisey 2008a, 2009); and (5) that material objects and arrangements of space can enforce compliance with norms even in the absence of internalized moral values (Latour 1987, 1992, 2002, 2005).

This study contributes to these recent developments by looking at the development of bird conservation in Great Britain and Germany from the late eighteenth to the early twenty-first century. Looking at a specific empirical case from a long-term and comparative perspective offers an opportunity to locate morality in the larger fabric of social life and to treat the five topics raised by the new sociology of morality as open research questions. The point of departure for this analysis is the theory of valuation of the pragmatist philosopher John Dewey and its adaptation by scholars belonging to what is known as the Chicago school of sociology. Key to this perspective is the insight that motives—moral motives included—are embedded in action and the institutional settings that facilitate this action. Motives are not fixed entities that trigger action from inside the actor, but they emerge within a given course of action. This insight has crucial implications for the way one looks at the five questions raised above: it implies that morality cannot be analytically separated from other social phenomena (as structural functionalism has done) but must be approached as a relational phenomenon, best studied in interaction with other dimensions of social life.

This study shows that the way valuations of birds are embedded in social life differs between Britain and Germany. In Britain, birds derived their meaning in the context of the practices and institutions of the game of bird watching. Birds are the most popular and best protected taxonomic group of wildlife in this country due to their particularly suitable status as "toys" on the "playground" of nature. Birds, in other words, are good to play with. In Germany, by contrast, birds were part of an economic set of practices and institutions in the world of work. They were initially protected as "labor birds" for their usefulness in agriculture and forestry. Actors who did not agree with this exclusively economic value attached to birds began in reaction to assign an equally exclusive ethical (i.e., moral) value to them. This study further looks at the interactions between conservationists in these two countries that over time made for a mutual adaptation of bird conservation policies and eventually led to its recontextualization within the legal framework of the European Union.

The remainder of this chapter has three aims: first, to outline Dewey's theory of valuation and the central distinction between work and play on which it is built; second, to extend the account of these two practices by a theory of institutions and to show how the conceptual distinction between work and play can be used to account for differences in the valuation of birds; and, last but not least, to show how these differences can be employed to explain the historical emergence of bird conservation and its transformation over time. In addressing these points, this chapter shows how a sociology of morality informed by the pragmatism of Dewey can enrich current debates on the topic.

PRAGMATISM AND MORAL VALUATIONS

The key to Dewey's theory of valuation is its conceptualization of the relationship between the means and ends of action. It is in fact the cornerstone of his entire philosophy, from aesthetics, ethics, religion, science, and epistemology to education and social reform. No single article, collection of essays, or monograph outlines the argument in detail; instead Dewey scatters its discussion throughout writings produced over more than five decades. The relation between means and ends is what the term *pragmatism* stands for in his case. As Josh Whitford has pointed out, Dewey's conceptualization of the relation between means and ends is also the part of his pragmatism that has the most far-reaching implications for current sociological debates on agency (Whitford 2002). Hans Joas has moreover demonstrated the importance of Dewey for a theory of creative action that is different from both theories of norm following behavior as envisioned by structural functionalism and theories of instrumental behavior described by rational-choice theory (Joas 1993, 1996, 2000).

Dewey develops his theory based on a critique of the teleological model of action according to which every act can be separated into means and ends (Dewey 1939). From this perspective, ends are the only carriers of value, and means have a merely instrumental function. In the folk model of action, ends have an intrinsic motivational force, while the means of action are entirely neutral or instrumental—hence the dualism of means and ends. Dewey challenges this dualism. In his view means and ends are neither analytically distinguishable nor empirically separable. Instead, he holds means and ends to constitute each other reciprocally in a given course of action. The very terminology of means and ends is thus misleading.

Dewey's argument unfolds from a rather simple yet crucial insight. He argues that what motivates action are not the ends of action as a state of affairs to be attained, which he calls ends-in-themselves, but the anticipation or idea of this state of affairs, which he calls ends-in-view. Since the ends or goals of action only motivate agency as long as they are unachieved, it is not the end result of action itself that motivates agency but the prediction or anticipation of this end. These ends-in-view organize a given course of action. As such, ends-in-view have the status of means for action, so-called procedural means: "On this basis, an *end-in-view* represents or is an *idea* of the final consequences, in case the idea is formed *on the ground of the means that are judged to be most likely to produce the end*. The end in view is thus itself a means for directing action—just as a man's *idea* of health to be attained or a house to be built is not identical with *end* in the sense of actual outcome but is a means for directing action to achieve that end" (Dewey 1966, 57).

Throughout a course of action, ends-in-view serve as procedural means; that is, they are tools or instruments for the attainment of a certain anticipated end result. Viewed a different way, means are intermediary ends. Means are never pure instruments, what Dewey calls means-in-themselves. They are instead intermediate ends with which we organize a course of action. This is to say that in a given course of action "means" as intermediary ends are on a par with "ends" as procedural means. At any given moment throughout a course of action every end is also a means and every means also an end.

> To grasp this fact is to have done with the ordinary dualism of means and ends. The "end" is merely a series of acts viewed at a remote stage; and a means is merely the series viewed at an earlier one. The distinction between means and end arises in surveying the *course* of a proposed *line* of action, a connected series in time. The "end" is the last act thought of; the means are the acts to be performed prior to it in time. To *reach* an end we must take our mind off from it and attend to the act which is next to be performed. We must make that the end. (Dewey 1922, 34)

Both means and ends are intermediaries in a chain or course of action. If we were to freeze a given course of action in time, what we are used to calling means and ends would be indistinguishable. It is only from the fixed vantage point of an arbitrarily chosen before or after that means as means and ends as ends can be distinguished. Yet this view must always remain a view from the outside, that is, a view ex ante or post hoc. It is an analytical abstraction that does not match any reality. Throughout the temporal unfolding of a course of action itself no such distinction can be made.

As long as action continues to proceed there is no beginning or end, no first step and no final goal. Ends are never final. Once a course of action comes to a close, new ends are designated. There are consequently no ends-in-themselves—the moment one line of action concludes, that which served as an end-in-view becomes a means for further action, that is, an intermediary for yet another end-in-view. In Dewey's view ends are in fact literally endless, forever coming into existence as new activities bring about new consequences. Once we have built a house we start furnishing it, once we have bought a car we start driving it, and so on. To talk about endless ends is a way of saying that there are no ends as fixed self-enclosed finalities. No end is forever final, unless it is indeed the final end.

Looking at the relation between means and ends as constituting each other reciprocally in the unfolding of a course of action turns the contrast between the two into an empirical rather than an analytical matter. The distinction between work and play captures this empirical degree of difference. Dewey conceives work and play as two different ways of relating means and ends to each other.

> The difference between them is largely one of time-span, influencing the directness of the connection of means and ends. In play, the interest is more direct—a fact frequently indicated by saying that in play the activity is its own end, instead of having an ulterior result. The statement is correct, but it is falsely taken, if supposed to mean that play activity is momentary, having no element of looking ahead and none of pursuit. . . . The point of these remarks is that play has an end in the sense of a directing idea which gives point to the successive acts. (Dewey 1916, 195–96)

Work and play are not two mutually exclusive concepts, but the difference between them lies merely in the length of the chain of intermediaries involved.

The difference between work and play is thus a matter of time-space extension—the longer the chain of intermediaries, the more real the difference between means and ends becomes. Play is an activity in which the result of each activity immediately motivates further action—that is, in which means and ends are in quick interchange—while work describes an activity in which a distant end-in-view motivates a course of action with many intermediate procedural means. A game of squash—in which every outcome of action (every ball hit) immediately affects the next action to be pursued—is play activity no matter how exhausting the task may be. A career in accounting, on the other hand—in which the only goal is to pay one's bills and make it safely into retirement—is a life of hard labor no matter how easy the task itself may be. The difference between the two activities lies in the length of the means-ends chains that are involved, not in the amount of effort involved or in the question of whether they are serious or fun.

This conceptualization of work and play as designating different degrees of time-space extension of means-ends relationships is the truly innovative step of Dewey's philosophy. Dewey applies this thinking in relational processes to his theory of value. The point of departure is the argument against the dualistic model of ends as self-enclosed finalities that motivate action and means as value-neutral instruments. In other words, a value is not an essence or an entity but rather the outcome of an act of evaluation. Value in this sense is not a noun but an adjective: "There is no peculiar class of things (much less of 'entities') to which value-qualifications can or should be attributed. The previous point indicates that 'value' is an *adjectival* word, naming that which is a trait, property, qualification of some thing—in the broad sense of *thing* mentioned. It is like, say, the words good, fine, excellent. . . . Anything under the sun may come into possession of what is named by 'value' as its adjective" (Dewey 1949, 66–67). There is no such thing as a value in and of itself—a value is a quality attributed to something else. Such an attribution is an act, and this act of valuation is part of a larger course of action. Being part of a process of action, no distinction between what other authors tend to call intrinsic and instrumental value can be made.

In the same way that means and ends cannot be empirically separated or analytically distinguished, means-value (i.e., that which is valued as a means) and ends-value (i.e., that which is valued as an end) cannot be empirically separated or analytically distinguished. In a course of action in which means and ends constitute each other reciprocally, the distinction between intrinsic and instrumental values becomes meaningless.

> It is impossible to form a just estimate of the paralysis of effort that has been produced by indifference to means. Logically, it is truistic that lack of consideration for means signifies that so-called ends are not taken seriously. It is as if one professed devotion to painting pictures conjoined with contempt for canvas, brush and paints; or love of music on condition that no instruments, whether the voice or something external, be used to make sounds. The good

> workman in the arts is known by his respect for his tools and by his interest in perfecting his technique. The glorification in the arts of ends at the expense of means would be taken to be a sign of complete insincerity or even insanity. (Dewey 1929, 279)

In a given course of action, that which is valued as means stands on the same level with that which is valued as end.

From this argument derives the assumption that a course of action once begun has an internal logic of its unfolding that cannot be reduced to a process of acting out prior calculations of how to relate given means to fixed ends. Every course of action, whether chosen habitually or reflexively, has an inbuilt transformative potential. Once the action has been initiated, the relation between what was believed to be an end and what was believed to be a means might be reversed. It is the internal logic of unfolding of a course of action that transfers value to people, objects, and events. Every course of action is thus a creative process in which value is produced and transferred. States of affairs do not attain value by sharing an essence with an abstract entity of valueness, be it located in yet another entity called structure or culture. An opera, a painting, or an academic degree is not valued because it contains traces of an essence called "musicness," "artness," or "scientificness" but because of its relational position in a course of action and that action's unfolding over time.

PLAY AND GAMES: THE VALUATION OF BIRDS AS TOYS

The pragmatist theory of value has a distinct blind spot when it comes to systematically investigating the conditions under which action falls on the side of play versus that of work. In Dewey's view, a separation between means and ends (and, thus, work-like activity) only exists when the process of action is interrupted—that is, when it is faced by environmental conditions that dissolve spontaneous action. Under such environmental conditions conscious thought sets in and seeks a new line of action. "Now at these moments of a shifting in activity conscious feeling and thought arise and are accentuated" (Dewey 1922, 179). In an unproblematic process of action, however, activity is play-like in that means and ends are mutually constitutive of each other. Dewey does not, however, engage in any theorizing on how problematic situations arise. For him these situations seem to be induced by outside events and are thus untheorized externalities in his theory of action.

I want to fill this theoretical void in Dewey's pragmatism with a theory of institutions. Institutions are arrangements of rules and resources that organize action. They represent ways to relate means and ends that can either foster or hinder the unity of the act. In this theory, problems (i.e., interruptions of a course of action) can be both internal and external; that is, institutions can contribute to their production as much as to their solution. To formulate a theory of games as institutions, a closer look at the specificities of the kind of situations that produce—and hamper—play is needed. As previously stated, play is an activity that takes place when means and ends constitute each other reciprocally in a course of action. In play, action follows upon action according to an internal logic that seems to need no conscious intervention by the actor. Mihaly Csikszentmihalyi coined the term "flow" for such creative processes of action (Csikszentmihalyi [1975] 2000). He describes play as an autotelic experience—an experience of complete involvement of the actor with his activity. The actor experiences this state as a unified flowing from one moment to the next, in which he is in control of his actions. Activities that reliably produce flow experiences provide opportunities for action that a person can act upon without being bored or worried.

This model is based on the assumption that, at any given moment, people are aware of a finite number of opportunities that challenge them to act; at the same time, they are aware of their skills—that is, of their capacity to cope with the demands imposed by the environment. When a person is bombarded with demands he feels unable to meet, a state of anxiety ensues. When skills are greater than the opportunities for using them, boredom follows. Flow (i.e., play) is experienced when people perceive opportunities for action as being evenly matched by their capabilities. This situation of action can be described as an equilibrium of forces joined with an uncertainty of outcome. Play thus comes to an end when uncertainty is too great and outcomes lie entirely beyond the control of players or when uncertainty is too slight and outcomes may become completely controlled by players. Play can be defined as a self-referential and self-renewing activity that is sustained by an equilibrium of forces and an openness or unpredictability of outcome. As such, play implies a potential to completely and utterly absorb actors in a stream of action. When, on the other hand, the equilibrium of forces is disturbed and the outcome becomes predictable, the play element of action comes to an end.

Games can be described as institutionalized forms of play. They provide the elements that will draw actors into play as they are structured systems of action that help to reciprocally adjust means to ends. Games can be defined as the institutionalized collection of rules and resources that foster the equilibrium of forces and openness of outcome that constitute play. The rules of a game put players in relationships with each other that produce and sustain an equilibrium of forces between both the preexisting differences in strengths and abilities of the players and the disparities of relative advantage that develop throughout the game itself. The handicap imposed on the superior or more experienced player is the most paradigmatic expression of a rule intended to ensure equilibrium of forces. The same can be said about the resources of a game: dice, cards, and fortune wheels are prototypical examples of resources that foster equilibrium and uncertainty. Rules do not introduce an element of routine and order into play, nor do resources add an element of extrinsic reward; they are not detrimental to but rather constitutive of the self-referential and self-renewing logic of play. Games in turn cannot be understood independently from play. Without games as institutionalized forms of play the success or failure of our propensity for play to translate into concrete action lies entirely at the mercy of the moment. Games stir and nurture our inclination for play; they make play transferable across spatial contexts and reproducible in time.

Games of skill and games of chance are both institutionalized assemblages of rules and resources that foster equilibrium and uncertainty—they are two different ways to attain the same result. Games of skill can be described as rule-centered, and games of chance as resource-centered ways of attaining equilibrium and uncertainty. In the former case, it is the selection of players by means of rules that guarantees equilibrium of forces. Tournaments, whether involving physical abilities or mental strengths, are the paradigmatic example. Here players are selected by an assessment of either formal characteristics or previous performance to match each other in ability and strength. In the most common form of tournaments, the contestants are matched in pairs, with the losers of each round eliminated and the winners paired anew until only one remains as the champion of the tournament. Despite a staggering amount of variation in procedure, all tournaments have in common that they arrange players by skill to allow for equilibrium of forces and uncertainty of outcome. This outcome can also be attained by introducing a handicap, which allows players of different proficiency to play against each other on equal terms.

Games of chance, on the other hand, attain equilibrium of forces and unpredictability of outcome by means of the resources that are employed. In a game of chance the outcome is influenced by a randomized device, such as dice, playing cards, roulette and fortune wheels, spinning tops, or numbered balls and billets. Unlike in the tournament, it is not the match of strength and ability of the players that ensures equilibrium but the surrendering to pure chance or luck of the potentially unequal players. Board games designed for the close-knit circle of the family are the paradigmatic example. Family members tend to differ in their abilities and skills due to their differences in age and thus dice, card decks, and equivalent devices can come into their own. The element of chance introduced by these resources guarantees an equilibrium of forces among otherwise unequal players. Without this element of chance the outcome of the game would be predictable, and the play element would fade away.

Viewed from a different vantage point, the difference between games of skill and games of chance is that in the case of the former it is the game as a fixed event that selects a group of players to match the requirements of the game, while in the latter case it is an already-established group that selects the game to match the abilities of the players. Accordingly, there is more involved in the choice between different games than individual preferences or collective norms—the relational logic of play as a self-referential and self-renewing process influences the number of possible choices. The more elements of the game are fixed in advance, the fewer options are left to choose from if play is to be attained. Games have to be adjusted to the players as much as players have to be adjusted to the game. Most games do contain elements of both skill and chance. In between the wide array of games commonly defined as pure games of skill, such as chess or the Olympic games, and pure games of chance, such as snakes and ladders or the national lottery, are games that mix elements of both—for instance, poker, blackjack, and bridge.

As institutionalized forms of play, all games assemble rules and resources in a way that allows for equilibrium of forces and uncertainty of outcome. If one element in this relational system is altered, all remaining elements must also be altered to preserve equilibrium and uncertainty. There is accordingly no specific rule or resource that all games share with each other. Yet there is, contrary to Wittgenstein's ([1953] 2001) influential assertion that the phenomena we call games are merely connected by family resemblance, something that all games have in common. They share the effect that they produce. Rules such as "count to a hundred before you open your eyes," "go to jail—go directly to jail—do not pass go, do not collect $200," or "play the ball as it lies and the course as you find it," and resources from race tracks and bungee cords to building blocks and tennis balls are not epiphenomenal to monopoly, hide-and-go-seek, bungee jumping, car racing, tennis, and golf. They are not the mere reflections of an essence put into physical shape and drafted as coded conduct, but they are what games are firstly and finally made of. In relation to each other these elements allow us to enter a course of action that is marked by equilibrium of forces and uncertainty of outcome. The procedural result of this relation is the common denominator of all games.

In games as institutionalized forms of play, the goals of action describe arbitrarily chosen obstacles. As such the activity of playing a game stands in direct opposition to a teleological model of action, whether in its utilitarian or normativistic variant. The game of chess provides a good example: the goal of the game is to capture the opponent's king. Yet the course of action that leads to this goal is not engaged in with the possession of the king as an actual finality of action in mind. If the activity were truly teleological, the player could with considerably less effort walk into a toy store, buy a game of chess, pull out the king, and walk home with it. Such an act would indicate a complete lack of understanding of what playing the game of chess is all about. One does not play chess in order to attain a set goal; rather the goal of the game is created as an arbitrary obstacle in order to organize a stream of action that is experienced as play. In an article in line with this argument titled "Means, Ends, and Galumphing," Stephen Miller gives an observational account of infants' play activity as awkward, exaggerated, ineffective, and deliberately complicated action. "An infant, who ordinarily can run with much more efficiency, galumphs in play. It is as though he were setting obstacles in the path of what he does in play" (S. Miller 1973, 90). Play activity or galumphing is not a means to an end but the circumvention of obstacles put there by the player or voluntarily accepted by him.

To identify the goals of play and games as arbitrary obstacles does seem to remove this mode of action and its institutional manifestation as far as possible from the phenomenon that we usually identify as morality. To call something a moral value is to say that it is experienced by the actor as beyond mere choice. As Charles Taylor pointed out in the language of analytical philosophy, a moral value is not a preference but a preference about preferences. It is not what the actor desires and wants to do but what the actor considers justified to be desired and thinks he or she should do (Taylor 1985). This kind of analytical distinction represents the central element of most sociological theorizing on morality. It is the kind of assumption that Talcott Parsons's structural functionalism took as its point of departure in distinguishing general value-orientations and their translation into situation-specific norms from economic preferences and utility-maximizing behavior (Parsons 1935, 1951, 1969, [1937] 1968; Parsons and Shils 1951). According to this theory, value-orientations determine the ends of action, while norms restrict the number of permissible means to attain these ends. "In the choice of alternative means to the end, in so far as the situation allows alternatives, there is a 'normative orientation' of action" (Parsons [1937] 1968, 44).

This normative dimension of action in the relation between means and ends is the centerpiece of Parsons's initial formulation of the theory of the unit act, which he elaborated and extended throughout his career.

It is at this point that the pragmatist theory of play and games can tell us something fundamentally new about morality. In the act of playing a game, arbitrarily chosen obstacles (i.e., instrumental values) can turn into deeply ingrained moral commitments (i.e., intrinsic values). It is only after analytically distinguishing the two that the entire hocus-pocus of fixing them back together empirically can begin. If we do not rip them apart in the first place, the entire mystery of which causes the other simply disappears. The central argument concerns the question of how the actor arrives at his or her valuations. From the pragmatist point of view there is something crucial about moral valuations as they emerge in play. The way the actor experiences these valuations is the opposite of those feelings and motivations that draw the actor into play and games. In a game the ends are arbitrary. Moral valuations, on the other hand, are deeply ingrained commitments. Arbitrarily chosen obstacles and deeply ingrained moral commitments are opposites on a spectrum of motives that makes a discrete analytical distinction between intrinsic and instrumental values. In the empirical reality of a given course of action and its institutional context, on the other hand, they are next to each other; that is, the latter develops from the former. The result of this transformative process is experienced as something that lies beyond choice. There is no necessity involved in such a process. The theory does not hold that every player in a game will arrive at a moral position. Not everyone is transformed in the process of playing a game—some actors simply play along. The argument holds the other way around that even those actors who possess deeply ingrained moral commitments did arrive at them in the process of playing a game. They talk about it as a transformative experience and couch their valuations in the vocabulary of the game that is played.

The practices and institutions of bird watching as a game account for why birds are the most popular and best protected taxonomic group of wildlife not only in Britain and Germany, but also in many other places in the world. What constitutes bird watching as a game is not the mere incidence of witnessing the presence of a bird, however far or remote, but the coordinated effort to spot, to identify, and, last but not least, to record wild birds on a regular basis. As such, bird watching is a form of field note collecting and record keeping. Bird records are collected with binoculars and with still and video cameras, and they are kept in the case of the professional ornithologist as "bird surveys" and in the case of the amateur bird watcher as "life lists." Bird watching is more than just one game. Efforts to collect field notes range from the far-flung attempt to "tick" as many additions to one's list as possible to the less ambitious but often equally demanding observation of one's "local patch" over a long period of time.

These two forms of bird watching mark the difference between what bird watchers call "twitching" or "tick hunting" as a game of skill and "birding" or simply "bird watching" as a game of chance. In both cases birds are observed in the natural and the built environment. Being part of a game, it is not the symbolic status of birds within a wider system of social relations but their peculiar pattern of diversity and distribution that gives meaning to them. As of 2010, the most recent edition of the British List, that is, the official inventory of bird species found in Britain, enumerates a total of 552 bird species that have been recorded at least once in their natural state since 1950, while the German List has 462 (Barthel and Helbig 2005; Dudley et al. 2006). An addition to the British or German List is a duly sought-after attraction and ensures that the lists are continuously updated.

The clear lack of anything overwhelming, astonishing, or extreme in these numbers is in fact what makes them so special. The passion for birds among bird watchers rests neither on a marveling at the diversity or abundance of nature nor on astonishment at the sublimity of its more archaic or aesthetically pleasing forms. What makes birds so fascinating to bird watchers is the fact that there are "not too few and not too many." Unlike the thousands of species of beetles that are estimated to live in the two countries, watching all or almost all British or German birds is a task that is possible to accomplish and yet not as easy to accomplish as spotting the few dozen species of mammals that are part of the countries' wildlife. It is in fact this somewhere-in-between status of the diversity and distribution of birds that fosters an equilibrium between action opportunities (challenges) and action capabilities (skills) and allows for the game of bird watching to stay in flow, which makes the birds so fascinating.

The ability of birds to fly contributes to the attraction of the game. Birds can potentially be observed everywhere, yet nowhere are they encountered with absolute certainty. Birds are the dice of nature that roll over the game board of field and forest. In addition to population diversity and distribution, weather adds a considerable amount of unpredictability. More than anything else, bird migration is an inexhaustible source of fascination. The question of which species will be seen in what location at what point in time creates the possibility of endless variation. Some species are regular visitors to some locations, others can be present only under favorable weather conditions, and yet others come as a complete surprise. For passionate bird watchers, autumn migration is a particularly exciting time of year. Websites and news services provide birders with relevant weather updates and the highlights of each month are reported in bird-watching magazines and ornithological journals.

Neither clear-cut cognitive categories nor the symbolic representation of social relations account for the fascination of wild birds. It is the twofold ambiguity of birds as "not too few and not too many" and being "neither here nor there" that produces equilibrium and uncertainty and constitutes bird watching as a game. The disproportional appeal of birds as compared to other forms of nature is based on their diversity and distribution in time and space, that is, the valuation of wild birds derives from their particularly suitable status as toys on the playground of nature. As stated above, birds are invested with meaning not because they are good to think with but because they are good to play with.

While birds as such are not a major topic of inquiry in sociology, the general insight that moral valuations can emerge from the activity of play has a long tradition within the discipline, albeit one that was temporarily superseded by Parsons's structural functionalism. The insight was central to W. I. Thomas's notion of the desire for new experiences, one of the four basic psychological desires or wishes that he listed alongside the desire for security, the desire for response, and the desire for recognition (Thomas 1923; Thomas and Znaniecki 1918–1920). In marked contrast to the desire for security, the desire for new experiences strives for uncertainty—that is, insecurity of outcome. It is precisely this insecurity or uncertainty that makes some activities appealing. The motivational force of these actions as described by Thomas is identical to Dewey's account of the motivating force of play, and it seems in fact to predate the latter's formulations on the topic. Thomas initially referred to the desire for new experiences as the gaming instinct (Thomas 1901). The gaming instinct is described as part of human nature and marked by a striving for contest and unpredictable situations. Hunting is the quintessential expressions of the gaming instinct and once described a way of life. Only after man's livelihood was provided by agriculture and regular occupations did making a living turn into labor and become experienced as drudgery. Yet even in contemporary society most people, according to Thomas, prefer those occupations that retain an element of unpredictability: "Some modern occupations are not irksome, and not all are irksome to the same degree; and an examination of them from this standpoint discloses a preference for those in which the element of uncertainty is pronounced, in which the problematical is present, or where, at least, the attention is intermittent" (Thomas 1901, 758). Industrial occupations are by contrast described as more or less regular, monotonous, and recurrent, with the same situation coming up again and again. Industrial labor is thus the opposite of play and the gaining of new experiences.

Deviant behavior, according to Thomas, is in most instances the direct outcome of the effort to escape such drudgery. Burglary, gambling, alcohol and drug use, and so on, have in common that they are driven by the play element of behavior and provide new experiences. The difference between these deviant activities and regular occupations that provide an element of new experience is not to be found in the motivation of the actors who engage in them but in the fact that society regards the one highly and the other not. "The gambler is distinguished by no particular psychic marks from other members of society. The gaming instinct is born in all normal persons, . . . but society justly condemns the exercise of the instinct aside from activities which create values" (Thomas 1901, 760–61).

This insight derived from the analysis of the desire for new experiences—that is, the gaming instinct—is a central element of what later would turn into labeling theory and what already represented a central analytical element in some of the Chicago ethnographies of the 1920s and 1930s. It figures prominently, for instance, in Nels Anderson's *The Hobo* (1923), Frederic Thrasher's *The Gang* (1927), and Paul Cressey's *The Taxi-Dance Hall* (1932). Following Thomas, these three studies show that the deviant behavior of those at the bottom of society, such as homeless men, gang members, and dime-a-dance girls, was driven by the same motivation for gaining new experiences as the behavior of those at the very top of society, whose motives are usually not questioned. A recurrent example of the discrepancy in the social evaluation of otherwise identical activities in Thomas's work and in that of other Chicago scholars is the comparison of the vagabond and the scientific field researcher. Arguing for the "psychological identity of a scientific quest with the pursuit of game," Thomas pointed out that both are migrant workers who travel to gather new experiences (1923, 10). The difference between the two is that society regards one highly and the other not. This insight that the activity of play can at times be found in what is commonly labeled as work also provides a link to the writings of scholars belonging to later generations of Chicago sociologists (Fine 1995). Michael Burawoy, for instance, analyzes the labor process in a factory as a game that is driven by the desire to escape the drudgery of industrial production (Burawoy 1979).

The link between the gaming or hunting instinct and the practice of scientific data collection—albeit officially labeled as work—is particularly pronounced in the case of ornithology. The process of data collecting in this scientific discipline is identical with hunting, which means that ornithology is a collecting game. The value attributed to birds derives from the practice of collecting. In a collection, objects, people, or actions are valued according to diversity and rarity. Collecting is thus different from the simple amassing of objects. Collecting in this sense is understood as the practice of obtaining, classifying, and storing objects and records of various sorts. All collections have in common that they aim for a full or complete set of collectibles. Within this full set special emphasis is put on those items that are difficult to obtain. Collectibles derive their meaning and value in relation to the notion of the full set of collectibles, whereby the exact specification of what such a full set consists of is derived from the classificatory principles of the collection itself. In a collection, an object or record is not valued because it is rare; it is rare because it is part of a collection. Everything that can be observed, measured, or procured can potentially become a collectible. Every place or space where it can be stored or displayed can in turn become a museum. It is not from the point of view of the experience of everyday life but from the point of view of the museum collection that an object is rare or common. Museum catalogues, ornithological handbooks, and field guides are a part of the rules and resources of the game that determine what qualifies as a collectible or, as it is called in the context of natural-history collecting, a specimen. They do not merely provide that institutional background that determines what is rare and what is common; they also represent the game scores of previous collecting enterprises. These publications list examples of notable specimens taken in the field, reference their current location, and enumerate previous owners.

Initially, ornithological publications were not limited to this kind of information. They frequently also reported on the practical or economic utility of birds, including the taste of their meat and the dishes they can be used for. Utilization in this sense is an instrumental activity. It is part of the world of work, rather than of play. Aspects of rarity and diversity do not matter when it comes to the utilization of birds as food. What counts instead is the taste of the meat, its abundance, and the effort needed to capture and prepare the birds. Those species that can serve as food are valued, while those that do not accordingly have no economic value attached to them. Not diversity and rarity but utility and abundance matter most.

Knowledge about the subtle differences between different species, let alone subspecies, the central focus of attention of the museum collector, is almost completely irrelevant to the utilization of birds in the kitchen. Tellingly, the most staple part of the traditional bird diet in German-speaking lands were the so-called *Krammetsvögel*, that is, properly speaking fieldfares (Wacholderdrosseln), although other thrushes were usually subsumed under this title as well. What counted was the similarity of the meat that could be derived from these species, not differences in feathers, song, or behavior in the field. The use of a composite label for several species reflected no lack of knowledge about birds but instead indicated the practical purpose to which this knowledge was put. It was an accurate classification for the use of these birds as food. The knowledge about birds produced by natural historians thus varied with the practice of which it was a part: while collecting produced detailed knowledge about the number of species and their geographical distribution, utilizing was prone to provide knowledge about the abundance of different species and their seasonal fluctuation.

WORK AND INDUSTRY: THE VALUATION OF BIRDS AS TOOLS

Bird conservation in Germany developed in the context of economic practices and institutions in agriculture and forestry. With the large-scale rationalization (i.e., industrialization) of both agriculture and forestry in the late eighteenth and early nineteenth century, habitats suitable for birds decreased on such a large scale that the negative effects of a diminished birdlife became apparent: it became increasingly recognized that songbirds played a vital role in the destruction of so-called insect pests. A century later, birds of prey, which were initially excluded from protection, became recognized as useful for the reduction of other vermin such as rats and mice. The creation of the concept of *ecology*—the nineteenth-century neologism for the notion of the household or economy of nature—reflected this. The transformation of the landscape changed the economic value attributed to birds. They turned from units of consumption into units of production. What caused this change was not the emergence of a new goal or ethical principle but an alteration in the calculus of costs and benefits. Protecting birds began to pay more than eating them.

Not all birds in the economy of nature, however, were perceived as useful. The benefits attributed to different species depended entirely on their feeding patterns and specifically whether they fed on harmful—considered from a human point of view—insects or on useful fruits and crops. Conservation efforts along this line were referred to as practical bird conservation (*praktischer Vogelschutz*) or applied ornithology (*angwandte Vogelkunde*). Which species devoured what kind of food was a question hotly debated within the emerging academic discipline of ornithology. Birds feeding exclusively on insects, such as blue tits, were valued by bird conservationists not for their beauty or rarity but simply for the fact that they helped to clear field and forest of dangerous insect pests. Sparrows, on the other hand, which as seed feeders were notorious for eating the grain of the fields, served as the example par excellence of harmful birds.

Birds were literally instrumentalized. They became means to an end (*Mittel zum Zweck*) in agriculture and forestry. Such separation between means and ends did not come naturally; it had to be institutionally reinforced and implemented in everyday practices. Bird conservationists ensured that people's playful urges did not get the best of them. Strict schedules and guidelines had to be observed when it came to putting up nest boxes and feeding birds, the two most common practices employed for their protection. These guidelines were part not of the world of play, as in the case of bird collecting, but of the world of work.

The rules and resources that constitute games can been described as institutions that foster the activity of play. Yet nothing intrinsic to rules and resources implies gameplay. They can in fact be arranged in ways that are highly detrimental to play. The prototypical example of an institutional arrangement that fosters play is the playground, and the comparable institutional setting that fosters work is the shop floor. While the former has been an object of social-scientific inquiry only marginally, the same can hardly be said about the latter. Social scientists have produced detailed analyses of the consequences of the system of industrial production for human labor. From the specific theoretical angle of this study, the system of scientific management set up by Frederic W. Taylor and the assembly line pioneered by the Ford Motor Company are the most clear-cut empirical examples. This form of industrialized mass production has been described as resulting in the deskilling of the worker (Braverman 1974) and the technical control of the work process (Edwards 1979). The separation of execution and conception under scientific management deprives the worker of any control over the production process, while the assembly line further dictates the rhythm of work and degrades the worker to a tool of the machine. Industrialized work in this form can be described as a systematic antiplay institution. The form of action that takes place on the shop floor is fixed in every single detail and entirely dictated from the outside; it is the very opposite of the self-renewing and self-referential logic of play.

Contemporary accounts of work on the shop floor are to a large extent informed by Karl Marx's theory of alienated labor, put forward in his early writings. In the view of the early Marx, who envisioned the world as entering a stage of history in which one could "hunt in the morning, fish in the afternoon, rear cattle in the evening, [and] criticize after dinner" (Marx and Engels [1845] 1998, 53), labor is a creative activity. Unalienated labor is identical to play, although what this unalienated labor really consists of is not fully elaborated and must be deduced from the negative—that is, from the four forms of alienated labor that Marx describes (Wolff 2011). It is through the productive (i.e., creative) activity of play as unalienated labor that man realizes his own species being. In the capitalist system of production, on the other hand, workers face the products of their labor as alien objects. The worker's labor is turned into a mere means for survival, while the bourgeoisie fetishizes commodities as absolute ends. In the capitalist economy both workers and the bourgeoisie are described as "a plaything of alien forces," albeit alien forces that are themselves a product of human action.

A clear conceptual account of play as a creative activity is one of the strengths of Dewey and the sociologists of the Chicago school. While this school had rather little to say about the organization of labor and the working-class movement, they had by contrast very much to say about the organization of leisure. Following Dewey's definition of play, Clarence Rainwater, for instance, who taught classes on recreation administration at the university, addressed these organized efforts as the playground and recreation movement, or as the play movement for short (Rainwater 1921). Marxist sociology and Chicago sociology supplement each other and show that both labor and leisure—work and play—can be addressed by organized social-movement action and result in permanent forms of institutionalization.

In the case of bird conservation, this process of institutionalization turned nature into a playground in Britain and into a shop floor—that is, a factory for the production of food—in Germany. Conservationists in Germany saw birds as pure instruments or, to use Dewey's terminology, as means-in-themselves. Once birds were reduced to absolute means, morality as the realm of absolute ends appeared on the horizon. The purely economic concern for conservation was paralleled by an equally pure moral concern for the protection of nature in its integrity. From the latter point of view birds were postulated to be ends-in-themselves (*Selbstzwecke*) and their protection a moral duty. Those who argued that nature should be managed to the best economic benefit were thus confronted by those who argued that it should not be interfered with at all. What took place in Germany was a complete separation between these two forms of discourse, one purely economic and one purely moral.

The fact that the emphasis on purely moral arguments was a reaction to a purely economic treatment of birds is highlighted by the absence of a comparable discourse from the British case. One of the crucial insights of this study is that actors in Britain and Germany used different accounts to argue for conservation. These accounts were not rooted in academic writings. Literally all actors I spoke to claimed not to have been influenced by academic scholarship on questions of ethics and morality, and they accordingly referenced no such writings in their own publications. To this extent it makes sense to address their accounts as folk philosophy, and folk philosophies consistently differed between Great Britain and Germany for the entire period studied. In Britain, the distinction between economic necessity (instrumental value) and moral duty (intrinsic value) was largely insignificant. British conservationists consistently did not think about conservation along lines easily decipherable as moral. They did not meaningfully distinguish mundane and moral principles of judgment: the enjoyment of birds as collectibles in bird watching was in fact the leading argument advanced for conservation. Birds were named in the same breath as objects of cultural preservation such as classical paintings or cathedrals. The enjoyment of birds was the main argument for their conservation, and while other reasons were also provided no effort was made to create a hierarchy among them; the more reasons for conservation one could give, the better. British conservationists did not evoke morality as a category. Although the deeply ingrained commitments that conservationists described are commonly referred to as moral feelings, using morality as a label to describe these feelings did not matter to them.

German conservationists produced narratives like those of their British counterparts when describing how they had become conservationists. It was only when asked why conservation should matter to other people that their accounts diverged. Enjoyment was quite distinctly not an argument for conservation put forward by actors in Germany; nor did they advance lengthy and unsystematic lists of arguments for conservation. Instead the idea that one must be able to formulate a single clear-cut principle for conservation prevailed, whether that principle turned out to be economic or ethical.

The way that wild birds are experienced in everyday practices—and the form in which this experience is institutionalized—is thus echoed in the arguments that actors give for conservation. Wild birds in Britain are part of a logic of play and games in which means and ends (instrumental and intrinsic values) constantly reinforce each other and thus are difficult to distinguish. In Germany effort is put into keeping means and ends separate, and morality therefore emerges as a distinct category. It is accordingly only in the German case that arguments for conservation are presented in a form that allows them to be referred to as moral arguments.

This observation has certain affinities with the hostile-worlds argument outlined by Viviana Zelizer (2005), which posits the existence of two distinct arenas for economic activity and emotional ties, with inevitable contamination and disorder resulting when the two spheres come into contact with each other. This argument can be found not only in popular discourse but also in the social sciences—for instance, in the form of Jürgen Habermas's notion of the colonization of the lifeworld (Habermas 1984–1987). While similar at first sight, there is a significant difference between the hostile-worlds argument and the case at hand. The conflict between ethical and economic arguments for bird conservation did not result from an infiltration or colonization by the market of a previously autonomous sphere of authentic moral values; instead it was the other way around: the formulation of a moral argument for conservation first developed in reaction to an already established economic valuation of birds. The causal arrow points in the opposite direction: the economic valuation came first, and the moral valuation developed later as a logical inversion of it.

It was the one-sided treatment of birds in Germany as tools or instruments within agriculture, forestry, and gardening that gave rise to the formulation of purely moral argument for their conservation and the effort to make certain species and their habitats exempt from economic use. Conservationists in this country failed to find a noneconomic language for their noneconomic feelings. In rejecting economic arguments for conservation, they reinforced the dualism between instrumental and intrinsic values. Lost to this dualism of economic necessities and moral duties is the sense of enjoyment that one can derive from birds as parts of a game. Instead, time and again German conservationists, both past and present, have pointed out that field ornithology is no child's play (*kein Kinderspiel*) and bird conservation not a game (*keine Spielerei*). The language of play and games that is so central to the entire concern in Britain was initially either absent in Germany or outright rejected. The centrality of a purely moral and a purely economic vocabulary in Germany and its absence in Britain reflects that in the latter country birds are part of the world of play and games, while in the former they are part of the world of work and industry.

THE EMERGENCE OF BIRD CONSERVATION

The difference between work and play as outlined by Dewey also helps us to understand the emergence of organized bird conservation. The most startling point about this development is its continuity. Those people who initially killed birds and contributed to their destruction and decline were the first to protect them. This seeming contradiction becomes understandable if one looks at the motivation to engage in conservation not as something that springs from internalized moral ideals or abstract ethical principles but as something that is embedded in concrete practices and larger institutional settings. Ends of action are only one part of a larger constellation or assemblage of practices and institutions. Only when we isolate ends of action, such as moral ideals, from the larger context in which they are embedded are we left with a need to search for the development of new ideas to account for changes in behavior. If we look instead at the entire constellation, we perceive manifold ways in which such change can be accounted for.

In Britain, a transformation of the practice of collecting birds in the field had the most decisive impact. The valuation of bird bodies as collector's items put on display in the museum was gradually replaced by a concern for the conservation of bird species living in nature. Collectors transformed their interest when binoculars and cameras became available for use in the field. While those who collected birds with the gun tried to get hold of their bodies, binocular and camera collectors tried instead to take sight records and pictures of living birds.

The process of change was one of continuity in discontinuity—the practice of collecting with binoculars and cameras was a transformation of the practice of collecting with guns. Early bird photographers were clearly aware of the fact that shooting birds with the gun and shooting them with a camera were in fact the very same practice. There was a rich vocabulary used both in gun and camera collecting that highlighted this continuity of practices. Bird photographers used, for instance, what was called a "gun-camera," that is, a camera fixed on a gunstock. Both the device itself and its name highlight the continuity of practices. The successful bird stalker had to know where to find birds, when to approach them, how to get close without disturbing them, and how long to stand still before an accidentally disturbed bird would eventually return to the nest. As far as practices are concerned, not much had changed in the transition from gun to camera, unless, of course, viewed from the birds' point of view. It was a transformation brought about by minute details of technological improvements, not by the invention of big philosophical ideas. Through the use of cameras and binoculars, bird collectors began to empathize with the well-being of birds through an alignment of the interests of self and other in the practices of collecting. The survival of birds was vital for success in collecting with binoculars and cameras.

Yet it was not the experience of looking through binoculars and carrying cameras into the field that alone made for the change. Both tools had long been available and could with some effort have been employed to the very same ends. What mattered was not the invention of these technologies, as such, but the shifting balance of forces that was achieved by their gradual improvement. The crucial improvements in the field of photography, as far as birds were concerned, took place in the closing decades of the nineteenth century. Faster camera shutters, improved lenses, and more sensitive plates, and, later on, roll films, flash photography, and telephoto lenses made it possible to photograph moving objects. These improvements in technology that were achieved throughout the 1880s and 1890s made it possible for the first time to photograph living birds in the field, rather than only bird bodies in the museum. Binoculars likewise improved in magnification and decreased in weight and, importantly, price. With these improvements it was now possible to carry cameras and binoculars into the field and to produce pictures and sights records of birds in full movement. From the point of view of the practices of field collecting a major shift in the balance of forces was thus attained. It was at that point in time—when to observe and record living birds no longer required exceptional effort and skill—that bird watching and bird photography became a game. Watching birds with binoculars or telescopes was technically possible long before, yet the effort required hampered its development as a game for any but the exceptionally motivated, talented, and wealthy. As long as the equilibrium of forces between skill and challenge described by Csikszentmihalyi could be attained by only a few, players remained scarce. When a critical mass of players was eventually reached in the first half of the twentieth century, bird watching and bird photography became a competitive game, institutionalized by bird-watching clubs and field-ornithological organizations. Once created, these organizations provided new methodological standards for collecting sight records—standards that constituted the rules of the game. The organizations further provided outlets where people could publish their sight records (i.e., compare game scores). Systematic field surveys and monitoring schemes followed, and it was this institutionalization of bird watching as a game that made all the difference. People's motivations had not changed—they were collectors before and they were collectors after. What had changed were the rules and resources of the game.

In Germany a link was forged between field ornithology and bird conservation as well. Field ornithologists in this country did produce data on birds in the field, but they did not count the same things that British bird watchers counted, and they did not do it with the same end in view. In Germany, the aim of observing birds in the field was to facilitate practical bird conservation for economic ends. Technological improvements mattered, too, but these were improvements in how to construct and install nest boxes and bird feeders. Field ornithologists' desire to increase the number of useful birds and decrease the number of harmful ones resulted in the mass production of nest boxes. These tools were means to already established ends. Field surveys were carried out at so-called Bird Protection Stations that focused on monitoring population levels of common species rather than on collecting sight records of rare species. What mattered most was conclusive knowledge about the increase in the numbers of useful bird species and the decline of harmful ones.

By the 1930s, important institutional ties had been forged between organized bird conservation and field ornithology in both Britain and Germany. In the

British case these ties bound conservation even closer to an amateur pastime of sight-record collecting; in Germany they joined conservation to the practice of economic ornithology. Both developments accentuated the differences between the two countries' relationships to bird conservation.

THE TRANSFORMATION OF BIRD CONSERVATION

The developments in the first half of the twentieth century provided the background for the further transformation of bird conservation that took place during the second half. More often than not, the period from the 1960s to 1980s has been identified as a turning point in the development of environmental awareness and Green movement activism. Within the larger historical process, however, these changes look far less seminal. While changes undoubtedly took place, all crucial aspects commonly credited to the "environmental revolution" of the second half of the twentieth century were already fully in place by the 1930s. In order to understand this period, we thus have to look at the entire historical trajectory, not just the more contentious events tied to Green movement activism in the time from the 1960s to 1980s.

What truly changed throughout this period was not a set of abstract attitudes and values concerning nature and the environment but the number of people who watched birds. The major British bird-conservation organization, the Royal Society for the Protection of Bird (RSPB), was not more successful during this period than the major German organization, the Naturschutzbund (NABU) because it was more progressive. It is true that the German organization used the decades after the war to continue the work it had engaged in during the decades before the war. Yet the British organization did just the same. The work of the NABU was focused on the distribution of nest boxes and bird feeders in gardens, fields, and forests, while the activities promoted by the RSPB centered on bird watching on nature reserves and across the wider countryside. The reason that the RSPB membership exploded while the NABU's membership stayed stagnant and even declined was not a change in direction, but the fact that bird watching became increasingly popular. Economic ornithology, on the other hand, lost importance after the war due to the large-scale use of pesticides such as DDT in agriculture. There was no homegrown field-ornithological tradition in Germany that could fill the gap produced by the lost importance of practical bird conservation for economic ends.

The transformation that took place in Germany after the 1960s was in consequence to a large extent an adaptation to the British model, not the beginning of something altogether new. Economic ornithology in Germany, with its focus on population levels of common species, gave way to bird conservation with a focus on rare and endangered species. Field-ornithological surveys to collect the appropriate data were modeled on the British example. This transition in Germany took place in the form of a generational shift that was most visible in the relation between the main bird-conservation groups and their youth organizations. In Britain, the relation between these groups was one of cooperation; in Germany, one of conflict. Accordingly, only the German case fits the model of an environmental revolution, and only imperfectly, since it was in fact not a complete departure from already established forms of conservation but merely a replacement of one established model by another—that is, a replacement of the German model by the British one. Looking at the development over time and in an international context makes this event indeed look much less eventful.

Conservationists in Germany did intentionally break away from their homegrown tradition of economic ornithology; in lieu of practical bird conservation of common species with nest boxes and bird feeders, they adopted natural-historical collecting's traditional focus on endangered birds. Lists of endangered bird species were first published in Germany in the 1970s. Once this transformation got underway, the membership numbers of the German organization began to grow.

The rise in membership numbers was even steeper in Britain. Yet no conflict-laden transformation comparable to that in Germany had taken place here. Instead the framework of already existing practices and institutions accelerated the rise in membership numbers. This is not to say that individual actions of British conservationists had no impact on this development. To the contrary, conservationists developed a set of strategies to nurture the growth of their organization. These strategies took their point of departure from actors' self-reflexive understanding of the experiences that had produced their own attachment to birds—that is, the experience of playing the game of bird watching.

Turning people into bird watchers remains among the main strategies for recruiting new members for bird conservation organizations. These organizations not only organize field-ornithological monitoring schemes but also publish field guides on garden birds and sell bird feeders and nest boxes for use in home gardens. The aim is to engage people in the game of bird watching and to make them fall in love with birds. Conservation organizations do not try to teach people abstract moral lessons; instead they seek to draw potential members into action and thereby induce in them the same kind of experiences that have proven vital to existing members. Their strategies exemplify the pragmatism of Dewey in action, operating on the assumption that moral valuations and emotional attachments do not motivate action, but that it is action itself that generates new valuations and attachments.

This strategic use of practices to produce emotional attachments raises the question of the relation between morality and instrumental action. One of the main debates in the sociology of morality in recent years deals with the question of to what degree morality or moral valuations can be described either as motivating action or as simply providing post hoc justifications for already established lines of action. The question has been addressed in particular in the work of Ann Swidler and Stephen Vaisey (Swidler 1986, 2001a, 2001b; Vaisey 2009). According to Swidler, culture influences action not by providing the ultimate values toward which action is directed but by shaping repertoires of habits, skills, and styles from which people construct strategies of action. Vaisey, by contrast, argues that agency has to be understood as a dual process. The fact that justifications for action are often disconnected from what actors do does not imply that morality is not a motivating force of action. It instead indicates that human cognition and motivation draw on two basic processes—one fast, automatic, and largely unconscious and one slow, deliberate, and conscious. Actors can be motivated by deeply ingrained moral commitments at an unconscious level, even if they fail to provide consistent or well-articulated justifications for their moral choices at a conscious level.

While differences between the two approaches resulted in a critical debate between Swidler (2008) and Vaisey (2008a, 2008b), common ground underlies them. Both approaches attribute weight to morality as both a motivating factor and a tool for justification: in Swidler's case by distinguishing settled and unsettled times, and in Vaisey's case by distinguishing practical and discursive modes of culture. A perspective that gives weight to both aspects of agency is further strengthened by acknowledging actors' ability to engage in strategic action to advance their moral commitments. Actors have a self-reflexive awareness of how their own deeply ingrained moral commitments developed, and they can use this awareness strategically to produce the same commitments in others, or to strengthen their own. Strategic action and moral motivation, in short, are not mutually exclusive.

The actors at the center of the bird conservation organizations who introduce and administer bird watching games as strategic tools can be described as moral entrepreneurs, to use the phrase coined by Howard Becker (1963). A moral entrepreneur is any actor who takes on the responsibility of persuading others to develop or to enforce rules that are consistent with his or her own moral beliefs. Moral entrepreneurs as described by Becker come in two distinct types: rule creators and rule enforcers. The prototypical example of the former is the crusading reformer, while the police are an example of the latter. The image of the moral entrepreneur as something akin to a puritan spoilsport who "operates with an absolute ethic" (Becker 1963, 148) owes much to the empirical case that Becker analyzed (marijuana consumption) when developing his concept. It can, however, be extended to fit other cases that are not about prohibitive rules. While the notion of the entrepreneur points to the world of work, it can also fit the world of play. In the world of work, moral entrepreneurs are akin to business managers; in the world of play, by contrast, they are coaches or game managers. The latter role is a significant one. A game manager is in the position to rewrite or even invent the very rules of the game. With these rules he or she shapes the experiences people make and the emotional attachments they develop. The actors who operate bird conservation and field-ornithological organizations are such game managers. They do not merely play the game of bird watching themselves but also have a major influence on how other people play the game.

The use of this strategy reflects that committed conservationists developed their own attachments to birds by playing the game of bird watching. Their moral motivations stand at the end of a process of action, not at its beginning. This study thus shows that in the case of organized bird conservation there exists no such a thing as the "moral dimension," to borrow an expression from Amitai Etzioni (1988), with morality understood as a self-contained entity that can be distinguished from other self-contained entities. Actors make moral valuations and have deep emotional attachments, but these valuations and attachments do not trickle down from a realm of intrinsic values. They instead develop over time in a relational process. The core insight of this study is that moral concern for the conservation of wild birds developed gradually out of the very practices and institutions of field collecting that had previously made for their destruction and decline. It developed out of a gradual reinterpretation and reappropriation of preexisting social forms to which no moral quality could initially be assigned. The following investigation shows that moral values were not the point of departure for this development. Instead, birds became the best-protected part of nature and the cornerstone of environmental legislation in Europe for the very profane reason that they are good to play with.

CHAPTER 2

Collector's Items and Viable Means

When the *Resolution* returned to England from Captain James Cook's third voyage in the Pacific, the many treasures it brought home included the bones of its captain, destined for burial in British waters. The great explorer was killed in 1779 at Kealakekua Bay on Hawaii Island, failing to live up to religious expectations after he was initially mistaken for the Polynesian god Lono (Sahlins 1981, 1985; Thomas 1991, 1994). The Hawaiians subsequently performed funerary rituals on his body in a fashion somewhat reminiscent of the treatment of European saints in the Middle Ages. His body was disemboweled, baked to facilitate removal of the flesh, and the bones were carefully cleaned for preservation as religious icons. Cook's remains turned out to be the last items on a long list of treasures, novelties, and exotica collected on the three voyages that had astonished the scientific establishment in Great Britain and Europe.

Cook's three voyages were accompanied by an entourage of medical doctors, natural historians, and artists, who were recording and collecting curious items discovered in the new lands, either on their own account or on behalf of wealthy patrons. Cook's journals are filled with references to such objects, and he noted that the artifacts offered by islanders in barter "generally found the best Market with us, such was the prevailing Passion for curiosities, or what appeared new" (Cook [1774] 1961, 531–32). The main collector during the first voyage (1768–71) was the botanist Joseph Banks, who would succeed Sir Isaac Newton as the president of the Royal Society. Banks, who joined the expedition at his own expense, was accompanied by the artists Alexander Buchan and Sydney Parkinson, the botanist Daniel Solander (a student of Carl Linnaeus), a secretary, four servants, and a dog. On the second voyage (1772–75)

Cook was joined by the German-born natural historian and ethnologist Johann Reinhold Forster and his son, Georg, along with the natural historian Anders Sparrman and the artist William Hodges. On the third voyage (1776–79) William Anderson and William Ellis engaged in natural-history collecting, assisted by the artist John Webber.

So pronounced was their curiosity for novel items that the collectors on board were at times mocked by the natives they encountered: "[It] was astonishing to see with what eagerness everyone catched at every thing they saw, it even went so far as to become the ridicule of the Natives by offering pieces of sticks stones and what not to exchange, one waggish Boy took a piece of human excrement on the end of a stick and hild it out to every one of our people he met with" (Cook [1774] 1961, 255).

Yet not all objects of interest could be carried home. Man-made artifacts, such as tools and religious objects, decorations, and precious metals, could easily be stored on a ship, as could natural artifacts that did not decay, such as shells and rocks. Exotic and curious fishes, birds, and mammals, on the other hand, had to be immediately preserved to prevent their decomposition, a task that was hampered by the limited amount of time spent on land and the absence of effective preservatives.

Birds were notoriously difficult to preserve, and their number among the curiosities on board was comparatively small. According to Banks, 500 birds were brought back from the first voyage, along with drawings of another 32. On the second voyage, Georg Forster made 139 drawings of birds, and the species represented are mentioned in his father's publications. On the third voyage William Anderson and William Ellis, assisted by the artist John Webber, yielded 96 drawings, and it has been estimated that about 120 species were found, including several new ones from the Pacific Northwest and Hawaii. The bird specimens were sent to the British Museum in London and to Banks.

The number of birds and other animals taken was limited not only by the lack of preservatives and space but by yet another obstacle, as Cook reported in his journal: "[The] reader will think the Ship must be full of such articles by this time, he will be mistaken, for nothing is more Common than to give away what has been collected at one Island for any thing new at another, even if it is less curious, this together with what is destroyed on board after the owners are tired with looking at them, prevents any considerable increase" (Cook [1774] 1961, 532).

A curiosity is curious only for a limited amount of time. If one shares close quarters with a novel item over an extended period of time it loses it novelty. It is this experience described by Cook that reveals one of the central motivational aspects of natural-history collecting.

COLLECTING, ACCUMULATING, AND UTILIZING

The value attributed to the items obtained on the Cook voyages derived from their status as collectibles. In a collection, objects, people, or actions are valued according to the criteria of diversity and rarity. Collecting understood in this sense is the practice of obtaining, classifying, and storing objects and records of various sorts. The institutional manifestations of this practice range from cardboard boxes, drawers, showcases, and museums to catalogues, checklists, and atlases. The exact features of these practices and institutions depend on the peculiarities of the collections themselves. Relics and memorabilia are collected differently from paintings, and plants differently from paleontological remains. Yet all collections have in common that they aim for a full or complete set of collectibles. Special emphasis is put on items that are difficult to obtain, such as exotic birds from remote locations.

Collecting is different from other modes of procuring objects of nature. I will refer to these other modes as accumulating and utilizing. For each, a different logic of valuation applies. In accumulating, the objective is to obtain large quantities of the same unit without paying any attention to variations among units. Accumulating trophies in a hunting bag, for instance, differs from collecting natural-history specimens, just as making money differs from collecting coins. Utilizing, likewise, does not entail any valuation of diversity or rarity. Money that is spent derives its value from the objects or services gained in exchange: its value is instrumental. Birds utilized as food derive their value from the satisfaction of culinary desires—they too can be instruments.

This chapter discusses the concepts of collecting, accumulating, and utilizing in relation to bird conservation. The practices and institutions of collecting proved to be the most decisive. With the rise of collecting throughout the eighteenth and nineteenth centuries, wild birds that had previously been valued (i.e., killed) as trophies and food were now considered specimens, and diversity and rarity became the standards of valuation.

Collectibles derive their meaning and value in relation to the notion of the full set of collectibles, whereby the exact specification of what such a full set consists of is derived from the classificatory principles of the collection itself. In a collection, an object or record is not valued because it is rare, but it is rare because it is part of a collection. The argument points to the fact that every object in the world is unique in some way, yet not every object is a collectible. It is only in reference to the classificatory principles of a collection that uniqueness turns into rarity—that is, into something that makes it desirable. Rarity is accordingly not a quality intrinsic to an object. With the appropriate classificatory principles in place, everything that can be observed, measured, or procured is potentially a collectible, and every place where it can be stored or displayed can become a museum.

A collection of this kind has an internal logic. The question of which object is most desired at any given moment can be answered only in the context of the objects already obtained and the full range of possible collectibles. It is with reference to the previously obtained objects that an additional object is seen as either redundant or as completing a set, and it is in anticipation of what might remain to be obtained that the commonness and rarity of each object is assessed. The value attributed to objects in a collection can accordingly never be fully reduced to either psychological longings or cultural scripts that predate the establishment of the collection.

The experience of discovery is a crucial element of collecting. Collecting is a creative process in that the value attributed to objects emerges from the process of collecting itself. The practice is more than just the execution of a set of existing values. The excitement and passion for collecting derive from the fact that the practice is open-ended, either because that which is to be collected is not yet fully known or because the success in acquiring it remains uncertain. Collecting in the context of an unknown universe of collectibles—as on the Cook voyages, which led to the discovery of new species—turns the collector into an explorer. This type of collecting is driven by an exploratory spirit, one that seeks discovery and hopes for surprise. The very openness of the outcome is part of the motivating force. It turns collecting into a game and collectors into players.

The practice of amassing large quantities of collectibles changed the way this collecting game was played. During the time of the Cook voyages, the persistent influx of collectibles led to the transformation of the cabinet of curiosities into the museum (*Museums* 1904; Arnold 2006; Asma 2001; Bennett 1995; Elsner and Cardinal 1994; Findlen 1994; MacGregor 2007; Pomian 1990; Shelton 1994; Stresemann 1923, [1951] 1975; Thomas 1991). The crucial difference between the two institutions is the way objects are valued. In a cabinet of curiosities, what is prized, cherished, and adored is that which stands out against the background of everyday experience, such as relics, treasures, antiquities, and exotica; in the case of the museum, it is the collection itself that determines what is common or rare and is valued accordingly.

The museum collection can make even the mundane exceptional. It disembeds and recontextualizes items that are meaningless outside of its context and bestows value on them. The valuation of each item in a collection is based on its abundance and availability relative to the abundance and availability of all other items in the collection. The standard currency of the collection is the collectible or collector's item. Collector's items are those that have value to a collector, even if to no one else. As collectibles, objects and records acquire a meaning and value that they lack from the perspective of everyday life. As far as the process of valuation is concerned, the collector does not carry the collectible from the field into the museum; he carries the museum into the field and creates the collectible. It is from the perspective of the museum collection—its catalogues, checklists, showcases, drawers, and so on—that the collector sees the world outside the museum.

The aspect of exceptionality so crucial to the cabinet of curiosity continued to be a significant aspect of the museum collection. It reappeared in the form of the so-called type specimen (Farber 1976, 1982; Warren 1966–1973). A type specimen is a single specimen or illustration used to describe and name a newly discovered species or subspecies. Type specimens exist even for commonly found species; as such, they have special status within a collection. On board the *Resolution*, some curiosities were thrown overboard once they lost their novelty. In the museum, such objects retained their exceptional status as types, even if they were not the only items of their kind in the collection. Bird specimens collected on the Cook voyages were in many cases the first of their kind to enter a collection. This added to their value, and they turned into central points of reference in systematic ornithology.

BIRDS AS COLLECTIBLES

Ornithology, like every other branch of natural history, developed alongside the invention and improvement of instruments for identifying, gathering, and preserving objects of nature (Allen [1976] 1994; Barber 1980; Haffer 2001b, 2006, 2007). In contrast to assemblages of curiosities, museum ornithology requires the long-term storage of large numbers of specimens. It was, among other factors, the availability of better preservation techniques that resulted in the decline of the cabinet of curiosities and inaugurated the museum era.

Only a very few of the bird specimens collected on the Cook voyages survived for long in British or European collections (Steinheimer 2005, 2006). Recipes for the preservation of mammals, birds, and fish had existed for centuries, but most were ineffective, and many virtuosi had accordingly to witness the decomposition of their new acquisition by insect pests. Birds had for a long time been preserved by mummification. Their organs (not the flesh or skeleton) were removed; wires were used to reconstruct the shape; and the bird was then salted or dried in an oven. This method of preservation is described in detail in Johann Reinhold Forster's manual of taxidermy (Forster 1771). All of the birds of the Cook voyages collected by Forster and his son were prepared in this way. Specimens prepared by mummification were not particularly long-lived and the majority of the specimens from the three voyages are now gone.

Toward the end of Forster's manual, however, a new recipe for a preservative appears, together with detailed instructions on its use:

> [The] inside of the skin may . . . be washed or brushed over with liquor, made of an ounce of Sal Ammoniac, dissolved in a quart of water, in which afterwards two ounces of corrosive sublimate Mercury must be put: or four ounces of Arsenic may be boiled in two quarters . . . : then the whole cavity must be stuffed with oakhum or tow, likewise imbibed with the above liquor, afterwards dried and mixed with a powder of four parts of Tobacco-sand, four parts of pounded black Pepper, one part of burnt Alum, and one part of corrosive Sublimate or Arsenic: Lastly, the whole is to be sewed with a thread dipt in the above liquor, and the skin thus stuffed must be gently dried. (Forster 1771, 35–36)

The use of arsenic soap, invented around 1755 by the French ornithologist Jean-Baptiste Bécoeur, enabled the first real advances in museum ornithology. With its use, the overall lifespan of birds in museums increased rapidly—although that of the taxidermists did not necessarily follow suit, since arsenic was highly poisonous even in the concentration used in the recipe (Browne 1896; Farber 1977; Frost 1987; Moss 2004; Schulze-Hagen et al. 2003).

Before the introduction of arsenic soap, collections had relied on a continued supply of new specimens to replace decaying ones. Acquisition of replacement specimens was particularly difficult for species obtained during long-distance endeavors. For many of the specimens from the Cook voyages that were insufficiently preserved, the only records were the rough sketches, drawings, and final, detailed paintings by the artists on the three voyages.

No fewer than eighty-five bird species have been named based on specimens or drawings of specimens collected on Cook's voyages (Stresemann 1949, 1950, 1953a; Whitehead 1969). The majority of these were first described by John Latham, a natural historian best known for his publication of a multivolume handbook on all known bird species. He had studied anatomy under William Hunter and was a practicing medical doctor at Dartford. His first publication on birds appeared in 1770 in the *Philosophical Transactions* of the Royal Society, of which he became a member in 1775. Latham did his research and writing during his free time, based on his own collection as well as on those of others that were made accessible to him. Joseph Banks was among those who supplied him with specimens and paintings from the Cook voyages.

In 1772 Latham met Ashton Lever, with whom he entered a friendly competition over their collection of specimens. Lever's collection contained a large number of specimens from the Cook voyages and was initially located at his estate near Manchester; it was transferred to Leicester Square in London in 1775 (Leverian Museum 1790). The sixteen rooms that were open to the public, for a fee, exhibited stuffed animals from around the globe, along with many thousands of birds, shells, minerals, weapons, and other artifacts from the South Pacific. An antique room displayed medals and casts. Lever's collection, with its wealth of exceptional items, stood in the tradition of the cabinet of curiosities and was lauded by many fellow naturalists.

Latham was a frequent visitor to Lever's collection. He also had access to the bird collection at the British Museum in London, which was not then notably different from private collections. Created in 1753, the British Museum originated in Sir Hans Sloane's bequest of his collection to the nation (E. Miller 1973; Stearn 1981). The national institution began to grow over time through the acquisition of private collections like that of the institution's founder. Birds from the Cook voyage came to the British Museum from the hands of Banks, who was initially the owner of the largest number of such items. He had not only procured items himself during the first voyage but also bought those collected by the Forsters and others on the two subsequent voyages.

With access to all of these collections as well as his own, Latham became known as the preeminent British ornithologist and the father of Australian ornithology. His most important work was *A General Synopsis of Birds*, published in multiple volumes between 1781 and 1785. Its opening page described Latham's ambitious goal: "The intent of the following sheets is to give, as far as may be, a concise account of all the Birds hitherto known; nothing having been done in this way, as a general work, in the English language, of late years" (Latham 1781–1785, i). The work placed particular emphasis on new species.

When it came to matters of classification, Latham made use of Carl Linnaeus's system for the organization of birds into genus and species, and of the scheme of the British ornithologist John Ray for the division into land and water birds. Latham was one of the early champions of Linnaeus and a founding member of the Linnaean Society in London in 1788. In the preface to his *General Synopsis*, Latham announced that his publication included four times as many bird species as were mentioned in the twelfth edition of Linnaeus's *Systema Naturae*, the last edition prepared by Linnaeus himself: "This

FIGURE 2.1 A perspective view of the grand saloon and gallery at the museum, late Sir Ashton Lever's, Surry end of Black Fryer's Bridge. Engraving by William Skelton after Charles Reuben Ryley, 1790. © Trustees of the British Museum.

we shall be enabled to do, from the numerous collections in Natural History, which have been formed of late years in England, and in which, of course, a multitude of new subjects have been introduced from various parts of the world; but more especially within these few years, from the indefatigable researches of those who have made so great discoveries in the Southern Ocean" (Latham 1781–1785, iv).

Latham was the first English author to give a name to many of the specimens collected on Cook's voyages. Among them was the Society parakeet, whose description was based on two specimens in Lever's collection. The specimens were procured on the island of Ulietea, now called Raiatea, which is part of the Society Islands in the South Pacific Ocean, so named by Captain Cook because they lay contiguous to each other (Medway 1979). The parakeet was about ten inches long, with a dark chocolate-brown head, a dark brownish red rump and upper tail coverts, and dusky yellow ochre breast and abdomen. A painting accompanied the description in the text. Within the handbook the species, of which only two exemplars had ever been obtained by natural historians, stood as one among many new discoveries.

The task of compiling a list of all known birds was not an easy one since the number kept increasing. Latham continued to revise, supplement, and expand his ornithology so that it included everything known, utilizing collections and published works. The *Synopsis* was updated with two supplements in 1787 and 1801, and a Latin *Index Ornithologicus* was published by him in 1790 (Latham 1787, 1790, 1801). Latham's own collection was growing throughout the process: even some of the specimens from the Cook voyages initially in Lever's possession found their way into his hands.

Lever's own collection continued to grow until 1783, when he ran out of money. When the British Museum declined to buy the roughly twenty-eight thousand collectibles, he decided to dispose of them by public lottery. The winning ticket was purchased by James Parkinson, who put the collection on display in London but, like Lever, was unable to profit from it. The collection was sold again at auction in 1806. Latham acquired some items from this sale, but the majority went to the goldsmith William Bullock, who owned a private museum in Sheffield. Bullock moved his collection to London in 1809. It became extremely popular after 1812, when he placed it in his newly erected Egyptian Hall at Piccadilly and enlarged the collection with Roman displays and memorabilia of Napoleon. Among the thirty-two thousand collectibles on display were about three thousand birds, a truly exceptional number for the early nineteenth century (Bullock 1808, 1813, 1819).

There was a crucial difference between Lever's collection and Bullock's, although many of the specimens were identical. The former emphasized the exotic, while the latter carefully labeled and displayed his specimens according to the Linnaean system. Bullock is in fact considered the first systematic collector in England. Yet the collections had one thing in common: Bullock's museum turned out to be unprofitable and was disposed of by auction in 1819 (Altick 1978).

The financial challenges not only of acquiring such collections but also of maintaining them were well recognized at the time. The riches from the Cook voyages that made their way through the hand of Lever, Parkinson, and Bullock were somewhat notorious in this regard. William Swainson commented on the matter in a manual on taxidermy first published in 1840. Although Swainson argued that Lever went bankrupt simply because he amassed more specimens than he could afford, he was quick to point out that financial failure in this field was in general not necessarily the result of mismanagement but a problem inherent to all natural-history endeavors. According to Swainson, there was no economic market for these kind of objects, regardless of their value to their owners.

> We have occasionally used the expression important and valuable; but these terms are only applied to subjects in natural history in a scientific, not in a commercial, sense. . . . [We] warn the commercial naturalist against indulging any hope, that the sale of his collections, even if he explore foreign countries, will at all remunerate him for the trouble, the anxiety, and the expense of their acquisition. If he collects on account of others, it is all very well; but he will be sure of disappointment, if he expects "the public" will give him such prices as will render it worth his taking the risk of remuneration upon himself. . . . Let no one, therefore, think of collecting for *profit*, or it is ten to one he finds himself a heavy loser. (Swainson 1840, 96–97)

Swainson himself was no stranger to the practice of amassing a collection with great personal efforts. He had spent time collecting in Sicily, Malta, Greece, Italy, and Brazil and produced a voluminous published output describing his exploits. His *Zoological Illustrations* (Swainson 1820–1833) established him as the first illustrator and naturalist to use lithography, a relatively cheap means of reproducing pictures that did not require an engraver (Natusch and Swainson 1987).

Illustrated natural-history handbooks became museums in their own right throughout the nineteenth century. The illustrations in these publications were arranged along the same classificatory principles as the actual physical objects displayed in a museum. The identity of museum collections and handbooks is particularly pronounced in those instances in which an illustration was the only record of a species available or the only record at hand for the majority of natural historians. With the decay and dispersal of many of the specimens from the Cook voyages, Latham's work became a crucial point of reference. The *General History of Birds*, published from 1821 to 1828, was intended as the second edition of the *Synopsis*. Its preface contained a comment on the current status of the Cook collectibles: "It will be observed in many cases, that birds, are said to be in various cabinets, not now in existence—but it must be remembered, that at the time of the first penning of the Synopsis, and long after, the Leverian Museum was in full preservation. Many subjects also, referred to in the British Museum, have since fallen into decay; and the very numerous and choice articles then in Mr. Bullock's noble collection are now dispersed. The reader has, therefore, to rely on the author only for the description" (Latham 1821–1828, ix).

Over time, the bird specimens collected on the Cook voyages became as difficult to trace in the museum as they had been to capture in the field. There is no lack of published notes on the topic of their whereabouts, including the remarks in Latham's handbook. These notes represent the art of natural-history writing in its most genuine form. They are historical records on who has taken which specimen at what time and place, cross-referenced with information on all subsequent owners and information on their current location. Yet the bird in the hand was by no means a more secure item than the bird in the bush. Specimens were difficult to obtain, constantly changed location, and almost inevitably decayed after a certain stretch of time.

Handbooks were a central element of museum ornithology. Virtually all collections were accompanied by libraries, and the catalogues of collections were largely identical to ornithological handbooks in that a handbook was nothing more than a descriptive account of species within a collection. The difference was that a handbook, unlike a catalogue, could reference more than one collection and draw on previous publications. Latham's *General Synopsis* and *General History* relied on multiple collections and various printed sources. Birds in the field, museum specimens, and printed records thus formed an interrelated system. Handbooks had the advantage of being reproducible at a somewhat higher rate than the specimens they described: there were, for instance, only two specimens of the Society parakeet in all of Britain and Europe, but many copies of Latham's handbook with an illustration of the bird.

In this system of representation, the ownership of the initial discovery of a species became an ambiguous matter. The credit for naming the parakeet did not go to Latham, who had first described it. Recognition of this kind usually went to the person who first assigned a binomial Latin name to a species. Johann Friedrich Gmelin used Latham's work and that of other authors to compile what he considered to be the thirteenth edition of Linnaeus's *Systema Naturae*. With the publication of Gmelin's two volumes dealing with birds (Gmelin 1788–1789), the Society parakeet was known as *Psittacus ulietanus Gm* (the latter two letters standing for Gmelin). The fact that Gmelin, a medical doctor at Göttingen in Germany who had no particular knowledge of ornithology, became the namesake of the birds described in the *General Synopsis* was not least among the reasons why Latham himself produced a Latin index of his work in 1790.

There was nothing special about the point in time when the Society parakeet was given a Latin name, although the date is referenced as the historical event of its discovery. The process involved a multiplicity of factors, including practices of collecting, techniques for preservation, cognitive scripts of classification, and so on. While the discovery of the species was not a discrete event, it did nevertheless have an impact. Once referenced in a handbook, the Society parakeet became one element in a larger collection of the birds of the world, which transformed the meaning of even the ordinary and well-known species. As part of a mass of records, it helped to establish the museum gaze that turned the birds of everyday experience into something exceptional. The Society parakeet was a curiosity when it was first collected. When it was set in the context of a wide range of collectibles, its own meaning and that of all other collectibles was gradually transformed.

BIRDS AS OBJECTS OF USE

This transformative quality of the museum gaze can be observed in the work of Johann Matthäus Bechstein, who translated Latham's *General Synopsis* into German between 1792 and 1812. The translator made no claims to novelty in the introduction to the first volume: his declared aim was to "deliver Latham complete and unaltered" (Bechstein 1792–1812, 2). Bechstein adopted the bird names already provided by other authors in previous translations of ornithological publications from abroad (in particular in those of Linnaeus); he further indicated at least two different local names for all German species, and formed novel German names for all those birds that did not yet have vernacular names. In the subsequent volumes, however, Bechstein went beyond his role as a mere translator and inserted footnotes with additional information. He added species that had only been described after Latham had published his handbook. Among the sources that he consulted for these additions were a natural-history collection in Darmstadt, an auction catalogue from Hamburg, and the literature available at the university library at Göttingen. He also added descriptions of specimens that he had procured himself in the forests of Thuringia, the region in Germany where he lived.

Bechstein's translation of Latham thus did more than introduce Germans to the birdlife of the South Pacific: it also transformed the meaning of the local birdlife, which suddenly became a point of reference for the most complete handbook of the birds of the world. The handbook turned the forest in Thuringia into a reference collection and thus into a place viewed with the eyes of the natural-history museum. Latham's publication thus provided Bechstein with a transnational perspective, although Bechstein himself had hardly ever travelled outside of his native Thuringia. All his own observations on birds were made near his home (Hildebrandt 1933; Pfauch 1966).

The translation of Latham's work was only the beginning of Bechstein's publishing career. He went on to produce eighty-three volumes on natural-history topics, the most famous of which was the four-volume *Popular Natural History of Germany* (*Gemeinnützige Naturgeschichte Deutschlands*) (Bechstein 1789–1795). Birds were the focus of this handbook as well as of most of his other publications, and he eventually became known as the father of German ornithology. In the introduction Bechstein pointed out that ornithology had from his early days been his favorite subject of natural history. Another highly influential publication, not only in Germany but also abroad, was Bechstein's *Natural History of Caged Birds* (*Naturgeschichte der Stubenvögel*) (Bechstein 1795). This practical manual on bird keeping was first published in 1795 and went through a second and third edition in his lifetime. Three further German editions appeared from 1840 to 1879 (Pfauch and Röder 1991).

Bechstein's handbooks were unlike those produced in the British Isles. In addition to the name, description, and distribution of each species, he provided information on reproductive behavior and feeding habits, instructions on how to hunt and capture each species, and, last but not least, assessments of their utility and harm. A separate chapter elaborated in greater detail the significance of the criteria of utility and harm in regard to birdlife. The criteria were presented as one of the most important aspects of the entire work, as he pointed out in the preface to the abbreviated edition of the *Popular Natural History*: "With this natural history I do not merely aim to advance the much needed factual knowledge, but I also aim to show the value that the objects of nature have for God's entire household on earth, and especially for our own basic necessities, comfort, etc., and for science at large" (Bechstein 1792, xv). The category was not to be found in any British natural-history handbook, whether popular or written for a learned audience, and least of all in Latham's work, which aspired to be a museum catalogue and nothing else. The innovative consideration of the economic benefit of each species set the standard for German natural-history handbooks for the next century.

Bechstein was largely an autodidact in natural history. He had studied theology at Jena but failed to gain a clerical position because he talked too much about botany in his inaugural sermon (Bechstein 1855). Instead he became known not only as the father of German ornithology but also as the founder of scientific forestry (Mey 2003). His popular handbook announced his intention to establish an institute for the education of young foresters and gamekeepers. It was eventually created in 1795, and Bechstein served as its director until the end of his life.

The goal of scientific forestry was to maximize the yield of wood from a forest. This entailed the extermination of all those species considered harmful to the trees. In his *Natural History of Insects Harmful to Forestry* (*Naturgeschichte der schädlichen Waldinsecten*) (Bechstein 1798), Bechstein argued that people should abstain from killing birds that feed on harmful insect pests. The task of the publication was to educate the public on ecological relationships within nature: "These pages are dedicated quite in particular to youth and to foresters and game keepers who are not familiar with natural history. They are the most likely to be able to prevent the devastation caused by the animals [insect pests] introduced here, partly by killing the eggs, larvae, pupae, and butterflies where they see them, partly through acquaintance with the interconnectedness of nature, and partly by refraining from the harmful habit of locating the nests of insectivorous birds and destroying their clutches or catching such birds at the wrong time" (Bechstein 1798, 3–4). At the same time, Bechstein encouraged efforts to increase the number of useful birds in a forest—that is, birds that feed on harmful insects. The accounts of bird species in his *Natural History* paid accordingly close attention to each bird's effect on forestry.

Bechstein's influence on natural-history writing in Germany was immense. The most famed German ornithological handbook of the nineteenth century applied the categories of utility and harm that Bechstein had introduced in his popular natural history: it was Johann Friedrich Naumann's *Natural History of German Birds* (*Naturgeschichte der Vögel Deutschlands*), published in twelve volumes between 1820 and 1844, with one posthumous volume following in 1860 (Naumann and Naumann 1820–1860). Naumann, too, was an autodidact.

A farmer with only a basic education, he worked his family's land and produced his ornithological writings during his leisure time.

Naumann followed in the footsteps of his father, Johann Andreas Naumann, who had published a natural history of the birds of northern Germany between 1795 and 1803 (Naumann 1797–1805). The handbook was not much of a success: it initially sold only seventy-seven out of five hundred printed copies. He also authored a book of observations of the natural world, *The Philosophical Peasant* (*Der Philosophische Bauer*), in 1791: uninformed by any writings on natural history, it was widely criticized (Naumann 1791). His son's work, by contrast, was a best seller that did more to shape German ornithology than any other title in the field. The decades following the publication of his handbook are commonly referred to as the Naumann period of German ornithology (Stresemann [1951] 1975; Thomsen and Stresemann 1957).

Naumann's interest in the study of natural history sprang in part from the excitement of acquiring something new for his collection, as he described in a letter to a friend in 1820.

> This spring I have made four journeys to observe the birds on their nests and have put life and health at risk to reach this goal. Wading through swamp and morass all day, up to the waist in water and mud, inadvertently starting to sweat through these efforts, and then getting soaked through a shower of rain from above is no fun! . . . Yet when I have happily ended a laborious day and feel that my efforts have been rewarded through a good find, then I am more satisfied and happy than a king and feel sorry for those who are not able to look into the face of nature in the same way. (quoted in Thomsen and Stresemann 1957, 97–98)

Naumann was assisted in his collection work by his younger brother, Carl Andreas, who obtained most of the specimens described in the handbook. By his own account, the latter shot 21,594 birds and trapped another 27,044 between 1816 and 1844. Only a small number of bird species, Naumann reported, "can claim never to have been shot or observed by one of us" (Naumann and Naumann 1820–1860, volume 1: vii).

Through these exploits Johann Friedrich Naumann assembled a large bird collection that he stored at his home. Birds were preserved with a tincture of arsenic that was becoming more and more common, as he described in a manual on taxidermy first published in 1815 and revised in 1848 (Naumann 1815; Naumann 1848). In 1819 the Duke of Anhalt-Köthen visited Naumann's collection and gave him permission to shoot birds throughout his principality, and Carl Andreas became the duke's personal huntsman. Chronically short of funds, Johann Friedrich repeatedly offered the bird collection for sale. In 1821 the duke bought it, established a museum in the nearby city of Köthen, and appointed Naumann as its curator. At the time of the sale, the collection comprised 691 specimens of 325 species and an additional 670 eggs of 263 species. Naumann continued to expand the collection while curator.

The main reason for the sale of the collection was financial problems. Naumann carried all costs for printing his works—including privately producing the plates—and sales never covered his expenses. In correspondence with other natural historians of the day Naumann repeatedly pointed to his limited funds as the reason why he did not travel much and largely collected locally. The statement turned out to be true. In 1831 he had to sell his house and land because of financial difficulties.

Despite this financial failure, Naumann's interest in birds was not entirely separate from his occupation as a farmer. His experiences in agriculture found their way into the handbook, particularly in his observations on the utility and harm of bird species. The role of birds as food stood in the foreground, and shooting was not the only way to obtain them. The Naumanns described themselves as passionate bird catchers, who trapped and limed birds in a specially prepared site, the so-called fowling floor (*Vogelherd*), and described the tools and techniques for catching the various species at this site in much detail. The *Natural History of German Birds* pointed out that "the flesh of most species is a tasty and nutritious meal" (Naumann and Naumann 1820–1860, volume 1, 140). The topic was not new: his father had written a separate manual on the practice some decades earlier (Naumann 1789).

In catching birds for food, abundance, not rarity, was prized. At the start of the eighteenth century, Naumann observed,

> all of ornithology and national ornithology in particular was advanced almost exclusively by uneducated hunters and bird catchers, people who connected the pleasure of the chase with its pecuniary benefit. Birds as game were always sought for the table of the gourmet, as they still are, for the greater part. The aim was not to distinguish the different kinds accurately, other than by the condition and taste of their flesh and by their monetary value. If a bird struck them as curious, they sometimes tried, as I heard from my father, to keep it alive insofar as they were capable back then, but they were much more likely to try if it would make for a tasty roast. (Naumann 1854, 150)

The group of species that formed the most frequent and staple part of local diets were the so-called Krammetsvögel. The term usually designated fieldfares, yet other thrushes were subsumed under it as well. The number of birds captured as Krammetsvögel was substantial. In the *Natural History of German Birds* Naumann estimated that the markets in the larger cities were full of them and that about six hundred thousand annually were eaten or sent to the market in East Prussia (Naumann and Naumann 1820–1860, volume 2, 309). The experience of catching large quantities of the same species proved vital to the assessment of the birdlife in the country. In Britain it was the distribution of rare species that mattered most. In Germany it was the abundance of common species. Perceived changes in the overall numbers of the birdlife in the country were duly reported. In 1849 Johann Friedrich Naumann described the declining capture of birds at his fowling floor:

> Wistfully do I remember the years of my youth, when sometimes after a quiet October night all the hedges in which berries grew were alive from the flapping and calls of the arriving thrushes and robins, hundreds of which got caught in the springes. When in late autumn flocks of fieldfares took over the tips of the highest twigs of trees, . . . and from there kept on traveling, chattering happily. Or in spring, hopping in our meadows by the hundreds, when the snow had just melted and they could hope to pull worms out of the molehills. Where have they gone, the immeasurable flocks of these and many other migrating birds? One hardly notices a dozen where previously hundreds showed up. (Naumann 1849, 134)

Without the institution of the fowling floor no such knowledge of population changes in nineteenth century birdlife—however incomplete, unsystematic, and biased—would have been available. The knowledge of and concern for the decline in the abundance of common bird species derived from the practice of catching birds for food.

POPULATION DECLINE AND EXTINCTION

No comparable lament for declining bird populations appears in the pages of early British ornithological handbooks. The most famous publication on British birdlife was William Yarrell's *History of British Birds*, which entered print in serialized form beginning in 1837 and was bound in three volumes in 1843 (Yarrell 1843). Second and third editions followed in 1845 and 1856. Like Naumann's handbook in Germany, Yarrell's work became the gold standard for ornithology in Britain and formed the basis for subsequent handbooks: many were hardly more than updates, revisions, expansions, or simple copies.

The handbooks reflected the state of the art in the two countries as much as they contributed to institutionalizing it.

The information that Naumann provided in his work reflected his experience as a farmer, bird catcher, and hunter, and the listing for each species included information on its utility and harm. Nothing of this kind was to be found in Yarrell's writing. He was the son of a bookseller and newsagent who inherited his father's business and used his leisure time to become one of the best shots of the period. A first paper, "On the Occurrence of some Rare British Birds," was published in 1825. The following year he was among the founding members of the Zoological Society of London.

Each of the three editions of the *History of British Birds* stated in the first sentence of the preface which species had been newly observed in Britain since the main text had gone to press. The third volume provided a total count of all British species. The handbook emphasized diversity and rarity. Yarrell offered scattered comments on the diet of certain of the more common species, yet it seems that he used this information to deflect any allegation that birds could be the enemies of agriculture, forestry, or gardening. In the entry for the house sparrow, for instance, commonly presented as the archetypal example of a harmful species in German handbooks, Yarrell included a rather defensive statement: "Their young are fed for a time with soft fruits, young vegetables, and insects, particularly caterpillars, and so great is the number of these that are consumed by the parent birds, and their successive broods of young, that it is a question whether the benefit thus performed is not a fair equivalent for the grain and seeds required at other seasons of the year" (Yarrell 1843, volume 1, 476). This passage on the economic impact of the species is presented as a mere afterthought. The main parts of the text comprised histories of rare occurrences and information on the distribution of each species. Utility was simply not a matter of much interest to the eminent British natural historian.

The knowledge of rare species produced by handbooks such as Yarrell's put the question of extinction on the mind of natural historians. For one thing, it was the information on rare birds and other wildlife that made it possible to assess if a species that once existed had ceased to do so. What is needed to assess extinction is not only a record of the present but also conclusive knowledge of the past. A species has first to be named and classified in order to become considered extinct at a later point in time. The Society parakeet *Psittacus ulietanus* provides an example. It is one species of which no living representative currently exists in nature. The two specimens collected on the Cook voyages that went into the collections of Joseph Banks and the British Museum are the only examples of the species in existence today. Had they not been collected on the Cook voyages they would be gone with no record left behind.

How many more species became extinct during the same period? We simply do not know because we can make no educated guess as to how many species might have become extinct before they were recorded. The list of 132 bird species that are known to have become extinct since 1600 is accompanied by a list of at least another 32 species for which no museum specimens remain, but which have been described by travelers and natural historians (IUCN 2010; Fuller 2001; Luther 1986). Since there is no conclusive proof that these species ever existed, they do not officially count as extinct. More than 50 percent of the birds that count as extinct were endemic to small islands, such as those visited by Captain Cook. It is thus an open question as to what degree our current knowledge about extinction rates is actual knowledge about extinction rather than about changing interest in natural-history collecting and record keeping.

Collecting was one of the major causes of wildlife extinction in the nineteenth century. Other causes included habitat destruction, competition from or predation by introduced species, pollution, and poisoning. No species was ever lost because of collecting alone, but once numbers had been reduced to a critical level, collectors pushed many species into extinction. Hunting for trophies and food were equally destructive in overall numbers, but these practices tended to stand at the beginning of the road to extinction, not at the end. The impact of collecting on extinction was hence particularly visible to contemporaries (Dixon 1888, 1898; Rothschild 1907).

A widely recognized report on the topic of extinction was produced by the natural historian Hugh Edwin Strickland together with Alexander G. Melville in 1848. Their work on *The Dodo and Its Kindred* was a systematic investigation into all records (historical, pictorial, and real) of the several species that made up this subfamily and provided an assessment of the reasons for its decline and eventual disappearance. They regarded the case of the dodo as "the first clearly attested instance of the extinction of organic species through human agency" (Strickland and Melville 1848, 5), as they put it, and pointed out that additional bird species might experience the same fate in a none too distant future. This insight, however, did not result in a call to action to reverse the trend.

> We cannot see without regret the extinction of the last individual of any race of organic beings, whose progenitors colonized the pre-adamite Earth; but our consolation must be found in the reflection, that Man is destined by his Creator to 'be fruitful and multiply and replenish the Earth and subdue it.' The progress

> of Man in civilization, no less than his numerical increase, continually extends the geographical domain of Art by trenching on the territories of Nature, and hence the Zoologist or Botanist of future ages will have a much narrower field for his researches than that which we enjoy at present. It is, therefore, the duty of the naturalist to preserve to the stores of Science the knowledge of these extinct or expiring organisms, when he is unable to preserve their lives. (Strickland and Melville 1848, 5)

In Strickland and Melville's view extinction was a side effect of the progress of civilization and as such inevitable albeit man-made. There was thus no reason for ornithologists to stop collecting rare species in the field. They saw it as their duty to collect specimens of species near extinction in order to preserve them in museums for study.

How could people who called it their duty to produce and preserve knowledge about rare species be almost entirely indifferent to their extinction? Was it outright hypocrisy? A pragmatist interpretation would suggest that it was not. One simply has to recognize that preservation (in the museum) and conservation (in the field) are two very different things. The exclusive focus on the preservation of bird bodies in the museum and the almost complete indifference to the extinction of living birds in the field highlights the validity of the pragmatist assertion that valuations are embedded in action. Value is an adjectival word, as Dewey put it, that is assigned in an act of valuation (Dewey 1949). It is such an act of valuation that matters, not any intrinsic quality of the entity to which it is attributed. With different lines of action—different practices—come different valuations. The valuation of birds reflected their status in the practice of natural-history collecting. Among the manifold aspects of birds that one could value, preservationists emphasized those aspects that related to existing scientific practices—the measurement and classification of bird bodies. Natural historians studied the anatomy, size, and color of birds, not their behavior, and all that was needed for this study could be preserved in the museum. From their own point of view, natural historians who shot birds did not destroy them, but they contributed to their preservation. Bird collectors were therefore neither hypocritical nor indifferent. The bird bodies in the museum were highly valued, and a substantial effort was spent on preserving them.

Such efforts at preservation were among the reasons why natural-history collecting was on the rise throughout the nineteenth century. One of the most important ornithological collections in Germany could be found at the Zoological Museum in Berlin, established in 1810 together with the University of

Berlin (Damaschun et al. 2010; Jahn 1985; Köstering 2003; Wirzberger 1970). Like the British Museum, the Berlin museum had the declared goal of acquiring specimens of all known species.

One of the natural historians working at the museum was Constantin Wilhelm Lambert Gloger. The son of a landowner, the young Gloger had already contributed to Naumann's *Natural History of German Birds* with information on the birds in his area. His first independent publication, in 1833, dealt with the invertebrate fauna of Silesia (*Schlesiens Wirbelthier-Fauna*) and was based on his own collections as well as on the available literature on the topic (Gloger 1833). He also made extensive use of the bird collection at the Zoological Museum in Berlin, which was administered by his former teacher Martin Hinrich Carl Lichtenstein. Gloger in exchange provided the museum with bird specimens from Silesia. He had ambitions to engage in expeditions abroad but never amassed the necessary financial means, despite support from Lichtenstein and other well-known natural historians.

The focus of Gloger's work remained the birdlife of central Europe, on which he aspired to compile a complete handbook (*Vollständiges Handbuch der Naturgeschichte der Vögel Europa's*). The material was largely copied from Naumann's *Natural History of German Birds*. Gloger's handbook was originally intended to be published in two parts for portability into the field, the first volume describing the birds of Germany and the second, which was never published, describing species in the rest of Europe: "It is designated for the purpose of in-depth lectures at institutes of higher learning, and particularly for the self-instruction of the practical ornithologist studying in the field, especially for educated foresters and gamekeepers seeking to extend their knowledge; for economists and cameralists it also aims to connect the amusing with the useful" (Gloger 1834, xiii).

When Gloger talked about practical use, he was referring exclusively to economic benefits, just as Naumann and other authors before him had done. The category of utility and harm was a key element in his account of the various species. His *Practical Hand- and Guidebook of Natural History* (*Gemeinnütziges Hand- und Hilfsbuch der Naturgeschichte*), published in 1842, made the point even more explicitly: "The utility of the study of natural history is as extensive as it is varied, that is, hard to calculate in its entire scope. Everything we own and everything that we need to sustain our bodies we receive from the hand of nature, to use in our own way according to our experiences" (Gloger 1842, 3).

Gloger's second handbook was, like the first, discontinued after the publication of the initial volume. Throughout the same period he was engaged in yet another project that was never completed or published: *Gloger's Natural*

System of the Animal World (*Gloger's Natürliches System der Thierwelt*), a system of zoological classification based on philosophical principles. In 1842 he was granted a three-year scholarship by the Prussian state for the project and gave up his secure position as a gymnasium teacher in consequence. At the end of the three years, no further financial support was forthcoming, and Gloger had to find another source of income (Glaubrecht and Haffer 2010; Haffer and Hude 2007). He eventually made a name and an income for himself with three treaties on the economic benefits of birdlife, published between 1858 and 1865, in which he attempted to provide guidelines for the increase of species of birds and other animals considered useful to agriculture, forestry, and gardening (Gloger 1858a, 1858b, 1865). In the preface to his *Brief Admonition Regarding the Protection of Useful Animals* (*Kleine Ermahnung zum Schutze nützlicher Thiere*), he observed: "There is only one sure way that will lead to a gradual but lasting prevention of the damage caused by vermin and mice: this is because there is only one way that is true to nature. In each and every case this can only be the complete and thorough protection of those animals that live on vermin and mice. . . . This goal can only be attained by means of their recovery, which is getting more important from year to year" (Gloger 1858b, iv).

This book, as well as the two subsequent publications, established Gloger as the leading promoter of practical bird conservation. He had previously published on beekeeping, meat certification, and the hunting of whales for human consumption, the latter being a concern that is currently not considered a mark of distinction for the aspiring conservationist (Gloger 1847, 1850, 1855b). His view on nature conservation was thus not about minimizing human impact. Quite the contrary, he saw conservation as making the best use of nature for human economic benefit. What was novel about his work was not the assignment of an economic purpose to birds and other animals (in which he followed virtually every earlier German ornithologist) but rather his treatment of birds as units of production rather than as units of consumption. Gloger was less interested in birds as food than in what the birds themselves fed on. This change in the column of natural-history bookkeeping marked the beginning of systematic bird conservation in Germany. The argument for conservation was a purely instrumental one and did not have any moral quality attached to it. No change in values was needed to bring the transformation about.

Bird conservation as promoted by Gloger did not emerge from an environmental ethic that stood in opposition to the approaching industrial age; instead it was the very application of the rules of industrial efficiency to nature that motivated this form of bird conservation. Nature conservation as a form of resource conservation is what Samuel Hays addressed as part of the industrial "gospel of efficiency" (Hays 1959). In the vocabulary of Dewey's theory of valuation, it meant that birds were part of the world of work, not the world of play. The countryside was turned into a factory and birds were used as tools, rather than as toys. The process of turning nature into a factory was not a metaphorical achievement, but required a real transformation of the countryside and of people's practices. It involved educating them on the economic value of birds and keeping them from catching birds for food. The practice of putting up nest boxes, as suggested by Gloger, came to be regarded as of national economic importance. The Prussian state commissioned 2,500 copies of his treatise for distribution among educational institutions.

PROFESSIONAL ORNITHOLOGY AND BIRD CONSERVATION

One of the most influential outlets for Gloger's systematic works on ornithology and for his practical considerations of bird conservation was the *Journal of Ornithology* (*Journal für Ornithologie*), established in 1853 by Jean Cabanis, the curator of the bird collection at the Natural History Museum in Berlin. The journal covered all fields of ornithology, which explicitly included bird keeping and bird conservation. Cabanis chose Gloger as the assistant secretary of the journal, and the position turned him into its most frequent contributor. From the first issue in 1853 until his death in 1863, Gloger published no fewer than 104 short articles and notes in the journal, more than any other author during this time. Among these articles were two calls by Gloger to put up nest boxes in the early 1850s, well before he produced his later, famous monographs on the topic. The articles gave appropriate dimensions for boxes and the address of a carpenter in Berlin who, for a small fee, would send model nest boxes for copying (Gloger 1853, 1855a).

In 1854 the *Journal of Ornithology* became the official publication of the German Ornithologists' Society (Deutsche Ornithologen-Gesellschaft). The organization had grown out of a section of the Society of German Natural Scientists and Physicians (Gesellschaft deutscher Naturforscher und Ärzte), formed in 1845. The independent ornithological organization was founded in 1851 in Berlin and headed by a six-member executive committee, with Johann Friedrich Naumann as president. By 1858 the society comprised almost 230 members (Bezzel 1984, 1988a, 1988b, 2000; Prinzinger 2001; Schalow 1901).

The stated aim of the German Ornithologists' Society was, like that of its journal, to cover all topics dealing with birds, including bird keeping and bird conservation. In this sense the organization was not much different from amateur bird fancier societies (*Lieberhabervereine*), which devoted themselves to practical issues of bird keeping. The German Ornithologists' Society was more professional only in the sense that it was devoted entirely to birds, excluding all other branches of zoology and the sciences. The majority of its members pursued ornithology as a pastime, and neither university employment nor a university degree was required for membership. Eligibility for election to the governing committee was open to all "personalities . . . who have achieved extraordinary merits as writers, travelers, collectors, or breeders" (Gründungs-Ausschuss 1867, 423).

The German Ornithologists' Society served as the explicit model for a British organization and journal. The British Ornithologists' Union (BOU) developed out of a series of meetings at the home of Alfred Newton, who held a chair in zoology at the University of Cambridge. At one of these meetings it was established that "an 'Ornithologists' Union' of twenty Members should be formed, with the principal object of establishing a new Journal entirely devoted to Birds" (Sclater 1909, 21). The rule restricting the number of members was dropped in 1864, and seven years later the BOU allowed for the nomination of ten honorary members from abroad. Three out of the initial ten—Eduard Baldamus, Jean Cabanis, and Gustav Hartlaub—were German.

The German influence was further evident when the first issue of the BOU's journal, the *Ibis*, was published in 1859. It listed among the reasons for its establishment "an experiment which in a neighbouring country, and among a kindred nation, had succeeded so well" (quoted in Mountfort 1959, 8). Yet beyond systematics—that is, questions of naming and classifying birds—the British and German journals had little in common. The attention paid to bird keeping and bird conservation in the *Journal of Ornithology* was virtually absent from the *Ibis*. The latter was grounded in the tradition of museum ornithology that was chiefly concerned with knowledge about bird bodies. Issues of the relation of birds to agriculture and forestry were likewise not discussed, and none of the leading members of the British Ornithologists' Union were known for achievements in these fields. Had it not been for the occasional travel reports, the work published in the *Ibis* would have been almost identical to museum catalogues.

Things were much different in Germany, where ornithology encompassed not only procuring birds for the museum but also catching birds for the kitchen. Yet by the time the German Ornithologists' Society was founded in 1851, the tradition of catching birds for food was already on the wane. Ferdinand Baron von Droste-Hülshoff, for instance, noted the decline of the practice in a report prepared at the request of the German Ornithologists' Society in 1872:

> Now everything has changed. In the city and in the country one has to search for the old-fashioned bird catcher with Diogenes' lantern. No longer do they stride out of the town gate by the dozen before sunrise, laden with nets and decoys, and if they did they would still not return with rich spoils, as they once did. Instead, every month all kinds of learned gentlemen and convivial regulars gather around a long table and discuss over a glass of beer, and with the proper seriousness, whether to build a house for this or that bird family. . . . Times have changed, they have changed a lot. (Droste-Hülshoff 1872, 37)

At the beginning of the nineteenth century, catching birds for food was still seen as making economic sense for both the individual bird catcher and the national economy. Over time, however, catching birds was regarded as harmful. Ornithologists now argued that bird catchers were driven by an irrational and destructive passion that did harm to the greater economic good of the nation (Barthelmeß 1981; Schmoll 2004).

One of the last outspoken proponents of bird catching was the Reverend Christian Ludwig Brehm, who published his *Complete Book of Bird Catching* (*Der Vollständige Vogelfang*) in 1855 (C. Brehm 1855). Brehm was a prolific author, contributing to many national and foreign journals and producing a number of major works (Haemmerlein 1985). In his *Contributions to Ornithology* (*Beiträge zur Vogelkunde*), published from 1820 to 1822, he minutely described 104 species of German birds, giving details of their plumage, internal anatomy, and habits (Haffer 1996). Other important books included works on European and German birds and guides to collecting and preserving birds (C. Brehm 1823–1824, 1831, 1842). He started collecting when he was a boy, and the last specimen was added to his collection only a few months before he died. His fifteen thousand birds were mostly from Germany, but he accumulated others through exchange and from his son Alfred Edmund, who collected in Egypt and Sudan as well as throughout Europe (A. Brehm 1855; Brehm Gesellschaft 1929; Schmitz 1984).

As the owner of one of the largest private collections, the older Brehm became one of the best-known and most highly regarded ornithologists of the era. Yet by the time he published his life's work, the *Complete Book of Bird Catching*, this reputation was in jeopardy. In the introduction to the book

FIGURE 2.2 Christian Ludwig Brehm among his treasures. Wood engraving by Carl Werner, 1859. © Victoria and Albert Museum, London.

Brehm repeated the familiar argument for catching birds: "Bird catching is profitable. It provides us with a very healthy, nutritious, and tasty dish, which in rural areas has to be esteemed the more, the less one has the opportunity to get fresh meat without great costs. . . . It is for this reason too, that bird catching is of importance to the farmer and at many places not a marginal line of business" (C. Brehm 1855, 1). It was an opinion that was no longer shared among ornithologists. Brehm was surprised to receive many unfriendly letters from fellow ornithologists who considered catching large numbers of birds for food as not just outdated but morally wrong.

Brehm's own son, who would ultimately become the most famous natural historian in the country with his popular *Illustrated Life of Animals* (*Illustriertes Thierleben*), first published between 1864 and 1869, was among those who had nothing good to say about the practice (A. Brehm 1864–1869). In his 1861 publication *The Life of Birds* (*Das Leben der Vögel*), Alfred Edmund Brehm went so far as to argue that bird catching indicates a dubious moral character: "A fowling floor for titmice appears to me not as the workshop of a noble gamekeeper, but rather as a playground [*Tummelplatz*] of bloodthirsty barbarians or merciless and irrational villains. Because titmice, the most useful of all birds in our forests, are too small to produce profitable spoils for the kitchen even in great numbers, they should be vigorously protected by all reasonable people" (A. Brehm 1861, 419).

In his view, nothing could justify the culinary use of birds. Yet his stance on keeping caged birds was completely different. Like some other ornithologists, he went to great lengths to point out that, despite the need to capture wild birds, bird keeping was ultimately beneficial to birdlife, not least because it bred bird lovers (Johler 1997). Alfred Edmund Brehm himself was not only a well-regarded traveling naturalist but also a bird keeper well known for his two-volume manual on the subject, published between 1872 and 1876 under the title *Captive Birds* (*Gefangene Vögel*). In the preface he argued that bird keeping is entirely consistent with bird conservation.

> To the thoughtful person it will be evident that the fancy for caged birds, now more widespread than ever, is something altogether different from a crude intervention into nature. The enthusiast [*Liebhaber*] turns, without conscious effort, into an observer, the observer into a friend and protector of these birds, and the latter finally into a researcher. . . . Protection for the birds! So much and so encompassing it can be—protection against the gluttonous bird catcher, the egg-pilfering boy, the bird of prey, the lack of nesting sites: but protection, or at least peace, also for a passion [*Liebhaberei*], which, if genuine, never harms, but seeks to preach and exercise protection with devotion and understanding" (A. Brehm 1872–1876, vol. 1, 5–7).

Among the species discussed as candidates for keeping in cages were those that had previously dominated the list of food species, such as fieldfares, blue tits, and skylarks. Brehm even named blue tits as among his favorite caged birds. He provided a justification for keeping birds in the *Journal of Ornithology* in an 1874 article on the economic benefits of birdlife: "In my view, catching with traps, nets, and lime, which aims solely at providing a caged bird for the fancier, or hunting for scientific purposes, should be allowed without concerns, because every bird fancier will turn into what every natural historian [*Naturforscher*] already is: a keeper and nurturer, a protector and vocal advocate for the birds. The craving of the stomach is harmful to them, but not the pleasure of eye and ear" (A. Brehm 1874, 37–38).

Bird keeping was nevertheless a contested issue. Bird fanciers had to defend their passion against the charge that it was at least partially responsible for the declining birdlife in the country. Moreover, controversy developed over the legitimacy of bird keeping as a professional scientific pursuit, as distinct from a mere hobby. This disagreement culminated in a personal battle between Alfred Edmund Brehm and Karl Ruß. The latter was the most prolific author on bird keeping in the country, with a special talent for rearing birds from abroad; unfortunately, he was also adept at making enemies among ornithologists at home. A bird keeper first and foremost, Ruß hardly ever ventured into the domain of museum ornithology. He started his career as a writer by contributing articles on housekeeping to women's magazines and only later turned to natural history. He drew no firm distinction between the study of wild and domesticated animals. His observations on animal life were made indoors as much as outdoors, and his writings included a book on homing pigeon and another on chicken (Ruß 1877, 1884).

The most famous of Ruß's many publications was the two-volume *Handbook for Bird Fanciers, Breeders, and Dealers* (*Handbuch für Vogelliebhaber, -Züchter und -Händler*), first published from 1871 to 1873 and translated into English in 1890 (Ruß 1871–1873). Its publication was one reason for his quarrels with the members of the German Ornithologists' Society. Ruß was initially a member of the organization and had several publications in the *Journal of Ornithology* to his name, among them *A Preliminary Report on the Breeding of Exotic Birds*, first presented at the annual meeting of the organization in 1868. Ruß took particular pride in keeping vast amounts of birds in his apartment in Berlin in a room specially prepared for this purpose where they could fly around freely. The overall number of birds in this room was by no means marginal: "The inventory of my bird chamber numbers in total 160 heads, among which 61 species in 1 to 3 pairs each are present" (Ruß 1869, 73). The annual meeting of the German Ornithologists' Society where he delivered the report was followed by a visit to his bird room by the council of the organization (Ruß 1870).

Ruß's relationship with the society began to turn sour soon after the initial recognition of his achievements in bird keeping. He vied with Brehm, who was a prominent member of the organization, over the claim to professional authority while they were both writing their magna opera on bird keeping. When Ruß founded his own journal on bird keeping, *Feathered World* (*Gefiederte Welt*), in 1872, it was seen as a direct challenge to the *Journal of Ornithology*. In subsequent years, Ruß's publications on bird keeping were largely restricted to his own journal, and he gained recognition from professional ornithologists only at international meetings.

In Britain, there was no comparable tradition of bird keeping (Birkhead and Balen 2008). Although publications on caged birds existed, they were not a part of the emerging scientific discipline of ornithology, and they tended to refer readers to experts in other countries. In 1891, for instance, W. T. Greene argued in *Favourite Foreign Birds for Cages and Aviaries* that "aviculture, long since carried to perfection in France and Germany, is yet in its infancy in Great Britain" (Greene 1891, 1). The most influential texts on the topic available in Britain were in fact a number of translations of Bechstein's *Natural History of Caged Birds*. A first English translation was available in 1835, and until 1905 a total of thirty different editions of this work entered the British market (Pfauch and Röder 1991).

The practice of translation indicates the state of ornithology in both countries. While handbooks on the birds of the world were translated from English into German (as in the case of John Latham's work that was translated by Bechstein), works on bird keeping traveled the other way. Around 1900, Ruß's manual on bird keeping was influential in both countries. One of the most prolific British authors on bird keeping, Arthur G. Butler, acknowledged his debt to Ruß in his *Foreign Finches in Captivity* (Butler 1899). Butler further pointed out that the practice of breeding foreign finches had for many years been a major preoccupation of German naturalists, whereas in his own country, scientific aviculture was still in its infancy.

Ruß was engaged in debates not only over bird keeping but also over bird conservation. A flood of articles on the topic appeared from the 1850s onward in the journals of foresters, fruit growers, gardeners, and agriculturalists, along with book-length treatises intended for a wider audience. Every ornithologist and every publisher in the country seemed eager to publish a work on the topic. In an official report commissioned by the German Ornithologists' Society that was published in 1878, Bernard Borggreve commented that "the favorable sales, the publishing trade, which has been made with Gloger's writings, sometimes maybe even without genuine fervor for the cause, has soon prompted the publication of yet some more dozens of variations about the same topic, mostly products of authors, who in regard to their previous publications appeared to have only little legitimacy to do so" (quoted in Ruß 1882, 7). The ensuing dispute concerned not the importance of bird conservation as such but the question of who had the professional expertise to declare which species should be considered useful and which harmful. Ruß's major contribution to the bird conservation literature consisted of a revised edition of Gloger's works. With Bruno Düring, he produced four volumes under the cumulative title *Protect the Birds!* (*Schutz den Vögeln!*) (Ruß and Düring 1876–1881). The fourth volume contained the editors' own contribution to the topic.

Ruß's was also the first author to produce a book-length discussion on the history of bird conservation, titled *On Bird Protection* (*Zum Vogelschutz*) (Ruß 1882). The publication of this work testifies to the fact that the practice of bird conservation was fully established by this time. In his review of the existing literature, Ruß listed several dozen monographs and pamphlets as well as a multitude of articles in influential scientific journals and popular magazines published in the fifty years prior to his review. He observed a bifurcation between purely economic and purely ethical arguments for conservation in this literature. His own position emphasized the ethical point of view.

> If we drive the dispute about this issue to its logical conclusion, we are left with the choice between the extreme materialistic point of view—which looks at birds just as at every other object in the world from the point of view of man's advantage, that is, from the point of view of utility and harm and thus teaches us to ruthlessly take advantage of everything, and to destroy everything that is opposed to us . . . —and the opposite point of view, which may be ridiculed as overzealous if it tells us that we should respect the right to existence of every creature and that we should value everything, even if it is not useful but simply beautiful, graceful, pleasurable to eye, ear, and heart. If a battle between both opinions ensues, I will be pleased to put myself on the side of the latter, and I will not be ashamed to be accused of sentimentality and zeal. (Ruß 1882, 16)

The matter was viewed somewhat differently when it came to the discussion of the individual species. Almost all authors concurred that blue tits were the most useful bird species and house sparrows the most harmful, and they were as vocal about how to reduce the number of sparrows as they were about how to increase the number of blue tits. Ruß was no exception. Despite his passionate defense of the intrinsic value of birds in the preface to his book, the body of the work advanced an economic argument: "All birds are worthy of nurture and protection, except for the small number about whose distinct harmfulness no further dispute can arise . . . ; those, by the way, are subject to the game law, which means that they can be shot at any time by those with a shooting license" (Ruß 1882, 17–18).

When it came to giving practical advice on how to treat birds, a tendency to ignore one's own pronouncement of their intrinsic value was not unique to Ruß. It could be found again and again in the literature on the topic. Economic reasons were not everything for Ruß, but what should one do in the case of conflict between the postulated intrinsic values and the instrumental ones already manifested in economic practices? All he could say was that a purely economic treatment of birds, at a complete disregard of all other considerations, did not match his personal sensibilities. The argument for the existence of intrinsic values had to fill in to capture these sensibilities. The notion of intrinsic values thus derived from a logical inversion (i.e., rejection) of the values embedded in the dominant economic practices for the treatment of birds. It was a mere declaration of faith, a personal pledge of allegiance to birdlife, that was not connected to any practices, neither any already established practices, nor any concrete guidelines for action.

Such free-floating moral ideals are according to Dewey prone to produce apathy—that is, a decoupling of discourse from practice (Dewey 1929, 280). By putting the intrinsic value of birds on a high pedestal, that is, above and beyond every established line of action, Ruß did not ultimately promote but instead undermined their noninstrumental treatment. The consequence was a complete ignorance of moral principles in action. He himself had advised ignoring the statement of the worthiness of protection of all birds when it came to dealing with harmful species. The postulate of an intrinsic value of all birds was of practical relevance only as long as no gun was at hand. More prone to produce practical conservation outcomes were statements that did not produce a stark opposition between intrinsic and instrumental values, as the example of Karl Theodor Liebe will show.

CHAPTER 3

Technology Comes to the Countryside

Organized bird conservation in the German Empire owed its mantra to Karl Theodor Liebe. *Liebe* means "love" in English, and Liebe was the most outspoken proponent of the ethical argument for bird conservation among natural historians in the Wilhelmine era. The way to encourage bird conservation, Liebe argued, was to make people fall in love with birds. He urged his fellow bird conservationists to "first get to know the life of birds thoroughly, if you want to protect them with proper success" (*Lernt erst das Leben der Vögel genau kennen, wenn Ihr sie mit rechtem Erfolge schützen wollt*). This sentence, part of a lecture Liebe delivered at the annual meeting of the German Society for the Protection of Birdlife (Deutsche Verein zum Schutz der Vogelwelt), became a canonically quoted expression in the field of bird conservation. It had the status of what in German is called a *geflügeltes Wort*, literally a "word on wings"—an expression that formulates proverbial wisdom in a poetic style (Marwinski 2004).

Liebe's mantra was a bit clumsy: its much better-known English equivalent is the expression "to know it is to love it," and this is what he had in mind. Concern for bird conservation, he argued, cannot be attained through teaching abstract moral lessons but only through the personal experience of birdlife: "With effusive, sentimental, and commonplace lectures alone we will not attain our goals: we have to seek for an intelligent bird conservation to gradually capture everybody's heart, of both young and old, in the way that everybody gets to know the birds more and more and learns how to enjoy them!" (Liebe 1894, 7).

A clear advantage of Liebe's mantra that seems to have made for its success was the fact that it turned out to be rather ambiguous. It spoke equally well to two groups of people: those who argued that passion for birds derives from knowledge about them and those who argued for bird conservation on rational economic grounds. Both ethical and economic bird conservation could be promoted with Liebe's word on wings, since both emphasized the importance of scientific knowledge. His own numerous writings on the topic mirror this ambiguity. Liebe's more theoretical and philosophical publications argued for the intrinsic value of birds, but, in making specific recommendations, he adopted the economic approach of distinguishing between useful and harmful species (Hennicke 1893, 1895).

A professor at the gymnasium in the city of Gera, in Thuringia, and president of the local natural-history society, Liebe exerted his greatest influence through his practical writings on bird conservation. Most of these were published in the *Ornithological Monthly* (*Ornithologische Monatsschrift*), the journal of the German Society for the Protection of Birdlife. The organization was founded in 1875 in the city of Halle as the Saxonian-Thuringian Society for Bird Study and Bird Protection (Sächisch-Thüringischer Verein für Vogelkunde und Vogelschutz) and was renamed in 1878. By 1883 it counted 1,112 members, predominantly rural dwellers. Liebe was a member of the society from its inception, serving a term as vice president, and was the editor of the journal from 1881 to 1894. The purpose of the organization lay somewhere between professional ornithology and practical bird conservation. Despite the existence of many comparable organizations in the German states, the German Society was the first to make a claim to represent a national rather than a local constituency.

The declared goal of the German Society was to raise interest in bird keeping and with it interest in birds and bird conservation more generally. It aspired to be to bird keeping what the German Ornithologists' Society was to museum ornithology. Its focus on captive birds included not only exotic birds and native songbirds but also poultry. Liebe, who had kept birds in his home since childhood (at times up to two hundred of them), saw this as the easiest way to generate support for bird conservation. The organization published extensively on the topic and gave advice on protecting birds, although it did not actively engage in conservation measures. In its first year the German Society published leaflets such as *Cultural Landscape and Bird Protection* (*Landeskultur und Vogelschutzand*) and *Hunting and Bird Protection* (*Jagd und Vogelschutz*).

The organization's major contribution to bird conservation, besides holding meetings, was the publication of its journal. The *Monthly Bulletin of the German Society for the Protection of Birdlife* (*Monatsschrift des Deutschen

Vereins zum Schutz der Vogelwelt) was founded in 1876 and renamed the *Ornithological Monthly* (*Ornithologische Monatsschrift*) in 1890. The inaugural article was written by Liebe and titled "On Breeding Experiments with Rock Thrushes" (Liebe 1876). The organization's leadership characterized the purpose of the journal as follows: "The Monthly, dear gentlemen, is the soul of the society, without which it would fall apart. It has to be kept in a popular style in order to reach the majority of the members, but still has to penetrate science, to emphasize bird conservation, to nurture bird keeping, to offer something to bird fanciers, and should not neglect poultry breeding (chickens, pigeons, geese)" (Vorstand des Deutschen Vereins zum Schutze der Vogelwelt 1884, 3).

Bird conservation was the undisputed main topic of the journal, with bird keeping a close second. Articles discussed not only the acquisition, habits, and health of captive birds but also, frequently, the legitimacy of the practice. An article by Georg von Gizncki titled "On the Moral Justification of Bird Keeping" in the first issue was followed by many similar articles (Gizncki 1876). Reports on curious observations or exceptional sightings were also popular. A regular section featured brief notes on rare species encountered in the field and curious behavior observed in captive birds. Travelers' reports and records of species shot abroad, on the other hand, were rare: more common were articles on exotic cage birds, both newly captured ones and those successfully raised in captivity for the first time.

The constituency of the German Society was predominantly rural. A statistic on the occupational distribution of the members presented in 1883 revealed that out of the overall 1,112 members 269 were landowners, economists, foresters, or gardeners. Of these four groups the majority had to be considered farmers, as the author of the study pointed out. "The farmers, of course, make for the largest group of the society—their occupation entails the highest exposure to nature" (Thiele 1884, 5). This group even outnumbered the fifty-two natural historians in the sample.

When it came to the causes for the declining birdlife, Liebe, along with the rest of the organization, was emphatic that the decline was due not to persecution and bird catching alone. Referring obliquely to the legal restriction on bird catching that had been introduced in East Thuringia and the surrounding principalities between 1850 and 1863, he declared that the overall trend of decline had not been reversed. From his field excursions in the area he could present quantitative data on the population change in 146 species that he had observed in the middle third of the nineteenth century: "More songbird species are increasing than decreasing; on the other hand, among the other species, which do not count as songbirds and small birds, many more have decreased in abundance than increased. Newly arrived in the area are seven songbird species and one species of duck. Five species have disappeared, i.e., become extinct, none of which belong to the songbirds" (Liebe 1878, 88).

Liebe considered changes in agriculture and forestry the major causes for the decline in birdlife throughout the century. Throughout the years he and other members of the German Society added to this list more recent changes to the landscape such as the installation of railway systems, electricity poles, and lighthouses. From the point of view of conservationists, birds died en masse when encountering these new technologies. They lost habitats due to the space taken by railway tracks, were electrocuted on telegraphic wires, and flew fatally into the windows of lighthouses whose lights attracted them at night. The transformation of the landscape through technology in its various forms was thus perceived as the major cause of species decline.

FORWARD TO NATURE, NOT BACK

Assessments like Liebe's have fostered the impression that nineteenth-century conservationists opposed all technological progress. In this chapter I argue that, although technology was perceived as a threat to birdlife by conservationists in both Britain and Germany, it was also perceived as a safeguard. I investigate the transformation of practices that led to the expansion of bird conservation efforts from the 1870s to the 1910s. Concern for conservation during this period developed not from a desire of the population in the urban industrialized centers to go back to nature but rather from a process of technology coming to people living in the countryside. Those who strove for the protection of birds were not the modern urban middle class but the traditional rural elite. Their concern for birds was not a nostalgia for an imagined golden past but a result of the transformation of the experience of birds inaugurated by the mass production and dissemination of binoculars, cameras, bird tables, and nest boxes.

Technological improvements were the driving force of this development. The observation resonates with the importance attributed to material objects within actor network theory as advanced by Bruno Latour, Michel Callon, and John Law (Callon 1987; Callon 1986; Latour 1987, 1999, 2005; Law 1987, 1992). Actor network theory emphasizes that material objects, just as much as human subjects, can produce causal changes in the world. In order to capture this causal equivalence between objects and subjects, the approach replaces the notion of actor with that of actant. "An actant can literally be anything provided it is granted to be the source of an action" (Latour 1996, 373).

Yet material objects as actants command no intrinsic causal powers. They are instead part of a relational arrangement of forces—a network—hence the name actor network theory.

Bruno Latour in particular has emphasized that technologies such as automatically fastening seat belts, door closers, and security keys have the ability to shape the motives of human actors and to interfere in a given course of action. Echoing Dewey's insight that means, and not just ends, can motivate action, Latour addresses this phenomenon as the morality of the means (Latour 2002). From this perspective it is not always clear where human agency ends and some sort of nonhuman agency begins. Social interaction must be studied by including interactions with nonhuman entities. A network analysis that restricts itself to merely social relationships will accordingly fail to recognize how technology and materiality mediate the character of such relationships. The more appropriate term for the kind of network envisioned by actor network theory is *assemblage*. A network analysis usually predetermines the entities that constitute the nodes and the nature of the ties between them. An assemblage, by contrast, although it shares many characteristics of a network, focuses explicitly on the heterogeneity of the constituent parts of the relational arrangement. The central argument is that such assemblages are hybrids between society and nature. They form what Wiebe Bijker has called a sociotechnical ensemble (Bijker 1995) that emerges over time in a dynamic interaction between society and nature in what Sheila Jasanoff has address as coproduction (Jasanoff 2004).

While actor network theory is fully consistent with pragmatism, it does not exhaust its full potential, since it does not distinguish between different forms of action. In this chapter I show that material objects as means of action produce different moralities depending on whether they are engaged in the practices of work or play. I show that material objects can be engaged either as tools (in the practice of work) or as toys (in the practice of play).

In Germany new technologies were employed as tools. Rational bird conservation, as it was promoted around the turn of the century, aimed for the discovery of ever more effective ways to increase the number of useful birds and decrease the number of harmful ones. Nest boxes and bird feeders were the preferred means to this end. Bird conservation organizations heralded these tools as the solution to the loss of natural habitats. The success of their work was in part measured by the number of nest boxes distributed throughout the empire. Simultaneously, public lectures and field outings aimed to introduce people to the economic importance of birdlife and educate them in the distinction between useful and harmful species.

In Britain, by contrast, new technologies were employed as toys. With the introduction of binoculars and cameras to the field in the second half of the nineteenth century, collectors turned their attention from collecting bird bodies to collecting field notes and photographs of living birds. The valuation of rare birds as collector's items displayed in private showcases and public museums gradually gave way to the concern for the conservation of living birds. The practice of collecting with binoculars and cameras was a transformation of the practice of collecting with guns. The survival of birds was vital for success in this new form of collecting. Through the use of these technologies collectors began to care for the well-being of their objects of inquiry. Concern focused on rare species as collector's items.

The diffusion of these instruments did initially not change the way they were used. When bird tables and nest boxes made their first appearance on the British Isles, they were not used as tools but as toys in the game of bird watching to enhance the diversity of species available for observation in one's garden. Conversely, once introduced in Germany, cameras and binoculars were used as tools and applied to already established ends. The possibility of using them as toys in a collecting game was not taken into account. The way these instruments were used would have a major impact on the development of bird conservation as an organized social movement.

BIRD PHOTOGRAPHY AND NATURAL HISTORY COLLECTING

Organized nature conservation in Britain derived its initial impetus from concern for animal welfare rather than from the study of natural history. The cause célèbre of British bird conservation around 1900 was the slaughter of birds for fashion (Gates 1998). Feathers decorating women's hats had become à la mode in Europe and the United States. The centers of the millinery trade were Paris, London, New York, and to a much lesser extent Berlin. Most popular were the long, plain white tail feathers of egrets, sold under the name of osprey feathers. The colorful feathers of exotic species such as hummingbirds and parrots were likewise in high demand. For the British trade, feathers were imported directly from the Cape Colony and the British East Indies or indirectly through France. The trade is well documented through commercial accounting records. During its peak, from 1891 to 1910, approximately twenty-five million pounds of feathers were imported to Britain (Doughty 1975).

Concern about birds in fashion provided the origin for the most enduring bird conservation organization in the country. The Society for the Protection of Birds (SPB) was founded in 1889, when Mrs. Williamson of Didsbury, Manchester, formed a group of ladies into an antiplumage association. The organization merged with the Fur, Fin and Feather Folk (FFFF), founded in the same year by Mrs. Phillips of Croydon and Mrs. Lemon of Redhill (both active campaigners against cruelty to animals), when it moved to the London area in 1891. The first meetings of the new organization were held at the office of the Royal Society for the Prevention of Cruelty to Animals (RSPCA) in London. Winifred Dallas-Yorke, the Duchess of Portland and a vice president of the RSPCA, became the president of the SPB, a position she would hold for sixty-three years until her death in 1954. Full membership was initially restricted to women, who managed the day-to-day business of the SPB, albeit usually using their husbands' first and last name together with the title Mrs. in their role as representatives of the organization (Clarke 2004; Lemon 1943; Rook 1966; Samstag 1988).

The organization's original purpose was to induce "a considerable number of women, of all ranks and ages, to unite in discouraging the enormous destruction of birdlife exacted by milliners and others for purely decorative purposes" (Society for the Protection of Birds 1891, 7). Members paid dues of two pence and pledged to refrain from wearing the feathers of any birds not killed for food (the ostrich being the only exception) (Lemon 1943).

Prevention of cruelty, rather than the threat of extinction, was at the center of the early campaigns. One of the major efforts of the SPB was to bring before the public eye the suffering the feather trade inflicted, which went largely unseen because most of the birds were caught abroad. The egrets were usually shot during the nesting season, when their feathers were the most fully developed and when they were easy to locate at their nesting sites. The orphaned nestlings starved to death. The underlying moral message geared toward women in their role as homemakers and mothers was difficult to overlook.

The SPB's first leaflet, *Destruction of Ornamental-Plumaged Birds*, written by Mrs. Phillips in 1890, made the point explicit. "Englishwomen, mothers with nurseries at home, wear these decorations even when engaged in the public worship of the Creator of the beautiful and useful life of which they are inciting the continued destruction" (Phillips 1890, 1). Up to the eve of World War I, the organization published slightly more than seventy such pamphlets, all of which could be ordered by the dozen or the hundred for distribution. The early titles included *Osprey, or Egrets or Aigrettes* (Hudson 1891), *Feathered Women* (Hudson 1893), *Shun Aigrettes and Save the Herons* (n.d.), and *The Trade in Bird Feathers* (n.d.).

In addition to educating consumers, the SPB lobbied for legislation to discourage the use of feathers at home and to limit hunting for fashion in the

British colonies overseas. Yet the practice remained legal until the Importation of Plumage (Prohibition) Act of 1921. Thus, for more than three decades, the organization devoted itself to lobbying and campaigning against the feather trade. Well-publicized activities included sandwich-board men parading along Oxford Street in London in 1911 with leaflets and pictures of egrets, both alive and slaughtered for their plumage.

Membership rose steadily during these years. By 1898 the organization claimed twenty thousand ordinary members in 152 branches and moved to its own premises, rented from the Zoological Society in London. In 1903 it founded a magazine, *Bird Notes and News*. Success was soon followed by recognition: the organization was given royal charter in 1904 and henceforth called the Royal Society for the Protection of Birds (RSPB).

Among the other organizations protesting the plumage trade was the Selborne League, founded in 1885 and named in honor of the famous natural historian Gilbert White of Selborne. One year later it merged with the Plumage League and was renamed the Selborne Society for the Preservation of Birds, Plants and Pleasant Places. It was largely a natural history field club that had adopted bird conservation as one of its causes. Although not a national organization, it had 1,712 members and fifteen local branches by 1907. The single issue focus on birds, however, seems to have been among the reasons why the SPB rather than the Selborne Society attained royal charter and managed to prosper in the long run. The latter society declined after the First World War and virtually ceased to exist after the second (Blackmore 1985; Clarke 2004).

Yet at the turn of the century it was the Selborne Society rather than the SPB that could best be described as a nature conservation organization. It was the individual animal that mattered to the early campaigners at the SPB, while nature conservation is concerned with the survival of the species, a concern that is sometimes advanced at the expense of individuals of the same or other species. The organization's initial focus on animal welfare in its early years has been largely forgotten today, even though it was this focus that made for its success.

One of the early members of both the Selborne Society and the SPB who facilitated the subsequent transition from animal welfare to nature conservation was Eliza Elder, who, under her married name of Mrs. George Brightwen, wrote several well-known books about British natural history. Brightwen was a member of the Zoological Society, among other natural history organizations, and the author of some of the early SPB pamphlets. Her most popular publication was *Wild Nature Won by Kindness* (Brightwen 1890). This was followed by *More about Wild Nature* (Brightwen 1892), *Inmates of my House and Garden* (Brightwen 1895), *Rambles with Nature Students* (Brightwen 1899), and other titles in the same genre. Brightwen opposed falconry and wrote pamphlets and essays deploring the use of feathers for millinery. She also rescued animals and lectured in local schools against bird trapping and the destruction of nestlings. Her concern for wild animals came from her observations of animals in the field, and through her writing she tried to forge a unity between humanitarian concerns and an interest in natural history. She believed that the study of natural history would foster a love of all living creatures. This study was to be pursued in a bloodless way—that is, through observing living birds and other wild animals in the field rather than their dead bodies in the museum.

Initially, Brightwen had been a museum collector herself, but this pursuit gave way to an interest in living birds and other animals when she began to use binoculars and cameras. Following its invention in 1839, photography was for many years limited to static objects in a studio. The first photographs of birds published in 1868 were of specimens in a natural history museum (Kennedy 1868). With the invention of faster camera shutters, improved lenses, and more light-sensitive plates in the 1880s, exposures short enough to photograph moving objects became possible. The move to dry plates and roll films, flash photography, and telephoto lenses had all been achieved by the mid-1890s. The improved tools prompted efforts to photograph living birds in nature (Guggisberg 1977). In Britain such efforts gathered momentum with the publication of the first photograph of a bird taken in the field in 1892. The first book entirely illustrated with wildlife photographs was the best-selling *British Birds' Nests* by the brothers Richard Kearton and Cherry Kearton (Kearton 1895).

Photography gained in popularity so quickly that by the beginning of the twentieth century it had become a standard tool of natural historians. In 1902 Granville Sharp observed that

> the great improvements in photographic apparatus . . . have made it possible for those who are fond of watching birds to press into the service a supplementary and interesting method of studying them. Indeed, it looks as if a camera, just as much as a pair of field-glasses, would come to be regarded as a part of a naturalist's equipment. If this is so, it will be a further indication of the increased interest taken in observing and recording the characteristics of the *living* bird, as against the desire to possess its skin or examine its bones. Nor need any one regret the decay of the old 'fowling-piece Theory' of a naturalist's work. (Sharp 1902, 1–2)

The invention that gave the greatest stimulus to amateur photography was the Kodak film camera, introduced by George Eastman in 1888. The Kodak reduced the ten or more operations hitherto necessary to make an exposure to three, reduced the weight and bulk of the equipment in the same proportion, and permitted the taking of one hundred pictures without reloading. The Kodak's greatest appeal lay in Eastman's processing service: the camera could be sent to the factory for the developing and printing of the film, after which it was returned with a hundred mounted prints and loaded with new film. It was captured by Eastman's marketing slogan "You press the button, we do the rest" (Gernsheim and Gernsheim 1969). The Kodak was launched in Britain in 1901 for a price of just five shillings and made photography available to a mass market (Moss 2004). Other manufacturers started to build pocket cameras as well, and by 1905 there were an estimated four million amateur photographers in Great Britain—roughly one camera user in every ten people. This development was accompanied by a rapid growth of photographic societies and camera clubs. Whereas in 1880 there had been only fourteen such societies in Great Britain, their number had risen to 354 in 1910. Most important among those societies concerned with bird photography were the Zoological Photographic Club (1899) and the Nature Photographic Society (1909).

The most famous bird photographer of the time was the already mentioned Richard Kearton (Chislett 1947; Hosking and Lowes 1947; Mitchell 2001). His gifts as a naturalist were discovered by the publisher Sidney Galpin, who employed him as a grouse-shooting guide. In 1883 he took Richard to London to work for Cassell Publishers, first in advertising and eventually as an author on the topic of natural history. Richard's brother, Cherry, followed him to London in 1887.

After procuring a cheap secondhand camera, Richard began to experiment with portraiture and landscape photography. The results gave him the idea of writing *British Birds' Nests*. Richard Bowdler Sharpe, the curator of the bird collection at the British Museum, commented in a review that it "marked a new era" in bird study (Kearton 1926, 111). A guide on wildlife photography, titled *Wild Life at Home: How to Study and Photograph It*, followed a few years later (Kearton 1898). The book emphasized not only the recent technological advances in photography but also the considerable amount of skill, bravery, and endurance needed to take pictures of birds in the field. Displayed were accordingly not only pictures of wildlife but also of the Keartons' own adventures.

Inspired by the success of their publications, the Kearton brothers spent the following decades touring the British Isles to photograph birds and give lantern slide lectures. Richard considered the impact of his photography on bird collecting to be his most important achievement: "[It] may be asked: What has photography done for natural history? In the words of a dead friend:

FIGURE 3.1 Descending an overhanging cliff. Illustration in *Wild Life at Home: How to Study and Photograph It*, by Richard Kearton, 1898.

'It has turned the destroyer into a preserver.' In the olden days the average Englishman used to say to himself: 'It's a fine morning; let us go out and shoot something.' Now he says: 'It's a fine morning; let us go out and photograph something'—very much to the benefit of birds especially" (Kearton 1926, 112).

Binoculars and cameras did not create new valuations of birdlife out of thin air and good intentions. The process of change was embedded in preexisting practices of collecting. As described above, prismatic binoculars and cameras had been around for most of the nineteenth century, the very stretch of time that experienced the rise of bird collecting with guns. What caused the change was a matter of degree, not of substance. It was a series of small improvements, not big inventions, that made all the difference. What changed technologically were exposure times, magnification powers of lenses, fields of vision, weight, and so on, and it was in relation to actual birds in the field—that is, in relation to their behavior, distribution, and size—that these changes made a difference. First, only stationary birds in a museum could be photographed. The next step allowed photography of immovable bird eggs in the nest; later followed almost motionless birds sitting on the eggs, and only in the very end birds in flight. Whether pictures were collected of bird bodies or of living birds was a function of these gradual changes.

Understanding this incremental technological shift provides us with a crucial insight into the motivation for conservation. It is only by making a normative distinction between preserving life and inflicting death and by attributing this distinction to different motivational forces in the actor that we end up with a need to explain the emergence of something altogether new. Yet a dualism between egoism and altruism, eros and thanatos, or ethically good and bad motives, does not apply to this case. Whether equipped with guns or with binoculars and cameras, bird collectors described themselves as bird lovers driven by a passion for natural history collecting. What took place was a redirection of this passion and love from bird bodies to living birds through changes in the tools of collecting; it did not involve a shift from violent and selfish interests to benevolent, selfless passions. The motivating force was the practice of collecting, and this practice stayed the same.

The continuity of practices is reflected in the language employed. Bird photographers talked of "stalking" birds with the camera in order to make a "snapshot" and to subsequently "mount" the exploits on the pages of their photo albums. This language was not merely metaphorical, but an expression and an indicator of a real continuity of the practices employed in the field. The knowledge that was needed to stalk birds with the gun was the very same knowledge that was needed to stalk them with the camera. The successful bird stalker had to know where to find birds, when to approach them, how to get close without disturbing them, and how long to stand still before an accidentally disturbed bird would eventually return to the nest. Knowledge of the tools was considered to be of secondary importance. As far as practices are concerned, not much had changed in the transition from gun to camera, unless, of course, viewed from the birds' point of view.

Binoculars and cameras can be described as actants in a larger assemblage of forces as outlined by actor network theory. The argument should not be mistaken with a position that has often been dubbed technological determinism—that is, the assumption that technological innovations can create new cultural forms. Technological determinism in its Marxist version, for instance, assumes that improvements in technology (i.e., the means of production) shape the relations of production and with them the dominant ideas in society (Marx and Engels [1848] 1988). Technology is part of the base of social life, and culture part of the superstructure that is causally determined by it. Innovations in technology, according to Marx, will ultimately lead to revolutionary ruptures in social relations and the ruling ideas in society. Actor network theory, by contrast, does not entail such a model of technological determinism. Technological objects as actants are always part of a larger assemblage—that is, of a constellation that is a hybrid of both social and material factors. The point made here is accordingly not that technology has the ability to produce radical social and cultural transformations akin to the social revolutions postulated by historical materialism but, to the contrary, that no radical transformation was needed to produce the change from bird preservation (in the museum) to bird conservation (in the field). What the introduction of binoculars and cameras into the field produced was not a revolutionary overthrow of natural history collecting but merely a shift in attention from bird bodies to living birds. The practice of natural history collecting stayed the same.

The transition from the technology of the gun to that of the camera itself was a fine-grained process that lacked any sign of revolutionary rupture. Oliver Pike, for instance, a vice president of the Nature Photographic Society and one of the noted pioneers of bird photography, used what he called a gun camera for his work and recommended the device to others. Pike was a self-taught photographer who began taking pictures in 1890. In the first of the more than twenty books written throughout his career, called *In Bird-Land with Field-Glass and Camera* and published in 1900, Pike recommended the device for the new sport of bird nesting:

> For taking occasional snap-shots of birds, I use what had been termed a gun-camera. This consists of a camera made on the "reflector" principle, fixed on a gun-stock. . . . To any one wishing to follow the new sport, as bird-nesting with a camera has been called, a gun-camera is a very useful acquisition, while it offers all the excitement of stalking a bird with a gun; the difference being that one's powers of woodcraft are taxed to the utmost, as a bird has to be approached much nearer than would be the case with any ordinary gun. (Pike 1900, x–xi)

Pike's invention was the model for the Birdland camera, which sold well among nature photographers. Later in life, Pike recollected how he had switched from the gun to the camera for collecting birds: "When I was a small boy I . . . thought that it was no crime to destroy. Then at the age of thirteen, I got a camera. It changed my whole life. Previously I had stalked the birds and mammals with a gun, but the thought came to me that it would be more fun to stalk them with a camera and obtain records of them and preserve life instead of taking it. Small are the incidents that will alter one's life for good" (Pike 1943, 88).

Arthur Radclyffe Dugmore was another natural historian who began collecting with the gun and was converted to the use of the camera. In his view, photographing birds was better sport than shooting them.

> If we consider the skill required for camera hunting, we must realize that more is needed than when the gun is used; for it is necessary not only to approach nearer to the animal, but, even when near, hours may be spent in trying to secure either a favourable place or a suitable attitude, and during all this time every precaution known to the hunter is called into practice. . . . A twig incautiously broken, the grazing of the camera against a dry branch, or any of the hundred and one accidents that may at times happen to the still-hunter, and where is your photograph? Gone! Whereas had you been using the rifle you might have easily have bagged your game. (Dugmore 1902, 73–74)

The shift in motives in natural history collecting engendered a new concern for living birds. Those who began to collect photographs of living birds developed a valuation of birds that was fundamentally at odds with that of classical natural history collectors who collected bodies. The amateur bird photographer John MacLair Boraston describes an encounter with a natural history field club on Puffin Island, an uninhabited island off the coast of Wales, in 1903, which demonstrates the difference in attitude:

> We had just made ready for photographing, when we were surprised to see a large party of visitors appear on the slope. This was, I learned, a Field Naturalists' Club on tour. . . . This horde of vandals swept the island, and we could mark their progress by the empty and disordered nests. . . . Did the sight of the sea-gull's nest—that one point of rest in a restless life—offer no charm to their eyes? . . . Let them have their Field Club, but, for God's sake, let it not be a club indeed, to beat and bruise and leave its bloody trail wherever they go. (Boraston 1905, 244–46)

It was this feeling for living birds that the RSPB wanted to instill in the population at large. The many pamphlets denouncing "feathers for fashion" were joined by publications explaining the use of binoculars and cameras for collecting, many of which were aimed at children and youth. A pamphlet titled *Appeal to British Boys and Girls*, written around 1900, exhorted children: "Throw away your catapults. . . . Sow the seed of kind deeds, and instead of hands smirched with blood, and ears hardened to the cries of wantonly-inflicted suffering, you shall reap 'the harvest of the quiet eye,' the gladness of heart which springs from 'loving and serving'" (1).

In the early RSPB campaigns that addressed a young audience, it was considered key to educate children in the habit of collecting birds and their eggs with pencil, notebook, and camera from the very beginning. In *Birds and Boys*, written in 1900, Linda Gardiner argued that this form of collecting is the better sport: "Even from a purely selfish point of view, killing animals is not the way to get the greatest enjoyment from them, but that we should obtain far more satisfaction if we were to treat them with kindness, so that they might approach us without fear, and we might have the constant pleasure of watching their winning ways" (Gardiner 1900, 1). Gardiner went on to encourage her audience to visit the countryside by bicycle, carrying field glasses and a camera. Collecting bird bodies was condemned, and photographing them encouraged and supported.

BIRD FEEDING AND AGRICULTURE

While natural history collecting dominated bird conservation in Great Britain, in Germany the decisive issue was the economic utility of birds. The closest German counterpart to the RSPB, the League for Bird Protection (Bund für Vogelschutz) was founded in 1899. The league followed the example of the ornithologist Rudolf Bergner, who had founded a bird conservation organization in Austria that at its peak counted around forty thousand members, many of whom lived in the state of Wuerttemberg, in southwestern Germany (Hanemann and Simon 1987; H. May 2003; Simon and Hanemann 1989).

When the Austrian organization collapsed in 1898, Lina Hähnle, the wife of the industrialist and liberal member of parliament Hans Hähnle, founded (for the Wuerttemberg members of the organization) the Swabian League of Bird Fanciers (Schwäbische Bund der Vogelfreunde) (Heger 1993; Helfer 1941; Wöbse 2003a). Its head offices were the two family residences of the Hähnles, in Stuttgart and in Giengen an der Brenz. The purpose of the organization was, according to its bylaws, to "work in a comprehensive way for the benefit of our useful birds . . . through the establishment of nest sites and winter feeding to maintain and multiply the useful native birds" (quoted in Hanemann and Simon 1987). The first pamphlet published by the organization in 1899 announced, "We should use every means to encourage the increase of the number of songbirds, that is, of all non-harmful birds, because only with nature's assistance can we control the menace of insect pests and the ensuing losses" (Bund für Vogelschutz 1899b, 1).

To the organization, it was self-evident that the primary reason for protecting birds was their economic utility. In its first annual calendar, it presented the costs of bird conservation measures as a long-term investment. "The small sacrifice is amply compensated by better harvests and larger incomes" (Bund für Vogelschutz 1899a, 31).

In its proposed measures, if not in their precise justifications, the league had much in common with the existing German Society for the Protection of Birdlife. The rationale for founding an additional society had more to do with difference in geographical location than with difference in purpose. The German Society had been founded in Saxony, and most of its members came from northeast Germany. Its name *German Society* was something of a misnomer: it also proudly listed members from countries such as Austria, Italy, and Switzerland, and in 1883, the organization had more members in each of these countries than it had, for instance, in Bavaria (Thiele 1884).

The league, by contrast, was based in the southwest. From the beginning, however, Lina Hähnle aspired to turn the league into a national mass-membership organization. Its foundation was clearly seen as a threat to the German Society's primacy. An editorial in the *Ornithological Monthly* for 1899 made the point explicit: "As much as such a fragmentation is to be regretted, it is a truly German approach. Everybody wants to found something new following his own will and exhibiting sovereign disregard for everything that is already established" (Vorstand des Deutschen Vereins zum Schutze der Vogelwelt 1899, 2).

The threat seems to have been real. Until the league's founding, the German Society had published membership lists and reported on the growth of the organization. In 1890 its membership peaked at 1,232. The number began to decrease the following year. Thereafter, membership numbers were harder to come by, and statements about stagnation became more and more frequent. It seems that the organization went into decline as soon as its major competitor entered the scene.

The German Society had never aspired or attempted to become a mass-membership organization: like many other natural history organizations of the time, it was largely a local bird fanciers' club. Elaborate recruitment strategies as the league employed were completely absent from the organization's agenda. It saw its major task as publishing the *Ornithological Monthly* and soliciting articles by well-known ornithologists. The organization's greatest expense, producing the journal, rose to new proportions in 1883, when the journal published four color prints: these four images, along with two black-and-white prints, accounted for roughly one-third of the organization's annual budget (Vogelwelt 1884). This was hardly the kind of investment that was needed to create a mass-membership organization.

Equally noteworthy is the fact that the investment in color printing inspired no sustained interest in the reproduction of photographic images of birds. The first photograph to appear in the *Ornithological Monthly* was a portrait of its president, August Wilhelm Thienemann, in 1884. Pictures of other council members appeared long before any images of birds. The attractions of bird photography as a sport seemed to have been fully lost on the German Society.

Bird photography had a different standing in Germany than in Britain. The very term *wildlife photography* demonstrates the point: there is no clear equivalent to the English term *wildlife* in German. The expression *Wildleben* would be a literal translation but it does not evoke any specific meaning. The English word *nature* certainly has an equivalent in the German word *Natur*, but this concept includes rivers, mountains, and landscapes, not just animals and plants, which are the declared object of wildlife photography. The concept of nature photography is not the same as wildlife photography. Photographs of birds in Germany were part of *Tierphotographie*, that is, "animal photography," and as such it included all animals, domestic and wild. Animal photography itself was part of *Momentphotographie*, meaning literally "moment photography," and being the branch of the art that specialized in taking pictures of moving objects. Here again all moving objects ranging from people, to animals, and physical objects were included. The matter was thus not simply one of translation. There was initially no German equivalent to the British branch of photography called wildlife photography.

The first German photographer to gain true acclaim for his pictures of animals was Ottomar Anschütz, the son of a decorative painter, who had studied photography before taking over his father's business. One of his most noted series of pictures was one of white storks arriving and departing from their rooftop nest. He chose storks because they are not particularly afraid of human beings and can be much more closely approached than other species. Otherwise, wild birds did not feature prominently in Anschütz's photographic oeuvre, which included photographs of pets and domestic livestock and a popular series on captive animals at the zoological garden in Breslau. A prime example of *Momentphotographie*, Anschütz took pictures of virtually everything that moved, such as walking horses, and his photographs of the Prussian army on maneuvers were exhibited by the War Ministry in Berlin. He eventually became the photographic advisor to the emperor and published a manual based on his experience of teaching the new art, titled *Indoor Photography* (*Die Photographie im Hause*) (Anschütz 1901).

The first manual on outdoor photography in Germany was written by Hermann Meerwarth in 1905—or, rather, translated, since the author took Arthur Radclyffe Dugmore's manual as his blueprint and merely adapted the text to continental European conditions. "It does no credit to us Germans that we are still considerably behind in the competition for good pictures produced by the camera. Foreign countries are very much ahead of us in this matter" (Meerwarth 1905, 3). Meerwarth intended to change this state of affairs. Together with Karl Soffel he served as the editor of a multivolume compilation of nature photography. *Portraits of Animal Life* (*Lebensbilder aus der Tierwelt*) was published in six volumes between 1908 and 1912 (Meerwarth and Soffel 1908–1912).

In marked contrast to the focus of British publications on the topic, Meerwarth argued that nature photography was not just a hobby or pastime but a serious contribution to the science of natural history. Meerwarth also hoped that observations made with the camera could contribute to economic ends, first of all by providing information on how to protect crops against their enemies in the animal kingdom.

The closest German counterpart to a British wildlife photographer at the time was Carl Georg Schillings, the son of a well-to-do landowner at Düren, in the Rhineland, who had traveled through British East Africa in 1896 and 1897 as a member of a scientific expedition (Schillings 1905a, 1905b, 1906). Schillings was a wildlife collector who supplemented the gun with the camera. On his return to Germany, he gave a vivid account of the decline of African wildlife. His photographs of threatened animals, mostly taken on an expedition in 1903–4, put "camera hunting" on the map in Germany. The publisher of his highly successful *With Flashlight and Rifle* (*Mit Blitzlicht und Büchse*) claimed to have sold the first eight thousand copies in less than a month.

Schillings called for immediate action to prevent the decline and extinction of African wild animals. "I want to raise my voice," he declared, "in order to involve all who have the power to do so, to save and to protect, what still can be saved." By this he meant "either the eventual preservation of the still existing treasures, or an immediate and intensive collection of exemplars of the single species for our places of popular knowledge, the museums." Schillings reported with pride that he had endowed museums with animals of species that had by then been "crossed out from the book of life" (Schillings 1905b, 10–15).

Schillings was one of the most prominent members of the League for Bird Protection. He became an outspoken and highly successful campaigner against the use of feathers for fashion, giving a talk on the topic at the first German Bird Protection Congress (Deutsche Vogelschutztag) in 1910 and writing a pamphlet illustrated with photographs of nesting birds of paradise killed for their feathers (Wöbse 2004) that was intended for a mass audience: "On all streets and squares in the civilized world one hears the cry of millions of birds descending from ladies' hats: 'You children of humanity, you ladies and girls, and—you gentlemen—protect our vanishing species, which are doomed to extinction through the feather trade'" (Schillings 1911, 15).

The efforts of bird conservationists in Britain and Germany on this issue looked very much alike, and the two leading organizations took note of each other. The league reported favorably on the campaigns in Britain, and the RSPB in turn pointed out that the German emperor and his wife abstained from wearing feathers. A law comparable to the British Plumage Prohibition Act was enacted in German New Guinea in 1911 (Föhr 2005).

To gain popular support, the league set its annual membership fee at the very low amount of fifty *pfennige* (cents) a year. The German Society for the Protection of Birdlife, by comparison, asked for an annual contribution of six mark—twelve times as much. A 1906 league recruitment leaflet, 106,000 of which were printed that year, was aptly titled *To Each and Everyone* (*An Alle und Jeden*). It had many well-known signatories, including the biologist Ernst Haeckel, the originator of the term *ecology*. The call was explicitly addressed to the entire people, or *Volk*. "Truly, those who love their fatherland and their *Volk* cannot be indifferent to the decline of birdlife. . . . Our homeland (*Heimat*) shall be preserved in its beauty and peculiarity, and our *Volk* shall be taught to love and to understand it" (Bund für Vogelschutz 1906, v).

The author of the call was Konrad Guenther, a professor of zoology at the University of Freiburg. In a book simply called *Nature Protection* (*Naturschutz*), Guenther emphasized that the experience of nature is open to everybody. "Here there is no difference between high and low, between poor and rich. Nature is the mother of all human beings in the same way" (Guenther 1910, 13). This reference to Mother Nature and the emotional tone in which it was presented was not an indication of a romantic longing for a golden past. Guenther was also the author of a book titled *Back to Nature?* (*Rückkehr zur Natur?*) that had a clear emphasis on the question mark, since the author strictly rejected any such goals. In his view nature conservation was about scientific progress—that is, about an improvement of the earth—and not about a return to an original state of nature that might have existed before human intervention took place. "Forward to culture, setting goals and attaining them—that is the purpose of the human mind! It is destined to claim dominion over the entire earth, and it will succeed. It will put an end to the horrible struggle for existence and wrest the forging hammer from the process of natural reproduction. Human reason is already guiding the life of mankind and that of many animals and plants" (Guenther 1907, 64).

Guenther looked at humans' relation to nature from an evolutionist point of view. Science was part of evolutionary progress, and it would lead to the improvement of both culture and nature. Contrary to the common wisdom of many contemporary commentators on what late nineteenth- and early twentieth-century nature conservation was all about, "Forward to nature" and not "Back to nature" would have been the appropriate slogan for Guenther's intentions when he wrote his pamphlet for the League for Bird Protection.

The pamphlet campaign was a complete success. In 1900, only a year after its foundation, the league reported having 3,500 members, more than the German Society had acquired in a quarter of a century. In 1909 the organization counted 18,300 members in 315 local groups. The organization's goal in expanding its membership was a dual one: on the one hand, the league was a political lobbying organization that wanted to boast mass support. On the other hand, it wanted to turn its mass membership into knowledgeable bird lovers who would undertake the practical work of bird conservation, such as feeding birds in winter and providing shelter for them during the breeding season. The organization thus lobbied its members as much as it used them for lobbying.

From its head office, which was still the residence of the Hähnle family, the league distributed nest boxes, bird food, and shrubs intended to facilitate bird conservation. Between 1899 and 1913 it claimed to have sent out as many as 75,387 nest boxes (Bund für Vogelschutz 1914). The tools were sold at cost. In fact, the Hähnle family paid a substantial portion of the running costs of the organization from their own pockets, because the token membership fee barely covered the costs of sending out the annual report. It is only a slight exaggeration to say that, in the early years, the membership functioned primarily as a mailing list Lina Hähnle could use to convince people to put up nest boxes and bird tables in their gardens at her family's expense.

In keeping with Karl Theodor Liebe's motto, which advocated getting to know birds in order to protect them, the organization aimed to educate its members about birdlife in order to turn them into bird lovers and conservationists. The goal was to make people "literate in birds" so that they could read the book of nature. In this insight the league did not differ much from the RSPB. The German organization as much as the British one distributed leaflets and other publications to awaken their members' interest in birds. These materials taught members how to identify various species and provided information on their distribution, habits, and food.

Yet not all birds were considered equal. For the league, the conservation of birds did not preclude assessments of their economic utility or harm. This point is illustrated by an ornithological handbook the league published in serial form, based on Johann Friedrich Naumann's famous *Natural History of German Birds*. It appeared in its annual reports from 1906 to 1910 and presented the sparrow as the archetypical enemy, a shameless proletarian with an unpleasant voice that flourished chiefly at farmers' expense. Ornithologists, the handbook argued, did not regard its benefits very highly. Only from May to July, when it was feeding on insects, did any utility derive from it. The same seasonal variation could also be observed for other species. Bird feeding had therefore to be restricted to winter. "Continuous lavish feeding without regard to the weather leads the birds to inactiveness and keeps the insectivorous ones from following their useful, diligent preoccupation" (Bund für Vogelschutz 1905, xi).

This effort to limit or discipline the practice of bird feeding highlights its localization in the world of work, rather than the world of play. According to Dewey, work or instrumental action is a form of action in which means and ends become separated through a lengthening chain of intermediaries. It takes a certain amount of organizational effort to turn action into work and keep actors from lapsing into play. In Germany this effort in disciplining action was accompanied by the attempt to downplay the significance of play. Conservationists were eager to distance themselves from the more playful aspects of the activity of bird feeding and to highlight that it was a serious economic enterprise and not "just" or "simply" a game, as the depreciative adverbs commonly used in this context had it. The involved instruments were accordingly tools, not toys or playthings (*Spielereien*). The enjoyment that can be derived from the activity of play was purposefully rejected as a justification for the pursuit of bird feeding. Instead, either economic or ethical arguments for conservation were presented as the only substantial arguments. If recognized at all, play was delegated to childhood, and described as a mere preparation of the child for adulthood.

Such a view of play is not unique to turn-of-the-century German bird conservationists. It finds reflection in models of agency that identify work as a productive or creative activity, while play and all leisure activities associated with it are identified as unproductive. In the words of Thorstein Veblen, leisure activity is the "non-productive consumption of time" (Veblen [1899] 2001, 34). From a pragmatist theory of action, by contrast, play is a creative activity per excellence. By not recognizing this creative potential of play, German conservationists deprived themselves of the possibility to make use of the joys of bird feeding and watching to attract people to their cause.

There was no equivalent to this neglect of play in Britain, since no comparable economic importance was attributed to birds to begin with and no ethical arguments were accordingly advanced in opposition to economic ones. The RSPB asserted that while some birds were without doubt useful to agriculture, the argument that others were harmful was at best anecdotal guesswork. The few British publications on economic ornithology that existed around the turn of the century tended to make a completely different argument from those in Germany. In *The Farmer and the Birds*, for instance, Edith Carrington argued that feeding on the farmers' grain and garden fruits is part of the bird's wage for services rendered: "It is neither generous nor prudent to reckon too closely with birds . . . after serving us faithfully for the greater part of the year. . . . The labourer is worthy of his hire, and birds, no less than men and women, will exact a wage sufficient to 'keep body and soul together'" (Carrington 1898, xii–xiii).

Although Carrington characterized the blue tit as one of the most useful birds and the sparrow as the most harmful, this assessment did not justify the sparrow's extermination. "Surely human intellect, which invented the steam engine and the electric telegraph, may find some less clumsy means of coping with a Sparrow than shooting him" (Carrington 1898, 158). Every species was considered partly useful and partly harmful, and no distinction between friend and foe was drawn.

This view was echoed in the RSPB's publications. Oliver V. Alpin, a member of the British Ornithologists' Union and author of *The Birds of Oxfordshire*, argued in a pamphlet titled *The Birds as Labourers*, published around 1900, that although birds did some damage to crops all year round, this was not a reason for extermination: "We ought not in times of insect scarcity to grudge a minute fraction of those crops which they have preserved for our use from their insect foes, crops grown on the land they have helped to keep clear and free from weeds. After all, the wages are very small. The work must be done; only the birds can do it. They will never strike; but do not let us try to 'dock' their well-earned wages, or to starve them out in hard times" (Alpin ca. 1900, 3–4).

Linda Gardiner, later the secretary of the RSPB, further argued in 1905, in *Our Ally the Bird*, that all birds are part of the economy of nature and that "the toll they take in man's crops of grain and fruit is only their just wages for destroying noxious weeds and destructive insects" (Gardiner 1905, 1). There seemed to be little desire on part of the British natural historians of birdlife to condemn any bird species tout court. Their arguments read rather like a rejuvenation of the medieval doctrine of just wages.

ECONOMIC ORNITHOLOGY

Systematic, firsthand investigations on the topic were almost completely lacking in Great Britain. The RSPB took this lack of knowledge as grounds for discrediting all calls for the destruction of certain bird species. This position was first challenged by Walter Collinge in *The Food of Some British Wild Birds: A Study in Economic Ornithology*, published in 1913. "The nature of the food of many of our wild birds has hitherto been largely guesswork. . . . It is only in, comparatively, recent years that the economic importance of birds has been realized and their habits scientifically studied" (Collinge 1913, n.p., 9). Collinge, a member of the British Ornithologists' Union and keeper of the Yorkshire museum, conducted postmortem examinations of 3,048 specimens and 312 nestlings as well as numerous observations in the field and took account of variation in feeding habits by habitat and season. His study was the most extensive and systematic on the topic ever initiated in Britain.

The RSPB duly reviewed the publication in its magazine and, faced with the evidence, used its well-worn line of wisdom on the matter, but in reverse. Previously the organization had argued that all evidence in economic ornithology was anecdotal rather than systematic—that is, based on scattered reports of farmers and newspaper clippings. Now it employed newspaper clippings and farmers' reports to contest the validity of Collinge's work (e.g., "Farmer" 1916). The organization was simply not prepared to alter its position that all bird species were to be protected without exception. Collinge, on the other hand, was very much opposed to the indiscriminate protection of all birds, as he spelled out in the revised and expanded second edition of his book: "Unfortunately, in this country, the question of the protection of wild birds is one that at times has fallen into doubtful hands. Actuated, no doubt, by the best intentions and the highest sentiment, bird lovers have clamoured for all-round protection, have refused to listen to reason, and so have stirred up a powerful antagonism that has done a considerable amount of harm" (Collinge 1924–1927, 355).

Collinge would have found more support in Germany, where the importance of birds for agriculture, gardening, and forestry turned ornithology into an applied science. One of its best-known practitioners was Georg Rörig, an entomologist by training. He was employed by the Imperial Biological Institute (Kaiserliche Biologische Anstalt), initially a division of the Imperial Health Office (Kaiserliche Gesundheitsamt) in Berlin-Dahlem, where he conducted experiments in applied ornithology (Aderhold 1906; Jaskolla 2005; Sucker 1998). Initially, Rörig had followed the standard practice of examining dead birds' stomach contents to analyze their feeding habits. In 1910 he reported having dissected the stomachs of no less than 11,846 birds (Rörig 1910, see also 1903b, 1905, 1909). Yet he considered the information yielded by this method as incomplete because birds could digest certain foods so fast as to leave no trace for analysis. Observations in the field were likewise incomplete because of the observer's inability to watch an individual bird around the clock. Rörig's solution was to install large outdoor cages at the office in Berlin-Dahlem. He offered the captive birds different foods and recorded the amount consumed by the various species on a daily basis, with attention paid to hourly and seasonal variations. He also computed the amount of food consumed relative to the size of the bird.

Rörig used the results from blue tits gained in this way to exemplify the practical significance of his work. "Twenty tits, for instance, that is one pair and its offspring, consume annually about half a hundredweight of dry matter, and therefore at least one and a half hundredweight of living insects, their eggs, larvae, or pupae" (Rörig 1903a, 34). This calculation turned into the most frequently cited quantitative evidence to support practical bird conservation in Germany. It was repeated over the decades, although not necessarily with any reference to Rörig or the experimental circumstances under which it had been produced.

Authors affiliated with the League for Bird Protection used data like Rörig's to call for increasing the number of useful blue tits and reducing the number of harmful sparrows. To this end the league explicitly recommended and distributed so-called von Berlepsch nest boxes, named after Baron Hans von Berlepsch, who invented them on his private estate in Seebach in Thuringia. The manufacturer used Rörig's numbers on blue tits in an advertisement to highlight the economic benefit of the product, and the advertising in turn was reproduced in the literature on bird conservation.

In the early years of the twentieth century, Berlepsch was the leading proponent of "rational bird conservation," which he distinguished from the many previous forms of "amateur play" (Berlepsch 1929a, 57). In a pamphlet, the league followed the inventor in his claim that "only so-called von Berlepsch's nest holes are suitable" (Bund für Vogelschutz 1906, 2) for the task. For many years the baron had experimented with ways to encourage bird

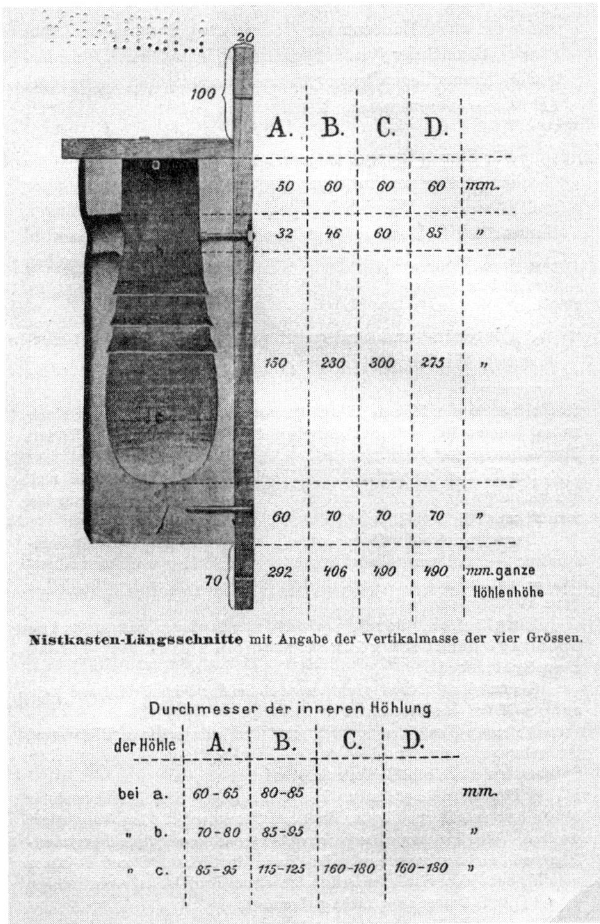

FIGURE 3.2 Original Berlepsch Nest Hole. Illustration in *Der gesamte Vogelschutz: Seine Begründung und Ausführung*, by Baron Hans von Berlepsch, 1899.

breeding, and he had devoted about five hundred acres of his estate to this purpose. For nest boxes to be successful, he believed, they had to copy nature as closely as possible. Since many songbirds nested in the abandoned nest holes of woodpeckers, the best way to substitute for the loss of such nesting sites in areas that were cleared of old trees was to imitate them. His nest box consisted of a wooden block with a hole carved in it that matched the dimensions of a typical woodpecker hole.

Berlepsch took substantial pride in the fact that it was he who had turned the enterprise of providing artificial shelter for wild birds into a real science—so at least he declared in his treatise *The Complete Book on Bird Protection* (*Der gesamte Vogelschutz*), which was first published in 1899 and went through twelve continuously enlarged and revised editions until 1929 (Berlepsch 1899, 1929a). The task of rational bird conservation was everything but to be belittled. The baron put special emphasis on the fact that he considered it to be a scientific means to an economic end and not a hobby, sport, game, or any other kind of amusement: "Bird protection is not simply a hobby, a passion derived from ethical or aesthetic motives—that is, not simply derived from the admiration for the song of birds, or from the effort to beautify and rejuvenate nature—but bird protection is first of all simply an economic matter, a matter of eminent importance. Bird protection is a matter from which derives for human beings a huge material and pecuniary benefit" (Berlepsch 1899, 1–2).

As proof of the scientific nature of his enterprise, Berlepsch pointed out that he had submitted his breeding research for publication only after years of experiment. He presented events in the summer of 1905 as the first large-scale proof of his measures. In this year, most forests of the area had been stripped by oak leaf rollers, a species of moth, while the trees on his own land, equipped with two thousand nest boxes, had been spared. The same experience was repeated in 1921. As the experiments show, Berlepsch's aim was not to go back to nature. He could have restored his own land to a natural state if he had wanted to, but his declared goal was, instead, a multiplication of insectivore birds "above and beyond the normal measure" (Berlepsch 1899, 4).

The Berlepsch nest box was a huge success. The model went into mass production by the company Hermann Scheid, which acquired the baron's concession to be the only company allowed to sell it under the name "original Berlepsch nest boxes." Scheid was a local wooden shoemaker who had helped the baron build the first models of his boxes and soon began to sell the boxes not only across Germany but also abroad—for instance, on the British Isles. This is not to say that German technology was more advanced than Britain's: the initial model of the Berlepsch nest box had in fact been made with specially made British drilling machinery. It was the demand for such a device, not the ability to design and produce it, that made for the difference between the two countries.

The success of Berlepsch's writings as well as of his private estate was tremendous. Seebach became renowned as a model for bird conservation. Officially recognized as an educational and research center ("Lehr- und Versuchsstätte") by the Prussian state in 1908, open to visitors and hosting lectures, it served as the example for many other German institutions in the early twentieth century. In its first twenty years the station counted 8,509 visitors

FIGURE 3.3 A pair of titmice and their offspring devour annually at least 120 million insect eggs or 150 thousand caterpillars. Illustration in Karl Mansfeld, *Vogelschutz im Garten*, 1934.

from more than one hundred countries. Berlepsch received an honorary doctorate from the University of Halle-Wittenberg in 1923 for his achievements in ornithology and bird conservation. “Looking back on my life,” he concluded his autobiography, published only a few years before his death in 1933, “I may find the truth of these words verified: Skill is knowledge, but knowledge is work” (Berlepsch 1929b, 85). This is a motto similar to the one expressed by Liebe that to know something is to love it, but relocated from the field of play into the world of work. There could have hardly been a better summary of the difference between the two approaches.

Bird feeders and nest boxes were the technological instruments or tools that allowed for the methodologically controlled experiments that Berlepsch had carried out throughout his life. The experience of nature that Berlepsch made in this context was mediated through the use of these instruments. Latour has coined the notion of circulating reference to indicate that the experience of nature is not unmediated but instead shaped by a chain of material instruments (Latour 1999). Nest boxes and bird feeders were such instruments used to produce knowledge about nature and involved in making a scientific distinction between birds as either useful or harmful. Conservationists themselves were eager to point out that these were serious scientific tools used for a rational economic enterprise. Yet, while bird feeders and nest boxes were without doubt used in this way by Berlepsch and other German bird conservationists, their use as tools was in no way intrinsic to these instruments. Whether a material object is used as a tool or as a toy is not determined by the object itself but depends on its relational position within a larger course of action, that is, on its position within the assemblage of which it is a part, to use the terminology of actor network theory.

When bird tables and nest boxes eventually made their appearance on the British Isles, they were used not as tools but instead as toys in the game of bird watching to enhance the diversity of species available for observation in one's garden. The practice of feeding birds in winter and providing shelter for them during the breeding season was perceived as a complete novelty by the majority of the population. In Britain the entire topic was somewhat of a curiosity, even decades after German agriculturalists, foresters, and gardeners had filled entire libraries with manuals on how to best engage in these efforts. Birds were fed and sheltered because it was pleasurable to have them around and watch them. A 1912 report in the RSPB magazine *Bird Notes and News* noted, "The appeal for 'Home for Birds' made at the Society's Annual Meeting . . . must have reminded his hearers how very recent is the growth of the popularity of the Nestingbox in Britain. When the R.S.P.B. first ventilated the idea it was generally regarded as a fanciful absurdity, and the boxes were often supposed to be for birds to roost and shelter in through the winter. We are still indeed behind the Germans in such matters" ("Homes for Birds" 1912, 14).

Berlepsch nest boxes were initially imported from Germany by the RSPB and offered for sale at exhibitions together with food tables, stands, trays, and bells ("Nesting-Boxes" 1913, 106). Baron von Berlepsch's ideas were first introduced to Britain through the translation of Martin Hiesemann's book on *How to Attract and Protect Wild Birds*, published in Germany in 1907 and appearing in English translation in 1908 and 1911 with an introduction by the Duchess of Bedford (Hiesemann 1907, 1911a, 1911b). Hiesemann was a faithful disciple of Berlepsch.

The resonance of the guiding principle of rational bird conservation seems to have been rather modest in Britain. A review of the second edition in the *Ibis* in 1911 pointed out that hedges and coverts are nearly absent in Germany but not in Britain, and there was thus no need for artificial replacements in the latter country. "We can cordially recommend this book to the attention of those who love birds, and are anxious to encourage them in their gardens and parks" ("Hiesemann" 1911, 388). This recommendation was that of a natural history collector, not of an economic ornithologist.

What is most striking about the British development is not its timing as such but the completely different ends to which these new means were put once they were introduced. In Germany nest boxes were economic means. On the other side of the shores of the North Sea, however, the often very same birds were attracted to local parks and gardens for very different reasons. In Britain nest boxes were employed as toys in the game of natural history collecting. The possible economic application of these instruments was so remote that when the first British manual on the topic appeared in 1897, it did not even acknowledge the voluminous literature in Germany. John Masefield's *Wild Bird Protection and Nesting Boxes* presented the devices as a new discovery. Masefield was the vice president of a natural history society, and his own manual "with illustrations of various designs of boxes, brackets, etc. that have actually been used by wild birds for nidification" was largely based on correspondence with bird watching friends regarding their own recent experiences in the field: "Of recent years many of my friends have corresponded with me on the subject of nesting boxes, and while some have expressed surprise that Wild Birds should nest in boxes provided for their use, others have kindly given me their experience and success" (Masefield 1897, 39).

Accordingly, Masefield went to some lengths to establish that birds do in fact nest in artificial places. The manual reported on about a dozen different designs to attract birds. It was silent on the various species' utility or harm to agriculture and instead emphasized the variety of species that can be attracted with such artificial breeding places. As evidence for his success, Masefield presented a list of thirty-six species that had nested in his bird boxes, including the sparrow, the species so reviled in Germany.

The reviews of Masefield's work displayed the same indifference to the possible economic utility of nest boxes. A writer in the *Ibis* attributed great value to them as a device for attaining a large list of collectibles, noting that Masefield had attracted thirty to forty species to his garden with the help of these boxes, and expressed the belief that "the list might be easily extended. We heartily commend this useful manual to all who are desirous of encouraging and observing birds" ("Masefield" 1897, 620). A review in *Nature* in the same year was equally outspoken on the point. The reviewer pointed out the list of different species as the true measure of success and concluded that descriptions and figures in the manual "will be found worthy of the best attention of those who may be inclined to devote their spare hours to this pursuit" ("Wild Bird Protection" 1897, 78). Any economic incentive was absent from the picture. In Germany nest boxes were part of the work performed in agriculture and forestry—in Britain they were perceived as part of a natural history collecting game. Even more tellingly, one of the first records of the mass production of nest boxes was reported in the *Irish Naturalist* to be engaged in by the "Ballycastle Toy Industry" ("Nesting Boxes" 1904, 140).

BIRD CONSERVATION AND THE WAR ECONOMY

What eventually brought the two countries closer together on the economics of bird conservation was the Great War. In Britain, as elsewhere, the war heightened the importance of domestic food production. Up to this point, the British Isles had been a net importer of agricultural products, a situation that was made possible by its numerous colonial possessions and easy access to global markets. The editorial of the magazine *Wild Life*, so far best known for its achievements in the field of nature photography, argued on the topic of feeding birds in winter with a stated benefit of birdlife to agriculture in view: "Looked at from any standpoint, the saving and protection of bird life is an act of national importance, recognized as such by many nations, and most of all, perhaps, by our greatest enemies, the Germans, whom no one would accuse of doing anything in which they could see no utility" ("Preservation" 1917, 269–70).

The editor went on to argue that, were it possible to express in terms of pounds sterling what the loss of birds during winter meant to the country, the figures would appear incredible. Yet, unlike the German bird conservationists, such as Baron von Berlepsch, British natural historians had no such data at their disposal. The war did nevertheless put an end to the import of the Berlepsch nest box as well as to the recognition paid to its inventor. The RSPB was reluctant to import goods from Germany and thus to weaken the war economy of the empire. An article in *Bird Notes and News* in winter 1914 accordingly reported on nest box making as a new British Industry:

> The difficulties encountered by the Royal Society for the Protection of Birds in providing Nesting-Boxes on the Berlepsch principle, but British-made, have now been happily overcome, and two enterprising firms have laid down

> special machinery by which boxes equal in every respect to those formerly imported in Germany are being turned out. The pattern devised by Baron von Berlepsch is undoubtedly most ingenious, strong, and attractive, and every effort will be made by the Society and by the collaborating firms to show that England can meet and beat the German manufacturer and to establish the business as a British industry. ("Nesting-Box Making" 1914, 57)

A report in the following spring on *British Homes for British Birds* went on to point out that it was satisfactory to know that the British-made nest boxes, manufactured for the RSPB, had proved equal in every respect to those formerly imported from Germany. The author argued that once the war was over, another industry would have been added to British production, and nest boxes would be among the thousand and one things for which German makers would never again pocket British money ("British Homes" 1915).

The war facilitated not only the domestic production of this device but also the recognition that it could be put to economic ends. It was during the war that the RSPB began to emphasize the economic value of birdlife, as a 1915 article on the RSPB in wartime indicated ("R.S.P.B." 1915). The society tried to raise awareness on the topic with the publication of a leaflet on *Birds, Insects and Crops*, which it supplied for free. It was distributed by the hundreds of thousands by summer 1917 ("Birds, Insects and Crops" 1917).

The same involvement of organized bird conservation in the war economy could be observed in Germany, where some of the technical advances achieved through wildlife photography were deployed in the production of war propaganda. Lina Hähnle's son Hermann was a keen bird photographer and an engineer. He had invented a stereo camera that produced three-dimensional pictures, which could be viewed through special glasses (Ecke 1964; Effenberger 1926; Helfer 1954). He presented these at public lectures of the League for Bird Protection, promoting them as a way to educate the eye for the perception of the natural world. "Those who deal with stereograms become aware of the effects of seeing with both eyes and enjoy with great appreciation the spatial composition of their environment, not only of its images!" (Hähnle 1955, 179). The league was thus involved not only in the use but also in the development of state-of-the-art technology (Schulz 1926, 1941; Steiniger 1955).

The league put this technological know-how to use in the war effort, as Lina Hähnle proudly stated in the annual report for the year 1915: "Since nobody could form a comprehensive idea of the tremendous experience of the war during the first months, we presented moving pictures mainly in the countryside and set the tremendous efforts that had been demanded of our troops into the right light. We have been among the first to recognize the positive aspects of the cinematograph" (Bund für Vogelschutz 1916, 9).

The use of technology for mass propaganda, called public education outside the war context, proved to be the true vehicle for the organization's success. Technology in its various forms was perceived as much as the reason for the decline in birdlife as it was used as a remedy for the situation. Bird conservationists were aware of this twofold role of technology and were accordingly not opposed to scientific and technological progress. Conservation itself was perceived as a progressive force. Rather than turning against established institutions, conservationists were by far more interested in creating their own institutions for the scientific study and conservation of birds.

CHAPTER 4

Field Ornithology and Practical Bird Conservation

Harry Forbes Witherby was a keen naturalist since childhood and the guiding light in the institutionalization of field ornithology in Britain. Through his work, sight records and photographs of living birds collected in the field slowly gained credibility as ornithological data. The transformation of bird watching from hobby to science took place outside the established institutions of ornithology, and it was Witherby's seminal accomplishment to create a new set of institutions alongside the already existing ones. When he joined his father's London printing firm at the age of twenty-one, he started to use the firm's publications as an outlet for his own photographs, sight records, and travel reports. When he later inherited his father's business he moved the firm into book publishing with a focus on birds and other natural history topics. His own *Bird Hunting on the White Nile*, published in 1902, was among the first, and by 1913 thirty titles, nearly all on birds and illustrated with photographs, had been issued.

One of Witherby's early contributions to his father's publishing venture had been the introduction in 1897 of a section in the magazine *Knowledge* called British Ornithological Notes, making the magazine one of the first to include bird photography. The innumerable submissions to this section led to the inauguration in 1907 of a separate journal called *British Birds* ("First Fifty" 1957; Ogilvie, Ferguson-Lees, and Chandler 2007). The publication sought to coordinate the work of amateur natural historians. The editorial in the first issue stated that "we hope, with the co-operation of our readers, to embark upon a series of more systematic investigations than have hitherto been attempted, with regards to matters concerning birds of this country" (Witherby and Pycraft 1907, 2). *British Birds* became the country's first journal of field ornithology. The use of photographs as scientific evidence was explicitly encouraged. "Of later years," the editorial stated, "photography has come greatly to the fore as an adjunct to science, but ornithologists . . . have used the camera more as a picture-maker than as a scientific recorder" (Witherby and Pycraft 1907, 3). Yet when Witherby made his editorial statement, he had no intention to replace the gun entirely. Initially, he had considered the camera as a mere auxiliary to the gun, as he stated in a book on forest birds: "Let us say a word for photography as an aid to the naturalist. A note book is indispensable, but minute details cannot be noted down, whilst a photograph, however bad it may be as a picture, gives a lasting impression of the position of the nest, eggs, and young, and their surroundings, which can always be referred to when the group is being cased" (Witherby 1894, 97–98). Photographs simply offered a visual tool to guide the arrangement of birds in museum showcases, not a scientific record considered valuable in its own right.

A few years later Witherby still assigned an equally auxiliary role to binoculars. "A good pair of binoculars is in every way the ornithologist's best friend," he wrote in *Bird Hunting on the White Nile*, "although a glass should not be relied upon to too great an extent in identifying birds" (Witherby 1902, 68). Only the measurement of a bird in the hand counted as genuine scientific data. The bookplates in the volumes of his private library accordingly displayed the then well-known motto "What's hit is history, what's missed is mystery" (Mearns and Mearns 1998, 7). As the editor of *British Birds*, he was initially known for his severe skepticism and scrutiny of observational records. Field ornithology as represented by the journal during its early years made use of the gun in the same way museum ornithology did. Even after the inauguration of *British Birds*, Witherby kept one foot in the world of the museum and the other in the field. On his travels he collected more than nine thousand study skins and mounted specimens, with a focus on variation of plumage by season and life stage. With this collection he became the leading British authority on plumage changes and molts. The phrase *systematic investigation* chosen by the editors did likewise not imply that the journal established any methodological standards for collecting data in the field. It simply meant that the journal drew on a larger pool of data than other ornithological publications.

This larger pool derived from the fact that *British Birds* accepted a plurality of approaches to data collecting. A series of articles, "On the More Important Additions to Our Knowledge of British Birds since 1899," written by the journal's two editors Norman Ticehurts and Harry Witherby, for instance, reported on, among others, an alpine accentor that had been "seen and carefully observed with glasses," on bearded titmouse sightings in Surrey, where one was "seen by an observer well acquainted with the species," while another observer "saw a party of five at Milford, near Godalming, and was confident of their identity." An adult cirl bunting was "watched at close quarters and its note heard," a little bunting in winter plumage was "caught with bird-lime," and a young bird of the same species was reported to have been "captured by a cat." A female ortolan was more conventionally "shot," and a male exemplar of the same species "entangled itself in some netting" (Witherby and Ticehurst 1907–1908, 109, 248–49). The practice of killing birds to obtain records was not eliminated but simply supplemented by additional ways to collect data.

Witherby's stance on photographs and sight records, however, was to transform throughout his career. Toward the end of his life he sold his bird collection to the British Museum (Natural History) and used the proceeds to fund field-ornithological survey work and to set up a bird reserve on his own land (Hickling 1983). He now preferred watching birds and collecting sight records to shooting them. While Witherby created a new institution that transformed the world of ornithology, the institution also transformed him. Once *British Birds* and the practice of sight record collecting were fully established, his own motives for collecting such data were no longer the same.

INSTITUTIONALIZING NATURE CONSERVATION

This chapter analyzes the institutionalization of field ornithology—that is, the practice of collecting records of living birds as they are encountered in the field. Just as museum ornithology was institutionalized differently in Britain and Germany, field ornithology also developed in different directions. In Britain this development resulted in the establishment of the British Trust for Ornithology, and in Germany in the creation of a network of state-supported Bird Protection Stations. In both cases the leading national bird conservation organizations allied themselves with the newly created field ornithological organizations. The contrast in bird conservation between the two countries increased through this process.

The analysis of this process of institutionalization is in large parts consistent with the major currents of neoinstitutionalism in sociology (DiMaggio and Powell 1983, 1991; Meyer and Rowan 1977; Scott 2013). A central assumption of this family of theories is often summarized in the formula that institutions are both enabling and constraining. They allow for actions that would not be possible without them, but at the same time limit the number of possible actions. Institutions thus lead action in a certain direction. The notion of institutions used in this study gives a slightly different touch to this account of institutions as enabling and constraining. Paired with a pragmatist theory of action, an institution cannot be regarded as a datum external to the actor. The very language of institutions as enabling and constraining conceives of them as means to an end, that is, as self-contained entities to which an actor relates in an instrumental way. Such an instrumental language ignores the creative or inherently transformative potential of both action and institutions.

In contrast to this view, I describe institutions as proactive and reactive. This is not meant to suggest that institutions are actors or even like actors but, instead, that institutions have an internal logic of unfolding, which under conditions of stable input can produce changing or unstable output. The consequences of the accumulation of wealth are the classic example: accumulated wealth, so the argument holds, whether found in family dynasties, monasteries, or empires, undermines the very personal qualities of courage, thrift, and hard work that made this accumulation possible in the first place. Such accounts highlight the inherently cyclical nature of institutions that is frequently indicated by the terms *rise* and *fall*. Institutions have the ability to produce such transformative effects under otherwise stable conditions. The current literature has paid most attention to institutions as factors of stability, yet institutions are equally crucial to processes of change. Andrew Abbott used the notion of turning points to refer to such changes. Turning points are short consequential shifts that redirect a process. "Turning points are thus best theorized as points at which the interlocked networks of relation that preserve stability come unglued and the (normal) perpetual change of social life takes over" (Abbott 2001, 259). Turning points thus mark the creative potential of institutions.

While institutions in the world of work can be described as enabling and constraining in the sense that an institution is a neutral means or instrument for the actor, in the world of play it would be more apt to refer to institutions as proactive and reactive, since institutional dynamics can change actors in the same way that actors can change institutions. Games as institutions are not just neutral means, but they have the propensity to produce or transform ends of action. To put it in the words of Hans-Georg Gadamer, while it is true that players can master a game, it is also the case that "the game masters the players" (Gadamer [1960] 1989, 106). Such a transformation of actors by institutions is what is meant by the proactivity of institutions, while the transformation of institutions by actors is addressed as their reactivity.

In the process of institutionalization described in this chapter, actors transformed in conjunction with the development of institutions, that is, the institutions formed the actors and their motives just as much as these actors and their motives formed the institutions. There was accordingly not a new spirit, consciousness, or creed of field ornithology at the beginning of the development that was merely translated into new institutional forms. As the example of Witherby shows, it was only at the very end of the process of institutional transformation that a new spirit could be found among the involved actors.

SIGHT RECORDS AND FIELD GUIDES

The amateur journal *British Birds* differed from the professional journal of the British Ornithologists' Union (BOU) the *Ibis* in its acceptance of sight records. Accepting such records enabled Witherby and his journal to collect an unprecedented amount of data. The journal was published every month and included articles contributed by various ornithological experts as well as submissions by amateurs. The total number of bird species known to nest or winter in Britain and information about their distribution increased substantially with the steady supply of data by sight-record collectors. The very first contribution to the new journal was an article by Howard Saunders, "Additions to the List of British Birds since 1899," which added accounts of twenty species to the second edition of the author's celebrated *Illustrated Manual of British Birds*, published in 1899 (Saunders 1899, 1907). Additions to the list of bird species on the British Isles were the dominant theme in the journal, represented by a section titled Notes, which published records from the field submitted by readers. It was the key feature of the journal despite its somewhat marginal-sounding title.

The journal turned out to be a success because it provided a growing army of bird collectors with an outlet for their sight records. The journal even explicitly encouraged amateurs to produce such records in the first place. As the editors stated bluntly in the preface to the first volume, the aim of *British Birds* was "to supply a need, and to make that need so real, that it will gain in intensity as the years roll on" (Witherby and Pycraft 1908, iv). With such a well-recognized outlet for field observation, owning collections of birds became less important. An individual's reputation as a bird collector (i.e., ornithologist) increasingly rested on the publication of sight records. A mention in the Notes of *British Birds* became the true collectible and thus the pride of collectors and the admiration and envy of their peers.

Natural history collecting is a competitive game, and with the establishment of *British Birds* it became possible to confirm one's game score and compare it to that of others without killing a bird. Without such an institutional outlet, the competitive collector would have had to resort to the established journals, in this case the *Ibis*, which at the time did not recognize sight records. It was this institutionalization of the practice of sight-record collecting through *British Birds* that made all the difference. Not everybody with access to binoculars or a camera automatically stopped collecting bird bodies. Witherby is a case in point. As outlined above, he initially promoted all forms of collecting, including that of sight records and actual specimens, without preferring one form of data over the other. It was only after the practice of sight-record collecting became fully institutionalized in the Notes section in his journal that the practice of collecting bird bodies became irrelevant.

A game is constituted by rules and resources that foster the practice of play; while binoculars and cameras provided new resources for the game of natural history collecting, the journal *British Birds* with its Notes section provided the accompanying rules to make the transformation of one game into another possible. Once the new field-ornithological collecting game was fully institutionalized, the old collecting game could be rejected tout court. This transformation from museum ornithology to field ornithology worked so well because the practice of collecting had not to be given up but simply to be channeled into a different direction.

Based on the growing supply of amateur field records, the editors claimed authority in determining the official number of British bird species. Witherby, together with a team of some of the more regular contributors to *British Birds*, prepared ornithological checklists and handbooks, beginning in 1912 with *A Hand-List of British Birds* compiled by Ernst Hartert, Francis Jourdain, Norman Ticehurst, and Harry Witherby (Hartert et al. 1912). The hand list described 469 birds, with an account of the distribution of each species in the British Isles and abroad. The list comprised breeding species, wintering species, onetime visitors, and one extinct species, the great auk. Its most noteworthy feature was the use of trinomials—that is, of subspecies as well as species names. The authors asserted that questions of nomenclature had been neglected in Britain since the last edition of Linnaeus's work about 150 years earlier, and they aimed to give a proper scientific names to "all those birds which in our opinion are entitled to a place on the British List" (Hartert et al. 1912, v). In doing so, the hand list and its authors simply ignored the previous efforts of the British Ornithologists' Union, which regarded such matters as its own purview. The BOU had published a *List of British Birds* as early as 1883 and had been working on an updated version since 1911.

There was no comparable German List to match the British List. The popular ornithological journals such as the *Ornithological Monthly* (*Ornithologische*

Monatsschrift) of the German Society for the Protection of Birdlife (Deutsche Verein zum Schutz der Vogelwelt), whose annual volume of printed pages far outnumbered that of *British Birds*, did not provide a total count of bird species that could be considered German. New sightings of rare species and occasional visitors were reported, but these were one matter among many others, and no cumulative list was provided. The German Ornithologists' Society (Deutsche Ornithologen-Gesellschaft) likewise did not provide any such count. This was certainly not for lack of interest in naming and classifying bird species. As it was for the British Ornithologists' Union, this was one of the primary tasks of the German Ornithologists' Society.

Something else made for the difference. The notion of a total set of collectibles present at a given territory that finds expression in a national list was not as central to German ornithology as it was to its British counterpart. In Germany the frame of reference for compiling data on birds was not a territorially bounded national entity but a place of belonging, or *Heimat.* The term *Heimat* is frequently translated as "homeland," although it does not refer to a particular territory such as the nation state: it represents the way one relates to a place. An ornithology focused on *Heimat* sought not to catalogue every observed form of birdlife but to compile data on those birds that were a part of daily routines and experiences over a prolonged period. Birds that were in Britain added to the list as "irregular visitors" were thus of little interest to those concerned with the birds of the German *Heimat*. Poultry and exotic cage birds, on the other hand, were included, as long as they were encountered regularly.

One of the most striking examples of the German approach to ornithology is the popular handbook by C. G. Friderich, published in four revised editions from 1849 to 1891 with two further revisions completed after the author's death by Alexander Bau in in 1905 and 1923 (Bau 1905, 1923; Friderich 1849, 1863, 1876, 1891). Friderich's aim was to provide a one-volume reference work. Much of his material was copied from Naumann's standard ornithological handbook, which remained unmatched by any other work in the country. Like Naumann, Friderich included exotic cage birds and poultry together with wild birds. The final edition prepared by Friderich himself carried the very informative but somewhat clumsy title *Natural History of German Birds, Including the Birds of Central Europe, Prepared Based on Many Personal Experiences and the Best Sources, for Fanciers of Aviary and Cage Birds, for Ornithologists and Egg Collectors, for Hunters and Hunting-Enthusiasts, as well as for the Instruction of Advanced Youths*. The author emphasized that the total number of birds in the world was not yet known with any precision and noted that "the birds of Germany described in our book only amount to the modest number of approximately 450 species, all of its sub-forms and races included" (Friderich 1891, 16). The notion of a total set of collectibles was not an organizing principle of the publication, and an approximate number was thus more than sufficient.

In part, this contrast to the British attempt at list making was a matter of geography. It simply was not as easy to define the boundaries of Germany as it was those of the British Isles. In most cases, neighboring territories were included for simplicity's sake. The first person to provide a count of the total set of German birds was Anton Reichenow, who had become curator of the ornithological collection at the Natural History Museum in Berlin (Berliner Museum für Naturkunde) in 1888. He was also named editor of the *Journal of Ornithology* (*Journal für Ornithologie*) in 1893 and president of the German Ornithologists' Society in 1894, and held all three positions until his retirement in 1921 (Stresemann 1943, [1951] 1975).

The expansion of the bird collection at the Berlin museum was one of Reichenow's goals. Before its move to a new and larger building in 1886, the collection had counted twenty-nine thousand mounted specimens. When Reichenow retired, the museum possessed an additional collection of sixty thousand study skins. As in Britain, the pursuits of collecting and colonialism went hand in hand. Germany had acquired its first overseas territory in 1884. When the influx of new species from the colonies arrived in Berlin, Reichenow became the leading authority on the birds of Africa, described in his monumental three-volume work *The Birds of Africa* (*Die Vögel Afrikas*) (Reichenow 1900–1905). Over his lifetime, he was able to assign his signum "Rchw" to no fewer than 950 species.

It was Reichenow who first took up the task of assembling a German List. He solved the problem of defining the boundaries of Germany by simply listing the bird species in all parts of Europe where German was spoken, including Luxembourg, the German-speaking part of Switzerland, the northern part of Tirol, Upper and Lower Austria, Carinthia, Styria, and Bohemia. The *Systematic List of the Birds of Germany and the Neighboring Central Europe* (*Systematisches Verzeichnis der Vögel Deutschlands und des angrenzenden Mittel-Europas*), published in 1889, was thus not a list *of* German birdlife but a list *for* German-speaking ornithologists. Reichenow arrived at a total of 396 species in these regions (Reichenow 1889).

The number of species was thus higher than the total of 376 species enumerated in the 1883 edition of the British List. This was soon to change: the second official edition of the *List of British Birds* published in 1915 gave a total of 475 species (British Ornithologists' Union 1915, xiv). However, the number of species reported to regularly breed in Britain had stayed almost the same:

180 in the first edition and 188 in the second. The number of winter visitors went from 31 to 46. The major increase, from 164 to 240 species, was in the category of birds of passage, occasional visitors, and rare visitors.

These additions to the list reflect an increase in the number of bird collectors. The higher the number of people collecting birds, the higher the number of vagrant or irregular species recorded. Regularly occurring species, in particular those that nest in a given territory, are the first to be detected and counted. Those that visit regularly in winter come second. Species that occur only irregularly are the most easily overlooked. Thus with a small number of collectors, the list of regularly breeding birds stays stable relatively early on, whereas the list of irregular visitors expands substantially with a rise in the number of collectors. The British Lists in these early decades are thus documents that chronicle changes in bird collecting with much more precision than changes in actual birdlife.

The rise of field ornithology also transformed the genre of the ornithological handbook. Early handbooks had been illustrated with lithographs based on paintings and wood engravings. Most bird artists were commissioned by wealthy natural historians to illustrate their ornithological works, working from specimens and study skins. A substantial number of the best-known handbooks were enlarged catalogues of museum collections. The collections most frequented by British bird artists were the Bird Room at the British Museum (Natural History) and a private museum at Tring established by Walter Rothschild.

The verbal descriptions and images of birds in these handbooks, based on museum specimens, were of little use for the identification of birds in the field. Almost every species was placed on a different plate and displayed in a slightly different posture, making comparison difficult. The descriptions in the text, moreover, did not necessarily match the details in the picture. They assumed that the bird to be identified was in the reader's hand and could be turned around and viewed from various angles. If the collector did not intend to shoot the bird, however, identification became a problem.

The first handbooks to explicitly aid identification in the field were produced by the same group of authors who published *British Birds.* In preparation since 1914, *A Practical Handbook of British Birds* was eventually published in two volumes in 1920 and 1924 (Witherby 1920–24). It was followed by an enlarged and revised five-volume *Handbook of British Birds*, published between 1938 and 1941 (Witherby et al. 1938–41). Both publications aimed to provide an up-to-date and easily accessible account of British birds for both professional ornithologists and amateur bird watchers. Like previous handbooks, it was a collector's manual and followed the same rules of arrangement. Each entry further contained a reference to the two standard ornithological handbooks: the fourth edition of Yarrell's *History of British Birds* (Yarrell 1871–1885) and the second edition of Howard Saunders's *An Illustrated Manual of British Birds* (Saunders 1899).

What made the two handbooks distinctive and their publication a much-celebrated event was the emphasis put on field characters—that is, the description of characteristics that allowed for the identification of living birds. A discriminating analysis was made of the bird's appearance, flight, notes, song, and other characteristics observable in the field. Comparative notes described how a species differed from other birds that might appear similar. As the authors noted, "It is often not possible to give and infallible guide to identification in the field: moreover, different observers note different characteristics, and it is often very difficult to put into words exactly how one identifies a bird in the field. The chief aim of the writers of this section has been to indicate the points on which attention should be concentrated" (Witherby 1920–24, 14).

Most of the accompanying illustrations were drawings of birds in their habitat, but these were supplemented with photographs and silhouettes of birds in flight. The majority of the field characters in the *Practical Handbook* were provided by Charles Oldham based on observations in the field. Most of the descriptions in the *Handbook*, however, were based on museum specimens at Tring and in the Bird Room at the British Museum, because there was simply nobody around who possessed sufficient field notes to replace museum observations. A section describing each species' sequence of plumages and molts from nestling to maturity and from season to season was based on Witherby's private collection.

The journal *British Birds*, published by the editors of these handbooks, facilitated the shift from museum observations to field observations. The journal made its reputation by requiring detailed field notes on every record submitted for publication. The editors stated in a 1930 issue that the worst types of submissions were those by observers who insisted on the certainty of what they saw in the field without supplying evidence. The editors recommended seven points every sight record should include, such as the distance of the bird and the direction of light, the nature of the ground it was on, whether it was seen from different angles, and whether it was in rest or flight. Potential contributors to the journal were asked to make field sketches showing the bird's color pattern, distinctive marks, and shape (Witherby, Jourdain, and Ticehurst 1930, 342–44).

This is not to say that field experience was considered irrelevant. To the contrary, an expanded version of these guidelines published by Horace G.

Alexander in a 1944 article highlighted that very point: "It seems likely that almost every separate species, and occasionally even separate subspecies, is in fact normally identifiable in the field, under favourable conditions, if you know exactly what to look for. But—and this is by far the most important proposition—such identification can only be achieved by observers who know how to observe" (Alexander 1944, 90).

The article gave credit to Witherby for securing the journal's claim to scientific credibility. Alexander, who was one of the pioneer bird watchers of the period, had frequently contributed not only to this journal but also to the section on voice in the Witherby handbook, and his name also appeared in the sections on field characters and general habits (Alexander 1974; Wood 2003).

A NATIONAL SOCIETY OF BIRD WATCHERS

British Birds was the central institutional locus of field ornithology from the 1910s to the 1930s (Lack 1937). One of the largest projects carried out by the journal during this period was the Oxford Bird Survey, inaugurated in 1926 by Witherby and his associates. Its designated organizer was a twenty-two-year-old upstart in ornithology, Edward Max Nicholson, who had just finished his first monograph on ornithology, carrying the not all too modest title *Birds in England: An Account of the State of Our Bird-Life and a Criticism of Bird Protection* (Nicholson 1926). The work aimed to channel the old-fashioned mania for mounted specimens and study skins into a form of collecting more beneficial to birdlife. Nicholson was one of the few who did not just play the game of bird collecting but consciously devised its rules. His success in this endeavor was based on a superb grasp of the motivational processes at work. In his view, museum and field ornithology were driven by the very same motive.

> A bird-lover . . . looks at birds as living beings and watches their movements, habits, and signs of life; so that for him only the living bird is a bird at all, after death it becomes merely a corpse. But the collector is always looking at even the living creature in terms of the glass case—he admires a fine specimen far above a fine singer, and to him the live bird is always a chrysalis ready to be transformed into the perfect imago which only a spell at the taxidermist's can make it. . . . I think that this, rather than the stage-villain type of man who slays for the lust of killing that we so often find painted, is the true psychology of the collector. When he shoots a bird he is not conscious of having *destroyed* anything, but only of having *secured* it. (Nicholson 1926, 258–59)

The remaining question for Nicholson was simply how to push this passion for bird collecting in the direction he wanted it to go. He can be described as a moral entrepreneur who turned into a game master, that is, he devised the rules of the game that other people played. Getting people to play the game of bird watching can transform their motivation in the sense that they develop an emotional attachment to birds. As described by Dewey, the activity of play is a creative process that produces new moral valuations. Nicholson was fully aware of this creative potential of play. By setting up the rules for the game of bird watching in the form of field surveys, he aimed to manufacture attachments to living birds in other people.

Looking at this possibility to purposefully produce transformative experiences has implications for how we conceptualize morality in relation to strategic action. All too often the two are discussed as mutually exclusive. Ann Swidler, for instance, has influentially addressed culture, moral values included, as a toolkit or repertoire that provides strategies for action, rather than motivations for action. While culture may have a motivating force during unsettled historical times, it provides mere strategies for action during settled times (Swidler 1986, 2001). Yet these two dimensions of action are not mutually exclusive—actors who are motivated by moral attachments can act strategically in order to produce the same attachments in others. The rules of a game are a cultural repertoire or toolkit that can be strategically used, but the emotional attachments that the process of playing a game according to these rules produces have nevertheless a genuinely motivating force.

The most important rules of the game that Nicholson tried to facilitate were scientific standards for field-ornithological survey work. His publications in the 1920s and 1930s focused on the use of censuses and other field survey techniques for more systematic observations (Nicholson 1927, 1931). The reviews of these publications in the ornithological journals reflected the ambiguous status of bird watching as somewhere between sport and science. An anonymous author in the *Ibis* denied Nicholson's *Bird-Watching as an Art* any claim to scientific credibility or originality: "Sometimes it seems that elaborate plans and statistics are made to prove what is commonplace knowledge to the mere collector" ("Nicholson" 1932, 188).

Divided opinions on the topic notwithstanding, field ornithologists continued their survey work. The experience of conducting the Oxford Bird Survey was translated into other endeavors. Soon after its successful inauguration, Nicholson ventured to create a survey that would provide data on the distribution and abundance of bird species throughout Britain. With Witherby's backing, Nicholson set up a national census of heronries (the breeding colonies of herons) in 1928. What suited the common heron, *Ardea cinerea*, to this undertaking was simply its size. Herons are to British birds what elephants are to African mammals: they are conspicuously big, not particularly afraid of human beings, and well known. The species appeared ideal for collecting accurate data on distribution and abundance for a large territory in a way that had never been attempted.

An appeal asking volunteers to participate was launched in *British Birds* and in several other natural history magazines and daily newspapers. The true aim of the survey lay in counting not herons but volunteer heron watchers. It was a test run to see whether or not sufficient numbers of bird watchers could be organized throughout Britain. This experiment was a true success. During the first year, 393 people participated and counted approximately four thousand heronries (Nicholson 1929, 273, 316).

As the survey work organized by *British Birds* grew more ambitious and the number of participants increased, the enterprise outgrew the journal's capacity to manage it. Nicholson was among the first to suggest the establishment of an independent organization for field ornithology. The novel organization was envisioned as a "National Society of Bird-watchers," as he stated in *The Art of Bird-Watching* (Nicholson 1931). An appeal supported by Harry Witherby, among others, was launched that year, and in 1933 the British Trust for Ornithology (BTO) was established, with Nicholson as honorary secretary. The trust's mission was to organize amateur bird watchers on a national basis for the benefit of ornithological science (Hickling 1983).

In order to gain financial support from the Ministry of Agriculture and the Empire Marketing Board, the trust was set up as an economic enterprise. Nicholson and his collaborators, with the additional support of the chancellor of the University of Oxford, the amateur ornithologist Viscount Grey of Falladon, convinced the ministry of the economic importance of field ornithology in inquiring into the utility and harm of the various bird species for agriculture. This economic rationale was, however, put forward solely for the purpose of gaining funds from the government, as Nicholson reported in retrospect: "We were up against problems of funding. . . . Our only hope seemed to be taking advantage of the then widespread, if scientifically dubious, belief that (before chemical pesticides had been heard of) birds could render great service in ridding farms of invertebrate pests" (Nicholson 1983, 16).

None of those who were involved in the trust's work were true admirers of economic ornithology. Nicholson had in fact argued in *Birds in England* that economic ornithology lacked a scientific basis. Commenting on one of the few British works in the field, Walter Collinge's *The Food of Some British Birds: A*

FIGURE 4.1 The heron was chosen for the first field ornithological census because its size makes it possible for almost anyone to recognize it. Illustration in Julian Huxley and Max Nicholson, *The Bird Population of Britain: A Plea for Co-operative Study*, 1935/36.

Study in Economic Ornithology, published in 1913, he argued that economic ornithology ignored basic ecological insights, such as the fact that feeding habits varied according to the availability of food sources and the competitors present in a given territory. If a species deemed harmful were artificially reduced in numbers, it was most likely that a competitor species would flourish and take over its feeding habits. Economic ornithology was enshrined in the trust's statutes but never formed any serious part of its work. The organization focused on surveying bird abundance and distribution as a collector's pursuit and not as an economic enterprise.

BIRD PROTECTION STATIONS

In Germany, in contrast to Britain, economic ornithology gained in importance and became firmly institutionalized in the first decades of the twentieth century. Unlike Collinge's *The Food of Some British Birds*, Baron von Berlepsch's *Complete Book on Bird Protection* (*Der gesamte Vogelschutz*) was highly influential, going through twelve editions, repeatedly enlarged and updated, until 1929.

Even more influential was the baron's experimental research station at his estate at Seebach. In 1926 it was reported that there were seventeen stations in Germany modeled on the Seebach example, several of which had, like the original, attained formal recognition by the state. Most of these places were known as Bird Protection Stations (*Vogelschutzwarten*). The name highlights their aspiration to scientific status: it was a slight alteration of the title of Bird Station (*Vogelwarte*) given to an ornithological research station established in 1901 by the German Ornithologists' Society at the town of Rossitten in Upper Silesia. The research at Rossitten had no economic component: following the example of Heinrich Gätke's work on the island of Heligoland, it was chiefly concerned with bird ringing and the study of bird migration. Heligoland became the second official bird station of the society in 1910, and a third one was established in 1936 on the island of Hiddensee in the Baltic Sea. Other facilities engaged in similar projects soon adopted the same name, even without formal recognition by the professional organization (*Vogelwarten* 1955; Bruns and Niebuhr 1961; Glasewald 1937).

In Britain, research on bird migration based on field observation was viewed, like all other field study of living birds, as unscientific by the British Ornithologists' Union. In the British Isles, the enterprise of bird ringing was initiated by Witherby in 1909 and carried forward by *British Birds*. It was one of the first tasks delegated by the journal to the newly created British Trust for Ornithology in 1938 ("Future" 1937). In Germany, on the other hand, the study of bird migration had the support of the professional ornithological organization. As early as 1875, Anton Reichenow had called for a committee on the study of the topic. The Bird Protection Stations, on the other hand, never gained official support from the German Ornithologists' Society. Economic ornithology became instead institutionalized within the ministries of agriculture and forestry of the German states.

Research on economic ornithology as carried forward by the Bird Protection Stations proved to be the central institutional locus for bird conservation. Inquiries were designed to investigate the feeding habits of the various bird species and to improve the means of protecting them. This work had a highly quantitative focus: practical bird conservationists cared about the number of nest holes that would be occupied in a given territory, how much offspring was produced in each, and what kinds of food the nesting birds would devour. When it came to the design of the nest holes, the central question turned out to be the specific size of the opening that would allow the useful but not harmful species to enter the holes. Baron von Berlepsch's original designs and measurements were continuously refined. Despite their inventor's claims, researchers concluded that his designs were not the only option. More and more designs claimed to represent the most scientific way to protect birds for economic ends (Föhr 2005).

The most competitive model was discovered by chance by the bird photographer Max Behr, who had experimented with nest boxes that would allow for easy photography of breeding birds. Behr had given up work as a farmer in 1902 to become a full-time nature photographer. He was one of the contributors to the multivolume *Portraits of Animal Life* (Meerwarth and Soffel 1908–1912), illustrated exclusively with photographs, and was best known for his pictures of the beavers of the Elbe River (Hinze 1937). In 1920 he rented a former vicarage in the area and put up nest holes to attract birds that would rid his garden of insect pests. The idea of photographing the birds nesting in those holes is told to have come to Behr as a mere afterthought. To pursue his envisioned goal, Behr had built simple rectangular boxes out of wooden planks whose front could easily be removed, since the Berlepsch nest holes that were carved out of a single block of wood had proven distinctively unsuitable for the purpose of photography. The new design was not much more than a wooden shoe box. What eventually made for its success was the fact that this design turned out to be completely sufficient for attracting birds. As many, if not even more, of Behr's boxes were occupied by nesting birds as were those made after the established design. In due course Behr's boxes were produced by the Scheid company, which was already manufacturing the Berlepsch boxes. Those made according to Berlepsch's design were known as *Nisthöhlen* (nest holes) and Behr's model as *Nistkästen* (nest boxes).

Behr was not only a bird photographer but also a serious economic ornithologist. Over the years he turned his land at Steckby into an experimental Bird Protection Station that carried out pioneering research in economic ornithology. Berlepsch's station at Seebach had gained recognition when the insect infestations of 1905 and 1921 had affected the land surrounding the estate but not the estate itself. At Steckby the insight that birds could prevent such calamities was tested experimentally. From 1925 onward, systematic inquiries were made into the exact relation between the numbers of breeding birds and insect populations. Some of the Steckby sites were equipped with nest boxes, while others were left unequipped for control purposes. Additional land on the nearby estate of Count Dürckheim was made available for the experiment, and the time-consuming inquiry was carried out with the help of Dürckheim's forester, Fritz Plate. The costs for this project, in particular the costs of the nest boxes, were carried by the League for Bird Protection (Bund für Vogelschutz), and Steckby was eventually recognized as an official experimental research station by the state of Anhalt in 1932. After Max Behr's death two years later, the station was run by the league.

The research at Steckby provided the work of the organization with scientific recognition. From the beginning, the league had claimed an ornithological grounding for its work, adopting Liebe's slogan "first get to know the life of birds thoroughly, if you want to protect them with proper success." Yet it was only with the work at Steckby that the league engaged in a research project of its own rather than drawing on insights from others. Starting in the mid-1930s, the results were published by Hermann Hähnle, the son of the organization's founder and president, Lina Hähnle. His analysis of the data was fairly basic, comparing changes in bird numbers with changes in insect numbers at the various sites but without calculating the statistical significance of the observed variations. Like the British bird surveys, it was an entirely descriptive exercise (Hähnle 1936).

The major difference between the British and German enterprises was the fact that the Steckby project focused on the numbers of birds breeding in a controlled set of artificial nest boxes rather than on a count of the population as it occurred in the wild, without intervention. In the first half of the twentieth century, the German Ornithologists' Society was much more receptive to such monitoring work than its British counterpart. This difference arose in part because in Britain, the old guard of museum ornithologists remained in office for much longer than in Germany. The turning point in Germany came in 1922, when the thirty-two-year-old Erwin Stresemann took over Anton Reichenow's role not only as the editor of the *Journal of Ornithology* but also as the president of the German Ornithologists' Society. Two years later he also succeeded Reichenow as the curator of the ornithological collection in the Natural History Museum in Berlin. Once in office, Stresemann began almost immediately to promote research in field ornithology (Haffer 1994, 2001a; Haffer, Rutschke, and Wunderlich 2004; Nowak 2003).

A systematic survey on the distribution and abundance of birds in the field was published in the journal by Gottfried Schiermann in 1930, with a follow-up article in 1934 (Schiermann 1930, 1934). At Stresemann's suggestion, Schiermann aimed to monitor the abundance of birds in a natural habitat free of nest boxes and bird tables. Finding such a habitat proved difficult, given the popularity of bird conservation in the country. Schiermann eventually chose a piece of forest in the Lower Spreewald near Berlin in the Mark of Brandenburg. From 1928 to 1929 he counted 2,729 breeding pairs of ninety-seven species. Most of his records were collected by sight or based on calls. Only in the case of an encounter with an unexpected rarity did he procure a specimen for the museum (Schiermann 1930).

Many other studies followed this example. From 1936 to 1938 Rudolf Berndt and Fritz Frieling, who had been involved in the work of the League for Bird Protection at Steckby, stocked a park with artificial nesting places to contrast its bird population with the density reported by Schiermann for a natural habitat. In terms of vegetation, the two sites were roughly comparable, but one site contained four hundred artificial nesting sites in approximately a quarter of a square kilometer. These were divided almost evenly between Berlepsch's nest holes and Behr's nest boxes in order to compare the effectiveness of the two designs. The author concluded, with apparent satisfaction, that "in the Park Prödel alone, about ten times as many birds are living, attracted by these devices, as in the entire Lower Spreewald" (Berndt and Frieling 1939, 607). The greatest contrast was observed in the case of the pied flycatcher, with 3.9 individuals per square kilometer on Schiermann's natural site and 281 on Berndt and Frieling's. The authors concluded that artificial nesting places have a measurable impact on the number of birds breeding in an area.

Most of the published work on density and abundance in subsequent years took these publications in the *Journal of Ornithology* as a model. Schiermann himself studied the density of birdlife at a cemetery in Berlin, Glasewald studied the birdlife of a pine forest in the Mark of Brandenburg (Glasewald 1933), Max Garling investigated the Berlin sewage farms as an example of an area with intensive agriculture (Garling 1940), and Georg Steinbacher counted the birds in a park and the zoological garden in Frankfurt (Steinbacher 1942).

These studies in the 1930s and 1940s notwithstanding, the old generation of museum ornithologists in Germany, just as much as their counterparts in Britain, initially dismissed field ornithology as a mere amateur pastime. Kurt Floericke was one of the unhappy pioneers of field ornithology who in consequence had to struggle to make a living (Franke 2009; Wegner 1919). He began his publishing career in the tradition of museum ornithology with a dissertation on the birds of Silesia (Floericke 1892). In the years that followed, he spent his time shooting, stuffing, and selling birds at Rossitten, located on the Curonian Spit on the Baltic Sea. At Rossitten he was the founding member of an ornithological field club called the Verein vergnügter Vogelliebhaber (VvV), a name that can be literally translated as "Society of Joyful Birdlovers"—or as "Blissful Birdlovers' Brotherhood," if one wants to retain the alliterative playfulness of the original. The club had a director of activities who was in charge of organizing field excursions and weekend trips.

Floericke eventually found a permanent position in 1907 at the Kosmos publishing house, which specialized in popular natural history and science. His output as a natural history writer was astonishing, comprising more than one hundred short books and eight hundred articles. Among them was *German Bird Book* (*Deutsches Vogelbuch*), published in the very first year of his engagement with the publishing house. His troubles finding gainful employment show up in the text:

> Ornithology is not a field of study to enter in the hope that it will provide a well-paid job! That should be said to everybody who is tempted by its charms to choose it as an occupation. For the full-scale study of ornithology one needs means that are for the most part available only to the millionaire or to the employee of a major museum. Among the contemporary circle of professional zoologists, where science only starts with a 300-fold magnification (these gentlemen are of course not able to identify a bird in the forest), ornithology is looked upon with reservations to begin with. This is the reason why somebody who chooses it as his main field of inquiry has hardly any prospect of receiving a professorship at university. (Floericke 1907, 122)

The best-known of Floericke's publications was his *Pocketbook for Bird Identification* (*Taschenbuch zum Vogelbestimmen*), a manual on field ornithology, albeit one that was equally geared toward museum ornithology—he had himself published a manual on taxidermy in 1897. Like other natural historians who had experience in both ventures, Floericke was aware of the peculiar difficulties faced by the field ornithologist:

> It's easier for the botanist and geologist, because the plants and fossils stay neatly in their place and do not run away. The student of entomology can likewise for the most part conveniently study his objects of inquiry at close range, capture them easily, carry them home and breed them as he pleases, and thus observe all their signs of life. . . . He certainly has access to lethal long-range weapons, which he can use to acquire these creatures [birds], but a shot fired immediately puts a bloody end to all further observations. The true friend of nature will not have the heart to point a deadly gun at an innocent songbird and to destroy with one stroke of his finger such a joyful bird's life. (Floericke 1912, 6)

But Floericke was also attentive to the needs of the museum ornithologist. The first part of the guide consisted of an expanded version of a systematic classificatory key produced by Reichenow (Reichenow 1902), which was intended for the identification of birds in the hand. The second part comprised a considerable variety of different keys to field identification, arranging birds by season, habitat, and size, and including information on voice, flight, and behavior as well as descriptions of nests and eggs.

FIELD ORNITHOLOGY AND ORGANIZED BIRD CONSERVATION

The work of the League for Bird Protection was put on a firmer footing in 1933 with the rise to power of the National Socialists. The Nazis' so-called *Gleichschaltung*—the synchronization of all aspects of social and political life—was an effort to centralize all organizations within the Reich and to exclude any that were not in conformity with their political leanings. Some organizations gained from this policy, while others lost. All bird conservation organizations were subsumed under the League for Bird Protection, now renamed the League for Bird Protection of the Reich (Reichsbund für Vogelschutz), which was then the largest and most successful organization in the field. The German Society for the Protection of Birdlife (Deutsche Verein zum Schutz der Vogelwelt) lost its independence in 1938 (Wöbse 2003b).

The League for Bird Protection itself was restructured. After receiving official acclaim for her work and the organization that she had founded, Lina Hähnle was replaced as its head in 1938. The Nazi Party member Reinhard Wendehorst, a holder of the golden party badge (*goldenes Parteiabzeichen*), took over the leadership, and the league's headquarters was officially moved from Stuttgart to Berlin, where Wendehorst was head of the National Socialist School for Engineers and was involved with the work of the Ministry for Education of the Reich. Unofficially, however, Lina Hähnle continued to run the organization. There is no indication that Wendehorst was significantly involved in its day-to-day business. The arrangement did not change much when Lina Hähnle died in 1941 at age eighty-nine: the Hähnle residence remained the unofficial headquarters, and the daily work of the organization was taken over by her son Hermann. With the *Gleichschaltung*, the league's membership rose from thirty thousand to fifty-five thousand between 1937 and 1941.

The leadership of the league had supported the Nazis' rise to power well before it came to the distribution of the spoils. In an article in the annual report for 1933, Lina Hähnle had welcomed the Nazis. "A miracle has happened. Germany has come together in an hour of inner need without outside force. Trustful, it follows its leader, Adolf Hitler, in its recovery, in which he wants to heal the German people through their connection to blood and soil. Love of one's *Heimat* and of nature are the strong roots from which Germany can gain strength, and for that reason we may count—small though our field of work may look—as fellow combatants. We stand joyfully behind our Führer and pledge to contribute all our strength to his high ambition" (Bund für Vogelschutz 1934, 1).

Yet the rise of the Nazis seems to have effected no fundamental change in the aims or agenda of the organization. Both before and after 1933, the league generally endorsed the political powers of the day. Earlier in the century it had counted members of the German aristocracy and the US president Woodrow Wilson among its honorary members. Now that the Nazis were running the show, the league emphasized its close collaboration with the Führer. Its annual report for 1939 noted that it had sent five thousand nest boxes to Hitler's domicile at the Obersalzberg in the alps. Lina Hähnle, still in office at the time, had administered the shipment personally.

It is difficult to tell from the few scattered documents that remain where to draw the line between true conviction and simple opportunism in the league's political stance. Beyond doubt, the organization strove to convey that its work had the full support of the new leadership of the Reich. In a leaflet written by Hans Schwenkel bearing the slogan "The Führer holds his protecting hand over the German hedgerows," of which twenty-one thousand copies were distributed between 1940 and 1942, the organization called for the preservation of shrubs and hedges (Reichsbund für Vogelschutz 1943, 2). The slogan was presumably based on a statement Hitler had made. Although the gesture in question can easily be pictured, it is not altogether clear whether Hitler had intended it to be understood in the way the leaflet presented it. It is more likely that the remark was no more than an offhand comment uttered in a very different context.

On the other side of the North Sea, organizational developments were running in the opposite direction. The Royal Society for the Protection of Birds (RSPB) had steadily lost members since the mid-1910s, and its leaders' competence had been called into doubt. Like the league, the RSPB continued to be run by its founding generation. The secretary, Linda Gardiner, who had been central in establishing the organization in 1889, was still in office when membership reached a low of 4,152 in 1929. Thirty years earlier the organization had reported as many as twenty-two thousand members. To a certain extent it was the organization's success that made for its failure. The RSPB had long been a single-issue organization focused on the plumage trade, and, when the Plumage Bill was passed by Parliament in the 1920s, it lost its cause célèbre.

At the beginning, then, the work of the RSPB was fairly far removed from ornithology as a field of study. This distance narrowed after ornithology made field observation its major method of inquiry. In the last decade of the nineteenth century and the first decades of the twentieth, a contrast between "science" and "sentiment" had loomed large, at least in the view of the British Ornithologists' Union. Although reflecting the positivist predilections of ornithologists more than anything else, this divide also reflected the fact that concern for animal welfare was incompatible with ornithologists' habit of shooting birds. The alignments that existed were to a large extent alignments in numbers. Animal welfare is concerned with the well-being of individual animals and nature conservation with the survival of species. Yet if one kills enough individuals of a given species as in the case of the feather trade the welfare issue turns into a conservation issue. In abstract principle, animal welfare and nature conservation are two different endeavors. In practice they are at times the same. To the extent that the RSPB condemned killing as a matter of principle, it diverged from the stance of the BOU and the early BTO. These three organizations subsequently found common ground not because they agreed in principle but because the RSPB gave up animal welfare and ornithologists largely gave up killing birds. Toward the end of this process, the RSPB had turned from an animal welfare society into a bird watching club.

The transition did not come about on its own but resulted from a palace revolution at the organization's annual general meeting in 1936, led by a group of field ornithologists who wanted to carry the organization into a new direction. At its head was Julian Huxley, the grandson of Thomas Henry Huxley and secretary of the Zoological Society of London, who had gained acclaim in ornithology for his observations of the courtship behavior of great crested grebes (Baker 1978; Burkhardt 2005; Clark 1960; J. Huxley 1970–1973). Huxley had been a proponent of field ornithology since his early days as professional biologist at the University of Oxford and the Rice Institute at Houston, Texas, when he published an article on bird watching and biological science (J. Huxley 1916). In an introduction to the posthumously published works of Edmund Selous, who in 1901 had coined the term *bird watching* (Selous 1901), Huxley argued that "the intensive watching of birds in natural surroundings has become a true branch of science, and has led to very great advances in our knowledge of scientific natural history" (J. Huxley in Selous 1927). He repeated this assessment in unaltered form over many years (e.g., J. Huxley 1930, 1946a, 1946b, 1950).

At the RSPB Huxley's aim was to move the work of the organization in the direction of the work of the BTO. Following the advice of a six-man committee under the chairmanship of Huxley, the matter was put to an election at the RSPB, and leadership passed from Margarette Lemon, who had administered the organization since Linda Gardiner's retirement in 1935, to Robert Preston Donaldson. After the restructuring, the membership numbers of the organization began to recover (Royal Society for the Protection of Birds 1936, 28–29). Huxley joined the governing council of the RSPB in 1936, and the annual report for the following year stated that his presence had a beneficial effect. Differences within the society had been happily put aside, and it was now engaged in cooperation with other societies. The BTO and the RSPB in particular, the report pointed out, "seemed to him [Huxley] to be doing complementary work: . . . study on one hand and protection on the other" (Royal Society for the Protection of Birds 1938, 43).

In both Germany and Britain in the 1930s, important institutional ties were forged between organized bird conservation and field ornithology. The nature of these ties, however, differed substantially between the two countries. In Germany a link emerged between the League for Bird Protection and the Bird Protection Stations, while in Britain a tie was established between the Royal Society for the Protection of Birds and the British Trust for Ornithology. (There was no true German equivalent to the British Trust for Ornithology; neither was there a true British equivalent to the German Bird Protection Stations.) On an institutional level, these new ties made the differences in bird conservation between the two countries even more pronounced. Although ornithologists in both countries produced data on birds in the field, they counted different things and for different purposes.

These new organizational ties were crucial for the future development of bird conservation. The British Trust for Ornithology and the German Bird Protection Stations were initially rather small enterprises. Yet they became the institutional settings that provided the larger context of organized bird conservation in the two countries. The same major events in subsequent decades, including environmental disasters that transgress national boundaries, would be experienced and interpreted differently within these two institutional contexts. Institutionalizing the experience of birds as work has different effects than institutionalizing it as play in the sense that culture (i.e., the way we perceive and evaluate the world) is shaped by the practices in which these perceptions and evaluations are embedded. Institutionalizing field ornithology as part of the world of work helped to produce data on species abundance, while institutionalizing it as a game facilitated the production of data on species diversity. Environmental problems, such as population decline or endangerment, cannot be assessed independent of these data. Institutionalizing the practice of data collection in these different ways determined which changes in the environment would be perceived as problems and which would simply remain unnoticed.

In Germany, organizational changes continued to be driven mainly by legal decree. Leadership of most of the bird conservation organizations was merged, and the League for Bird Protection and the established Bird Protection Stations, above all Seebach, were the winners. The Decree for the Protection of Wild Plants and Nonhuntable Wild Animals (Nature Protection Decree) of 18 March 1936 (Verordnung zum Schutze der wildwachsenden Pflanzen und der nichtjagdbaren wildleben Tiere [Naturschutzverordnung]) declared in paragraph 2(1) that "the titles 'Bird Station' and 'Bird Protection Station' may be used only by permission of the federal Nature Protection Agency" (1936). Out of the dozens of research stations that existed under these and similar titles and that had received federal support, only three bird stations (Rossitten, Heligoland, and Hiddensee) and six Bird Protection Stations (Seebach and Steckby among them) remained. The three bird stations continued to be administered by the German Ornithologists' Society. The oversight of the six Bird Protection Stations recognized by the decree was assigned to a Working Group of Bird Protection Stations (Arbeitsgemeinschaft Vogelschutzwarten) led by Karl Mansfeld, who had already organized the work at Seebach since 1925 (Mansfeld 1938).

The German Society for the Protection of Birdlife was hit the hardest by the *Gleichschaltung*. Not only did the organization cease to exist, but Mansfeld also took over the editorship of its journal, the *Ornithological Monthly*, which was renamed *German Birdlife* (*Deutsche Vogelwelt*) in 1938. It became the joint publication of the Working Group of Bird Protection Stations and the League for Bird Protection of the Reich. Karl Hennicke, the author of some of the most comprehensive works on bird conservation (e.g., Hennicke 1912, 1917a, 1917b), lost both his position as the vice president of the German Society and as the editor of the journal. The League for Bird Protection, on the other hand, gained by the merger and was put into closer conjunction with the economic aspects of bird conservation through its cooperation with the Working Group of Bird Protection Stations.

The similarity in titles between the newly created journal *German Birdlife* and the well-established *British Birds* calls for a comparison. Despite similarity in name, there are striking differences in content. The German publication was primarily concerned with bird conservation from an economic standpoint, while the British journal paid no attention to economic ornithology whatsoever. Conversely, the collecting and list-making pursuits of *British Birds* received some attention in *German Birdlife* but were far from central to its content. The new German journal placed even more emphasis on economic ornithology than its predecessor. The previous editor, Karl Hennicke, had endorsed Liebe's ethical approach to bird conservation that highlighted the intrinsic value of birds, while Mansfeld was a disciple of Berlepsch and paid most attention to economic ornithology. The war led to the consideration of additional topics, such as the selection of bird food to replace seeds that were now rationed.

Mansfeld's most important publication was the booklet *Bird Protection in Forest, Field, and Garden* (*Vogelschutz in Wald, Feld und Garten*), which went through six editions from 1934 to 1944 (Mansfeld 1934, 1944). It was an abbreviated and slightly updated version of Berlepsch's well-known work on the same topic. If anything, Mansfeld's statements about the economic value of birdlife were even more straightforward than those of his predecessor. The booklet contained descriptions of sparrow-proof bird feeders (called *Antispatz* or *Kontraspatz*) that were constructed to allow blue tits but not the slightly larger and heavier sparrows to use them. Mansfeld further recommended reducing the numbers of sparrows by destroying their nests and putting up sparrow traps. He estimated that twenty to thirty sparrows could be caught with such traps every day. "The captives are sedated and killed through smashing their head with a hammer on a hard surface or through quickly tearing their heads off" (Mansfeld 1944). Mansfeld was also the author of several leaflets, issued by the Bird Protection Station in Seebach, on how to feed birds in winter and how to build nest boxes (Mansfeld 1935, 1937a, 1937b).

An even more systematic effort at population monitoring and control was carried out by Otto Henze during the same period. Henze had studied forestry and written his dissertation in the field of applied zoology at the University of Munich. In 1936 he issued the first edition of his *Checklist for Nest Boxes in Agriculture and Forestry* (*Kontrollbuch für Vogelnistkästen in der Land- und Forstwirtschaft*), published by the Reichsnährstand, the statutory corporation of farmers in the Third Reich (Henze 1936). The brief text consisted largely of a reminder about the basic principles of practical bird conservation. The booklet's true purpose and novelty was a table for record keeping. Henze explained that from the experience of checking tens of thousands of boxes over a decade, he had concluded that artificial nesting places were useless if not properly managed. He urged his fellow bird conservationists to check their nest boxes at least three times a year and record the results in his checklist. For consistency and efficiency, Henze proposed using a system of symbols to represent the most common species. The prime reason for monitoring the boxes was to remove any unwanted species. Sparrows, of course, headed the list. "To prevent the sparrows from reproducing in the nest boxes at the expense of the tits, the young sparrows are taken out of the boxes at particular points in time and serve a purpose in chopped-up form, as chicken feed" (Henze 1936, 15). He also emphasized the importance of cleaning the boxes at the end of each nesting season to prevent infestation by parasites such as fleas. Even without parasites, uncleaned boxes would become useless after the third year of use because dirt, rotten eggs, and dead birds would accumulate.

Henze's checklist allowed users to keep a record over several years, depending on the rigor of the controls and the nature of the habitat. Nest boxes placed in gardens needed more monitoring, those in a forest less, yet every owner of nest boxes was urged to maintain such records. Henze himself used them in the preparation of his *Bird Protection against Insect Infestations in Forestry* (*Vogelschutz gegen Insektenschaden in der Forstwirtschaft*), published in 1943, the most extensive field study on the economic utility of birdlife in the country to date (Henze 1943). He recorded young birds in the nest boxes and observed what kind of food the parents fed them. His aim was to put the calculus of the benefit and harm of practical bird conservation on an even firmer scientific basis. The author was aware that a species could be useful in one habitat while harmful in another, yet he stuck to the idea that the task of practical ornithology was to determine which species are overall more useful. To those species considered useful he assigned the term *labor bird* (*Arbeitsvogel*): "For the terminology to be more comprehensible and easy to use in practice, we use the term 'labor bird' to refer to those birds that are successfully introduced in many breeding pairs to combat a certain pest insect within a limited territory of a district" (Henze 1943, 31).

FIGURE 4.2 Labor birds at work. Illustration by Franz Murr, in Otto Henze, *Vogelschutz gegen Insektenschaden in der Forstwirtschaft*, 1943.

Birds that did not work for human benefit were excluded from protection. The neologism *labor bird* clearly indicates what German bird conservation of the time was all about: it was the very opposite of a game or pastime. Bird conservation, in Henze's view, was a serious economic enterprise, and the birds were anything but toys.

FIGURE 4.3. Checklist for Nest Boxes in Agriculture and Forestry, Otto Henze, 1936.

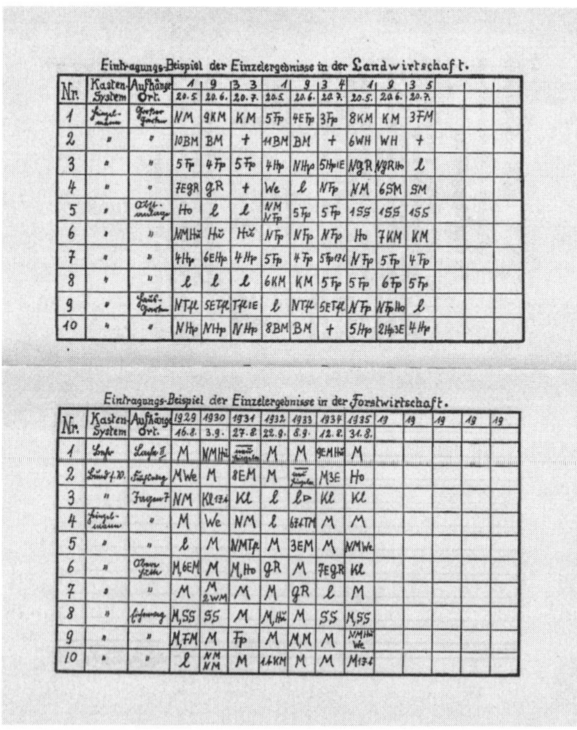

Eintragungs-Beispiel der Einzelergebnisse in der Landwirtschaft.

Nr.	Kasten-System	Aufhänge-Ort	1	9	3 3	1	9	3 4	1	9	3 5			
			20.5.	20.6.	20.7.	20.5.	20.6.	20.7.	20.5.	20.6.	20.7.			
1	[illegible]	[illegible]	NM	9KM	KM	5Fp	4EFp	3Fp	8KM	KM	3FM			
2	"	"	10BM	BM	+	11BM	BM	+	6WH	WH	+			
3	"	"	5Fp	4Fp	5Fp	4Hp	NHp	5HpIE	NgR	NgRHo	l			
4	"	"	7EgR	gR	+	We	l	NFp	NM	6SM	SM			
5	"	[illegible]	Ho	l	l	NM NFp	5Fp	5Fp	1SS	1SS	1SS			
6	"	"	NMHü	Hü	Hü	NFp	NFp	NFp	Ho	7KM	KM			
7	"	"	4Hp	6EHp	4Hp	5Fp	4Fp	5Fp1Tl	NFp	5Fp	4Fp			
8	"	"	l	l	l	6KM	KM	5Fp	5Fp	6Fp	5Fp			
9	"	[illegible]	NTfl	5ETfl	Tfl1E	l	NTfl	5ETfl	NFp	NFpHo	l			
10	"	"	NHp	NHp	NHp	8BM	BM	+	5Hp	2Hp3E	4Hp			

Eintragungs-Beispiel der Einzelergebnisse in der Forstwirtschaft.

Nr.	Kasten-System	Aufhänge-Ort	1929	1930	1931	1932	1933	1934	1935	19	19	19	19	19
			16.8.	3.9.	27.8.	22.9.	8.9.	12.8.	31.8.					
1	[illegible]	[illegible] II	M	NMHü	[illegible]	M	M	9EMHü	M					
2	[illegible]	[illegible]	MWe	M	8EM	M	[illegible]	M3E	Ho					
3	"	[illegible] 7	NM	Kl1Tl	Kl	l	l▷	Kl	Kl					
4	[illegible]	"	M	We	NM	l	6T1TM	M	M					
5	"	"	l	M	NMTfl	M	3EM	M	NMWe					
6	"	[illegible]	M,6EM	M	M,Ho	gR	M	7EgR	Kl					
7	"	"	M	M 2WM	M	M	gR	l	M					
8	"	[illegible]	M,SS	SS	M	M,Hü	M	SS	M,SS					
9	"	"	M,FM	M	Fp	M	M,M	M	NMHü We					
10	"	"	l	NM NM	M	1tKM	M	M	M1Tl					

Yet Henze also seems to have recognized the element of excitement and exploration that could be involved in observation and record keeping. Among his permanent records, some findings stood out as curiosities. Common bird species were recorded through the use of a special sign key. Items, however, that he had come across only once throughout the years of his investigations were presented in a separate illustration on "Rare Curiosities from Nest Boxes." Henze was positively excited about these finds. "Downright detective studies can be conducted with bird nests taken from nest boxes" (Henze 1943, 211). Included in the picture were a mole, a bone of a deer gnawed on by mice, a six-inch-long piece of root, a blue piece of a New Year firework, and a piece of coconut fiber. There is of course nothing special or curious about these items other than the fact that they stood out against the mass of Henze's records. In any other context these would have been quite mundane items, yet to this day-to-day recorder of nest boxes they were exceptional.

Persistently promoted by ornithologists and bird conservationists such as Henze, nest boxes and bird feeders could be found throughout the entire German Reich. Shortly after Germany's defeat in 1945, Andrzej Zaorski accompanied a team of doctors sent to Auschwitz by the Polish Red Cross to save surviving inmates. He found, to his astonishment, that the entire camp and the staff living quarters were equipped with nest boxes. In the office of the camp commander, Rudolf Höß, he found an ornithological article titled "Observations on the Bird-Life of Auschwitz" (*Beobachtunge über die Vogelwelt von Auschwitz*), published in 1941 in the *Annals of the Natural History Museum* in Vienna (Niethammer 1941). The author was the SS officer Günther Niethammer, who thanked Höß in the article for his support. Niethammer had worked as a guard at the main gate of the camp from October 1940 to October 1941, a point in time before the gas chambers were installed but well after the first mass executions inside and outside the camp had begun (Nowak 2010).

According to Niethammer's own account after the war, "It was sufficient to be at that place for a single moment to notice this great crime" (quoted in Nowak 2010, 78). Niethammer was certainly a devout Nazi when he became an SS officer, but it is not clear from the historical record or from his own account what effect his time at Auschwitz had on him. It seems that when he realized the full implications of being an SS officer at Auschwitz, he took refuge in bird watching. This escape from his duties was possible because of his cordial relations with the camp leaders, whom he supplied with wildfowl, mainly ducks, shot on his ornithological excursions in the area. "I am a kind of concentration camp SS game-keeper [*Jägermeister*] around here," he stated in a letter to Erwin Stresemann in 1941 (quoted in Nowak 2010, 78).

Before and during his tenure in the SS Armed Forces (Waffen SS), Niethammer worked with Stresemann at the Natural History Museum in Berlin, where he was in charge of editing the *Handbook of German Birdlife* (*Handbuch der Deutschen Vogelkunde*), published in three volumes between 1937 and 1942 (Niethammer 1937–1942). The last volume, like the bird watching article found in Höß's office, was signed by Niethammer in his capacity as an SS officer. By the time the first volume was published, the handbook had been in preparation for almost twenty-five years. The initiative for the work came from Ernst Hartert, who had coauthored the *Hand-list of British Birds* with Witherby, and was supported by Stresemann, but came to a temporary halt due to the Great War. The plan was taken up again some years after the war, with Harry Witherby's *Practical Handbook*, published in two volumes in 1920 and 1924, as a model.

Stresemann assigned Niethammer the task of editing the handbook in 1934. In the introduction to the first volume, the twenty-nine-year-old ornithologist did not hesitate to dismiss the German handbooks available at the time as outdated and declared to follow the example of the more recent British publications. Like its British models, the German handbook began the entries for each bird species with a reference to the standard ornithological handbooks in the field. Niethammer had personally taken the measurements of the various species at the Natural History Museum in Berlin, under the general supervision of Stresemann. The descriptions of each species further contained sections on field-ornithological markers, general distribution, distribution in Germany, migration, habitat, breeding habits, nests, eggs, and feeding habits. Peculiar to the German version was a section on the parasites of each species. Also different from the British example was the fact that the question of which bird species could be observed within the boundary of the nation repeatedly changed over time. Editorial decisions did not always reflect the currently existing boundaries. Included in the first two volumes were bird species of Alsace-Lorraine and those parts of Poland that were part of Germany until 1919 and had been ceded to Germany's neighbors after the previous world war. By the time the third volume was published in 1942, these territories had become German again, along with other territories that had been occupied by the Reich during the war. Additional bird species were accordingly incorporated into the text. "Wherever possible, the borders of Greater Germany as established by the victorious campaigns in the new Europe have been taken into account" (Niethammer 1937–1942, vol. 3, iii). Right after the annexation of Austria to the Third Reich in 1938, Niethammer had published a list of species now newly German in the *Ornithological Bulletin* (*Ornithologische Mitteilungen*). A postscript to the third volume contained additions to German birdlife for those taxonomic groups already dealt with in the previous volumes, encompassing the birds of the now German territories in Eastern Europe.

It is one of the more striking features of this publication signed by an SS officer and appearing on the market during World War II that it had more in common with the emerging British science of field ornithology than with the economically oriented German ornithological literature. The *Handbook of German Birdlife* was an inventory of birds for a clearly demarcated territory and as such in large part a collector's list. It did not mention the economic benefit or harm of birdlife, which was previously the outstanding feature of German publications in ornithology. A withdrawal from economic considerations and a heightened focus on rare and endangered species following the British example would in subsequent decades dominate not only German ornithological handbooks but also the work of practical bird conservationists.

CHAPTER 5

Endangered Birds and Indicator Species

Nobody knows for sure when the birds first arrived on the Minsmere Level on the Suffolk coast in the east of England, but the story was recounted many times in autobiographies, ornithological journals, and newspapers. The pied avocet, *Recurvirostra avosetta*, or simply avocet, a wading bird about the size of a goose, with long legs and a long, thin, upwardly curving bill, had been extinct in Britain since the mid-nineteenth century because of persecution by skin and egg collectors and habitat loss: the marshy Minsmere Level was drained and reclaimed for grazing by 1890. In 1940, to defend against a possible landing by German tanks, part of the level was flooded and the seashore planted with steel pikes and massive blocks of concrete. The tanks never arrived, but the flooding led to the formation of a series of shallow freshwater meres and a new and very large reedbed (Axell 1992; Axell and Hosking 1977; Dougall and Axell 1982).

Recolonization of the Minsmere Level by the avocet began after the transformation of the site. The story was recounted by Lieutenant-Colonel John Keith Stanford, who with his brother, Brigadier H. M. Stanford, discovered the birds in 1947.

> I could hardly believe my ears, and I felt curiously like a burglar as I waded out through the warm yellow-brown water over mud which was pocked with mortar-bomb holes from the battle school, so that I had to be careful not to go in over my waist. The avocets kept piping shrilly overhead. And there on the mud islet were the three nests within a few yards of each other, made of dead rushes and two already with full clutches. It was one of those high moments

of discovery which come to all birdwatchers, if they are lucky, once or twice in a lifetime. Words fail you; you feel at once very proud and humble, as if at some undeserved accolade. (Stanford 1954, 193).

Together with the local landowner and a small number of others, the Stanfords formed a group to protect the Minsmere birds against egg collectors, rats, and other predators. The men often spent entire days and nights guarding the nests (Stanford 1950; Stanford 1954).

In 1947, the former tenant farmer of the Minsmere Level insisted that all the land should be reclaimed for agriculture under the Agricultural Act. In response, the Royal Society for the Protection of Birds (RSPB) leased around fifteen hundred acres of the area in 1950 and turned it into a nature reserve. John Keith Stanford commented on the benefits of such a conservation effort: "It is to me a most striking instance of what bird-protection can, and should be, not a matter of Acts of Parliament and schedules hung on police-station walls, but a combined effort by people living in their own countryside to save, it may be for one year only, or it may well be for posterity, some portion of the beauty in our marshland which our forefathers too often allowed to perish" (Stanford 1954, 165). Minsmere was a huge conservation success. In 1952, an RSPB leaflet counted as many as 171 bird species observed on the reserve over a period of four years.

Yet the society soon realized that the restriction on agriculture and the absence of shooting alone would not suffice to maintain Minsmere as a bird reserve. About a quarter of its open water was lost under a cover of reeds in the first ten years. To maintain large sections of the wetland, work began in 1962 on the construction of a fifty-acre shallow lagoon system, subsequently christened "the Scrape" because it was scraped out with a bulldozer. Minsmere turned into the RSPB's model nature reserve and was used to advertise the work of the organization. Today Minsmere is the most famous bird reserve in Britain. It is not only a mecca for bird watchers but also a model for conservationists the world over.

The RSPB had begun to purchase the land at Minsemere in 1977, expanding the reserve with further land purchases over the years. Although the organization had already acquired land at the beginning of the century, it was not until the 1950s that the society adopted the creation of nature reserves and habitat management as its major conservation strategy. Previously, the main aim of land ownership had been to keep the birds from being shot by sport hunters and museum ornithologists. In keeping with the then predominant understanding of the causes of declines in bird populations, these efforts were

FIGURE 5.1 Leaflet advertising the Minsmere Reserve of the Royal Society for the Protection of Birds. Illustrations © Robert Gillmor. Photographs © RSPB.

all about preventing bird shooting, not about habitat protection. In some cases, the RSPB did not even buy the land or a freehold but simply acquired the shooting rights (e.g., "Brean Down" 1912). A watchers fund was established by the organization in 1905 to make sure that these rights were enforced. With the help of watchers the RSPB could keep the birds from being shot, although the land might still be used for agriculture by its owner.

At Minsmere, by contrast, conservation focused for the first time on habitat, and with this new focus came the realization that conservation required more than just leaving birds alone: it required active reserve management, as a decline in avocet numbers in the 1950s made evident. After the war, rising membership dues were invested in more land purchases. This reserve- or habitat-based form of conservation practiced by the RSPB and other nature conservation organizations proved remarkably successful. *The Breeding Birds of Britain and Ireland*, a survey that covered the years 1800 to 1967, noted that "in recent years gains have outnumbered losses: since 1940 only one species has ceased to breed, while ten have become established or reestablished. Indeed, there are currently more species known to be breeding in Britain and Ireland than at any other time since 1800" (Parslow 1973, 187).

Most of the reestablished species, the author pointed out, nested on reserves, including the avocet. Following its success with the conservation of this species at Minsmere, the RSPB chose the avocet for its logo in 1970.

ENVIRONMENTAL (R)EVOLUTION

The decades after World War II are frequently credited with the rise of environmental movement activism. An integral part of the current sociological debate on Green issues is the classification of this activism as a "new social movement" (Buechler 1995; Rucht 1994). Despite a multiplicity of theoretical models, two central assumptions are commonly shared by all new social movement theorists. First, it is assumed that a radical break or rupture within modernity took place in the decades following the 1960s, generating the need to distinguish between an old modernity and a new modernity. Second, it is argued that this radical rupture is the consequence of the internal logic of the process of modernization itself. That is, the new modernity is considered to be a reaction to the sometimes intended yet more often unintended consequences of the old modernity.

Theories differ profoundly in whether they characterize the consequences of modernization as negative or positive. The negative evaluation is represented by theories of need defense, which interpret environmental values and environmental movements as attempts to contain or overcome the undesired side effects of modernity. Examples are the production of health risks resulting from new biological, chemical, and nuclear technologies (Beck 1992; Beck, Giddens and Lash 1994; Giddens 1990), as well as the increasing colonialization of an authentic lifeworld by a capitalist economy and a bureaucratic state following a logic of economic growth and scientific progress (Habermas 1981, 1984–1987; Offe 1985).

Theories of rising demands, on the other hand, interpret these movements not as deriving from the production of new problems but rather from the solution of old ones; that is, they provide a positive characterization of the consequences of modernity. By providing economical security and political stability in the decades following World War II, modern societies are held to have shifted from materialist to postmaterialist values, qualifying them as postmodern societies (Clark and Inglehart 1998; Inglehart 1977, 1990, 1997). It has been argued further that contemporary postindustrial societies have moved from a struggle over the means of production toward a struggle over a growing capacity for self-production and self-control based on information and knowledge (Touraine 1981, 1985; Touraine et al. 1983), replacing economic conflicts with symbolic issues of identity and self-expression as the leading social forces (Melucci 1985, 1989).

Theories of need defense and of rising demands both argue for the existence of a radical rupture between old and new social movements driven by overall changes in social structure. The nature conservation of the nineteenth century is identified as a phenomenon of the past and contrasted with the environmentalism of the twentieth century. Yet the membership numbers of the relevant movement organizations show that bird conservation, as an integral part of nature conservation, is not a thing of the past—its high time is the present. Bird conservation for the most part outstripped environmentalism in membership numbers throughout this period. Moreover, this trend was not driven simply by changes in social structure but to a significant extent by the work of traditional conservation organizations.

The different ways in which bird conservation had been institutionalized in the first half of the twentieth century shaped this development in the second half. Bird conservation in Britain continued to be a collector's pursuit, with a special value placed on the preservation of rare species. With the shift from museum ornithology to field ornithology, natural history museums ceased to be the major institutional loci for collecting, giving way to nature reserves such as Minsmere for the collecting of rare birds. The only difference between the two forms of collecting is that while a museum preserves bird bodies, a nature reserve conserves living birds. The increasing popularity of bird watching as a leisure pursuit contributed to an explosive growth in RSPB membership in the late 1960s and early 1970s. Membership fees provided the bulk of the funds for the expensive business of acquiring and managing reserves.

Developments in Germany in the same period, by contrast, marked a radical rupture with the past. Membership in the League for Bird Protection collapsed after the war, and the organization came close to dissolution. The league continued its work with the old generation of bird conservationists, most of them now in their seventies and eighties, and its aims were the same as before the war: the protection of useful birds for economic ends and the destruction of harmful species. Given both the financial situation of the league and the lack of primary interest in rare bird species, the organization acquired no noteworthy reserves during this period. Bird extinctions in Germany actually increased after 1945.

In the 1970s and 1980s, conflict arose within the organization through a process of generational succession. The new generation taking the reins saw conservation as driven by larger political causes. The relationship between the main office of the League for Bird Protection, now renamed the German

League for Bird Protection (Deutsche Bund für Vogelschutz, or DBV) and its youth organization, the DBV Youth (DBV Jugend), remained conflicted for a long time. No comparable conflict could be observed in Britain. Many of the RSPB's leading members had been members of the society's youth organization, the Young Ornithologists' Club (YOC). Youth work was seen as central to the organization's success. The relation between young and old generations was harmonious. Except for a new emphasis on lobbying that came with a growing membership, bird conservation in Britain remained unaltered. The period was one of growth, not of change in direction.

THE POSTWAR BOOM OF BIRD WATCHING IN BRITAIN

RSPB membership increased during World War II. In 1939, it stood at 4,825; by 1945 it reached 5,535. A commentator in the annual report for 1944, apparently puzzled by this wartime success, speculated that bird watching must have become a means of coping with the horrors of war. "In spite of, or possibly because of, the times, more and more people are turning to Nature and the beautiful side of life" (Royal Society for the Protection of Birds 1945, 5). Yet membership numbers continued to rise after the war. The increasing success of the RSPB was due not to the war but to its new focus on amateur bird watching. When bird watching as a leisure activity flourished in the mid-twentieth century, the RSPB flourished with it. This increasing popularity was both reflected in and catered to by the publication of numerous bird watching guides (Cocker 2001; Fisher 1966).

One of the best known of these guides was James Fisher's *Watching Birds*, first published in 1940. It contained descriptions of the more common bird species and detailed instructions on how to observe them successfully. Fisher left it to the reader to decide the purpose of the activity. "The observation of birds may be a superstition, a tradition, an art, a science, a pleasure, a hobby, or a bore; this depends entirely on the nature of the observer. Those who read this book may give bird-watching any one of these definitions" (Fisher [1940] 1953, 16). The aim of the publication was to increase interest in the practice and encourage new bird watchers to join the major organizations in the field. The book included full-page descriptions and the addresses of the RSPB and the British Trust for Ornithology (BTO).

Fisher was the assistant curator at the Zoological Society of London from 1936 to 1939, and he had played a leading part in establishing the amateur-based British Trust for Ornithology, of which he was honorary secretary from

1938 to 1944. His guide thus promoted the organization he had just started to administer. He wrote several other books that were instrumental in stimulating popular interest in birds, among them *Birds as Animals* (Fisher [1939] 1954), *The Birds of Britain* (Fisher 1942), and the three-volume *Bird Recognition* (Fisher 1947–1955). Fisher's *Watching Birds* was the most successful of them. It went through several editions and soon inspired further publications of a similar kind.

Another member of the BTO who promoted amateur bird watching with the publication of a field guide was Bruce Campbell. In his youth he was one of those bird lovers who shot birds and collected their eggs. He eventually abandoned these practices after he took part in a bird survey in the 1920s and started to publish sight records in the journal *British Birds*. In 1943 he joined the BTO after he met James Fisher, and he became the first full-time secretary of the organization five years later. Campbell not only took part in ornithological survey work but also kept private lists, assembling in his diaries a "fun list," a "walk list," and a "European list," together with notes on "first" sightings of a species. He also frequently contributed writings to the RSPB membership magazine *Bird Notes*, the best known of which was a 1951 article on the competitive motivation of field ornithology, aptly titled "birdmanship" in analogy to sportsmanship (Campbell [1951] 1962). He did not exclude his own motivation from the analogy. In his autobiography Campbell confessed that "birdwatching to me has always been a sport rather than an art or even a science, though, of course, a scientific structure can be built on it" (Campbell 1979, 91). In 1952 he published a field guide titled *Bird Watching for Beginners*, with the intention of converting museum ornithologists into field ornithologists. An appendix advertised the work of the BTO and RSPB (Campbell 1952).

The undisputed classic among the British field guides, still in print today, was *A Field Guide to the Birds of Britain and Europe*, prepared by Roger Tory Peterson, Guy Mountfort, and Peter A. Hollom in 1954. The introduction, written by Julian Huxley, described the niche the book aimed to fill: "Knowledgeable ornithologists who already possess one or several of the encyclopaedic works, such as Witherby's five volume *Handbook of British Birds*, ask for a similar authoritative and complete reference book 'small enough to carry in the pocket'" (Peterson, Mountfort, and Hollom 1954a, ix). The publication was innovative in that its illustrations were prepared for the specific purpose of identification in the field. It included pointers noting the birds' distinctive field marks, thumbnail distribution maps, and a concise, clear text, rather than the exhaustive treatment offered by handbooks. In addition, the book was small enough to be carried into the field.

Peterson, a US ornithologist and painter, had pioneered this approach in his *A Field Guide to the Birds* (Peterson 1934). The plan for a British equivalent had been made when Mountfort and Peterson met in Pennsylvania in 1949 (Mountfort 1991). Peterson was also friends with Fisher, with whom he made a birding trip in 1953 from Newfoundland to Alaska via Mexico. The journey resulted in their classic book *Wild America* (Peterson and Fisher 1955).

Like its American predecessor, the British and European field guide turned into an almost immediate classic. The preface to a revised edition in 1966 claimed that with "the single exception of its earlier companion volume, *A Field Guide to the Birds (of North America)* it has now sold more copies than any book on birds yet produced in any country" (Peterson, Mountfort, and Hollom 1966, xi).

Field guides such as this consolidated the rise of field ornithology and helped strengthen the link between science and conservation. A large number of them were written by bird conservationists (e.g., Dobinson 1976; Ennion 1963; Ferguson-Lees, Hockliffe and Zweeres 1975; Gooders 1975; Hammond 1978; Lockley 1968) and used to promote the work of bird conservation organizations. The RSPB itself was engaged in publishing field guides. Among the titles was, for instance, the *RSPB Guide to Birdwatching*, written by Peter Conder, the director of the organization from 1963 to 1976 (Conder 1978).

One of the most prolific guidebook authors who had helped to establish the genre was Richard Fitter, who in 1952 published the *Collins Pocket Guide to British Birds*, illustrated by Richard A. Richardson. Designed to allow even the inexperienced to identify birds, it was innovative in arranging the species by habitat and size, and in including a silhouette of a sparrow in each picture in order to indicate the relative size of the depicted bird. In the introduction Fitter echoed Fisher, whom he also mentioned in the acknowledgments, in describing bird watching both as a science and as a sport:

> Though bird-watching is becoming an increasingly popular pastime, many are still deterred from taking it up by the initial difficulty of telling one bird from the other . . . But there is no reason at all why anybody with a pair of sharp eyes, a reasonably good ear and a capacity to sit and wait should not get a great deal of enjoyment from bird-watching in his leisure hours. If he also has the integrity to record only what he is sure he has seen, he may also have the satisfaction of adding to the stock of known facts about birds. (Fitter and Richardson 1952, 1)

The book sold more than one hundred thousand copies over the years. The success inspired Fitter to produce about a dozen other field guides, including

FIGURE 5.2 *RSPB Guide to Birdwatching*, by Peter Conder, a former director of the Royal Society for the Protection of Birds, 1978.

the *Collins Pocket Guide to Nests and Eggs* (Fitter [1954] 1968) and the *Collins Guide to Bird Watching* (Fitter 1963). In 1972 he published his best-known work, the *Birds of Britain and Europe with North Africa and the Middle East*, with illustrations by Herman Heinzel and maps by John Parslow (Heinzel, Fitter, and Parslow 1972b). The guide introduced the practice of placing the pictures and text description of each species on facing pages rather than in separate sections to aid identification. Fitter promoted interest in bird watching and conservation not only through his publications but also through his involvement in founding the Berkshire, Buckinghamshire, and Oxfordshire Naturalist Trust and through serving in different capacities in both the BTO and the RSPB, including as a member of the councils of both organizations.

An even better-known conservationist who used publications of this kind to raise interest in birds was Peter Scott (Courtney 1989; E. Huxley 1993; Kear 1990; Scott 1980, 1966). His father, whom he never met, was the Antarctic explorer Robert Falcon Scott. In a last letter to his wife before he died on the return journey of his 1912 South Pole expedition, Robert Scott had advised her to "make the boy interested in natural history," and so she did. She brought her son into touch with some of the best-known naturalists of the day, and he became a member of natural history societies early in life. His main interest from the very beginning were ducks and geese—he published among others *A Coloured Key to the Wildfowl of the World*, illustrated with his own paintings (Scott 1957b). Scott, who had already provided some of the illustrations for the celebrated Witherby *Handbook of British Birds* (Witherby et al. 1938–41), became known as both a painter and a conservationist. After obtaining a degree in zoology and art history at the University of Cambridge in 1930, he spent a term at the Academy of Visual Arts (Akademie der Bildenden Künste) in Munich and then two years at the Royal Academy Schools in London.

Scott was also a sportsman: he excelled at sailing and won a bronze medal in the 1936 Olympic Games for single-handed yachting. Later he developed a passion for gliding and won the British gliding championships in 1963. In his view, bird watching and ornithological record collecting was a competitive sport as well. "Bird-watching has become a delightful game, and what fun to read how others have scored" (Scott 1957a: 229). As with some other conservationists, it was the practice of bird watching that convinced Scott to give up the sport of bird shooting.

Scott became a vice president of the RSPB in 1950 (Royal Society for the Protection of Birds 1951, 1), but he made his major contribution to bird conservation by founding his own organization. While still a bird shooter, Scott visited the River Severn at Slimbridge in Gloucestershire in 1945, seeking the rare lesser white-fronted goose among the wintering greater white-fronted geese. During this trip he chose the location for the establishment of a research organization that he had planned for many years, devoted to the study of the swans, geese, and ducks of the world. The Severn Wildfowl Trust, later renamed the Wildfowl and Wetlands Trust, was set up at Slimbridge in 1946 (*Wildfowl Trust* 1968). The ground at Slimbridge illustrates the institutional continuity between natural history museum and nature reserve particularly well. As in a museum, Scott aimed to acquire examples of every wildfowl species from around the globe.

His collecting trips led him to the Perry River region of northern Canada in 1949 (an area that he mapped while searching for the breeding grounds of the Ross goose), and to Iceland in 1951, where he studied pink-footed geese in their breeding grounds. Stocked with species captured on these and other excursions, the reserve at Slimbridge turned into the largest known wildfowl collection—which Scott could enjoy from his living-room window. His private residence was built on the reserve at Slimbridge and overseeing an artificial pond that became home to a large proportion of his collection. From a collector's point of view, the wildfowl reserve and the pond served the same function as an exhibition floor and a showcase in a natural history museum. The only difference was that Scott's collectibles were not stuffed, mounted, and preserved, but very much alive.

In the early 1950s Scott became involved with the International Union for the Conservation of Nature (IUCN). With two friends, he founded the World Wildlife Fund (WWF) in 1961 (later the World Wide Fund for Nature), initially as a mechanism to raise the money to finance IUCN conservation projects. Scott became the first chairman of the WWF and designed its panda logo. He traveled extensively on behalf of the fund, establishing national appeals, advising on conservation issues and areas for reserves, lecturing, and fundraising. At about the same time he helped establish the Species Survival Commission (SSC) of the IUCN and served as its chairman from 1962 to 1981. His most influential contribution to the work of the SSC was the invention of the International Red Data Books (later known as the Red Lists), "a register of threatened wildlife that includes definitions of degrees of threat" (Scott, Burton, and Fitter 1987, 1). Starting in 1964, for fifteen years these lists were issued by the organization as loose-leaf, ring-bound volumes (Fitter 1974; Fitter and Fitter 1987; Jarvis, Leon, and Oldfield 1981; King 1981).

The Red Data Book was not an entirely new invention: inventories of extinct and endangered animals and plants had been published decades earlier. One of the best known was provided by William T. Hornaday, the director of the New York Zoological Park (now the Bronx Zoo) in *Our Vanishing Wildlife: Its Extermination and Preservation* (Hornaday 1913). The book included accounts of species already extinct and those in danger of extinction, such as the passenger pigeon, which was officially declared extinct one year later. In 1931 the American Committee for International Wildlife Protection was established with the aim of gathering information on endangered mammals and birds. The committee raised funds to support research on this subject, which culminated in the publication of *Extinct and Vanishing Mammals of the Western Hemisphere* by Glover M. Allen (Allen 1942), *Extinct and Vanishing Mammals of the Old World* by Francis Harper (Harper 1945), and *Extinct and Vanishing Birds of the World* by James C. Greenway (Greenway 1958, 1967).

The IUCN Red Data Book, which was aimed chiefly at an academic audience, was accompanied by publications geared for a wider audience. Most of these volumes were in the format of coffee table books and illustrated accordingly. Richard Fitter (who succeeded Scott as chair of the SSC at the IUCN) published *Vanishing Wild Animals of the World* (Fitter 1968); James Fisher, Noel Simon, and Jack Vincent wrote *The Red Book: Wildlife in Danger* (Fisher, Simon, and Vincent 1969); and Prince Philip and James Fisher produced *Wildlife Crisis* (HRH The Prince Philip and Fisher 1970). Most of the authors and contributors to these publications were involved in the work of the Species Survival Commission, and Peter Scott wrote the foreword to each of these volumes.

The discourse on endangered or threatened species was one of the most consequential innovations in communicating the concern for conservation. Yet its development shows that this discursive tool was not an invention in the literal sense of the term. The notion of endangerment was nothing more than a label attached to already established practices and institutions, in this case the practices and institutions of rarity collecting. An endangered species is the same as a rare species with an additional sense of urgency and call to action attached to it. As such the notion of endangerment, unlike rarity, has not merely a descriptive but also a prescriptive (i.e., normative) connotation. Where did this normative connotation come from? One will search in vain for any moral philosophers among the innovators. The Red Lists of endangered species were created by natural historians and field note collectors. It is for the record collector of living species in the field that a rare species qualifies as an endangered species. The new vocabulary or label of endangerment used by Scott in the 1960s was not the outcome of a newly emerging moral discourse on environmental risks but the consequence of the transformation of the focus of bird collecting from bird bodies to living birds. For a museum ornithologist rare species were not endangered because their bodies could very well be preserved in the museum. It was only after collectors made the shift from the museum to the field that rare species could be described as endangered. It was accordingly the already established practice of natural history rarity collecting that fueled the concern for endangered species.

More than this, without such already institutionalized practices it would have been impossible to describe a species as endangered or threatened.

Classifying a species as such presupposes a substantial amount of data; a bird has to be named as a separate species, its distribution in the field known, and its population total estimated. Of all the labels that a species can be assigned, the label "threatened with extinction" is one of the most data intensive. It stands at the end of a process of data collecting, not at the beginning, and its assignment is related to a large number of observations made in the field. The initial Red Data Books thus did not offer exhaustive assessments of every known species, either for birds or for any other taxonomic group. Information on the population level of each species was limited, and it was usually left to ornithological experts to judge whether a species should count as threatened.

Data on birds proved to be the most abundant in the compilation of the Red Lists. In Britain, information of this kind had been systematically collected since the establishment of the journal *British Birds*. Its regular Notes section contained records of rarities observed in Britain, and the number of records submitted rapidly increased with the postwar boom in bird watching. Eventually the editor of the journal could no longer deal with the volume of records submitted, and in consequence a special committee was formed in 1959: the British Birds Rarities Committee (BBRC), comprising ten amateur bird watchers and field ornithologists. Its main task was to determine the validity of the submitted records and to create a unified set of standards for their assessment. All validated records were published in an annual report in *British Birds* (Dean 2007).

These rarity records were eventually included in the British List, the official list of wild birds records in Great Britain. An updated version of the British List, now called *Check-List of the Birds of Great Britain and Ireland*, had been in preparation since 1946 and was completed in 1952 by a subcommittee of the British Ornithologists' Union (British Ornithologists' Union List Sub-Committee 1952). The committee had close ties to both the BTO and *British Birds*. Bernard Tucker, for instance, who was a member of the council of the BTO and the editor of *British Birds*, served on the committee until his death in 1950. The 1952 *Check-List* was the first British List issued by the BOU to make extensive use of sight records, a shift that reflected the gradual professionalization of field ornithology. In 1954, the BOU went one step further and formed an independent committee, the British Ornithologists' Union Records Committee (BOURC), to advise on the authenticity of field-ornithological records. Since its formation, the committee has been the ultimate authority on the addition of birds to the British List. The records published by the BBRC in *British Birds* form a basis for the work of the BOURC.

The data also form the basis for the conservation work of the RSPB. In 1968, the organization created a special committee dealing with rarities, the Rare Breeding Birds Panel (RBBP) (Spencer et al. 1992). It records the abundance of breeding birds in Britain and can be considered the British version of the Red List. As such it is a tool for conservation. The RBBP eventually became an independent body funded by a multitude of conservation organizations, including the British Nature Conservancy, the BTO, and the RSPB. The reports of the panel are published annually in *British Birds*.

The popularity of rarity collecting was on the rise throughout the period in which these institutions were formed. A 1976 publication, *Rare Birds in Britain and Ireland* by John T. R. Sharrock and E. M. Sharrock, which compiled all available records on the topic, noted that the number of rare birds reported annually in Great Britain and Ireland rose from approximately two hundred per year in the late 1950s to about five hundred per year in the early 1970s: "Most birdwatchers are fascinated by rarities. To some, the occasional rarity is an unexpected but welcome excitement in the course of the normal year's birding, but to others the finding of rare birds becomes the raison d'etre of their hobby. Once known as tally-hunters, tick-hunters or tickers, these rarity seekers are now known colloquially as 'twitchers' " (Sharrock and Sharrock 1976, 9).

The publication contained information on 221 rare bird species based on almost eight thousand sight records collected from 1958 to 1972. The heightened attention paid to endangered species throughout this period was not a reflection of a rise of some abstract form of environmental awareness but a result of the institutionalization and increasing popularity of rarity collecting.

THE POSTWAR DOOM OF BIRD CONSERVATION IN GERMANY

The systematic observation of birds in the field had been institutionalized in Germany at the same time as in Britain. In both countries a link had been forged between field ornithology and organized bird conservation by the mid-1930s. Yet the focus of field investigations differed between the two countries: in Britain it was rarity and diversity that mattered most, while in Germany the focus was on the abundance of common species. In Britain the institutional context for this work was the British Trust for Ornithology; in Germany it was the network of Bird Protection Stations. Both continued to be the loci for large-scale programs on bird observation after 1945. For a long time, the employees of the German Bird Protection Stations were as prolific in their published output as the volunteers of the British Trust for Ornithology.

One of the most noteworthy authors on bird conservation in Germany was Sebastian Pfeiffer. Born in 1898, he began publishing on field ornithology in the 1930s. His first monograph was a 1936 field guide, *The Birds of Our Homeland* (*Die Vögel unserer Heimat*), which went through three editions until 1952 (Pfeiffer 1936, 1952). In 1937 Pfeiffer participated in the founding of the Bird Protection Station in Frankfurt, and in 1944 he became its director and chief research scientist. Its work focused on the economic utility of birds. In 1953, Pfeiffer, together with Kurt Ruppert, published a study on the increase in population density of common breeding birds achieved through the use of artificial nest boxes (Pfeiffer and Ruppert 1953) . The study concluded that the number of birds could be increased to ten to fifty times the natural level, much more than even the best-known proponent of these devices, Baron von Berlepsch, the founder of the first Bird Protection Station at Seebach, had estimated (see also Pfeiffer 1955a, 1955b; Pfeiffer and Keil 1958).

Pfeiffer took this research as a point of departure for the publication of a practical guidebook on bird conservation. *The Pocketbook on Bird Protection* (*Taschenbuch für Vogelschutz*), which he edited in 1957, went through an additional four continuously revised and enlarged editions until 1980 (Pfeiffer 1957, 1980). Pfeiffer asserted in the introduction that bird conservation was first and foremost an economic enterprise. His focus was on promoting useful bird species to the explicit exclusion of harmful ones. The pocket guide spent many pages describing what kind of nesting boxes and feeders to choose for various species, where to buy them, and how to use them.

Not much had changed since the time of the publication of Berlepsch's *Complete Book on Bird Protection*, to which Pfeiffer's *Pocketbook* was the successor. Added to the agenda in the 1930s and 1940s was the use of poison for eliminating harmful birds, in addition to the more traditional nets, traps, and guns. Pfeiffer was not a strong advocate of this practice, but he conceded that it was sometimes necessary: "It is a deep insult, in particular to the bird lover, when birds, even so-called harmful birds, are simply put on the same level as rats and mice and the larger pests of the insect kingdom. . . . [Yet] for the prevention of substantial economic damage, the upper levels of the nature conservation agencies can allow the use of poison for the reduction of house and field sparrows in particularly urgent cases" (Pfeiffer 1957, 88–89).

Experiments on the most economically efficient way to poison sparrows were carried out at most of the Bird Protection Stations throughout the 1950s. Large-scale poisoning had first been made legal by an Act for the Protection of the Agricultural Crops (Gesetz zum Schutz der Landwirtschaftlichen Kulturpflanzen) in 1937 (extended, in slightly altered form, in 1949) and had been practiced ever since. If anything, efforts at killing sparrows had been increased over time, not least because of the shortage of food after the war.

The League for Bird Protection (Bund für Vogelschutz) endorsed such measures in the years immediately following the war, and it cooperated closely with the Bird Protection Stations. Many of the council members of the league were full-time employees at the Bird Protection Stations. Sebastian Pfeiffer eventually became the president of the league, or what was left of it, after his retirement from Frankfurt in 1965. In the first two decades after the war, the organization continued to be led by the generation that had dominated its work before the war. Hermann Hähnle, the son of the league's founder, Lina Hähnle, had become the official president of the organization in 1945 and continued in this office until succeeded by Pfeiffer in 1965. In other words, since its foundation in 1899, the organization had been led by two people, mother and son.

Hermann Hähnle never retired from his office as president: he died in office at age eighty-six. When Pfeiffer took over the role, he was already in his late sixties. The choice of a man of this age for the office cannot be attributed simply to the organization's inherent conservatism: it was due to the circumstance that the league continued to be a volunteer organization without any full-time employees. Pfeiffer had the time to do the job because he was retired. He was also willing to move the headquarters of the league from the Hähnle residence in Stuttgart to his own home in Frankfurt. In the move, a substantial amount of the paperwork of past decades was discarded, including all membership files and correspondence. It was the year in which the organization came considerably close to being dissolved.

Pfeiffer was moreover hardly a visionary. In the 1967 annual report, he describes as one of the year's achievements that "all books have been provided with a dust jacket, and a catalogue has been produced" (Deutscher Bund für Vogelschutz 1967, 1–2). The report mentioned almost no progress or future plans for bird conservation, but it must have been soothing to the president of the foremost organization dealing with the topic that at least the bird books were fully protected. The organization had simply failed to recover from its collapse at the end of the war. While organized field ornithology and bird conservation was growing rapidly in Britain, it was in decline in Germany.

The membership of the organization reached a postwar low during the 1950s and 1960s. The lack of reliable records probably helped to conceal how low it really was: for many years, the league simply published the same number of members. When the organization again tried to make an accurate count, the membership turned out to be much lower. In 1941, during its first peak, the organization had reported as many as fifty-five thousand members; in 1966 the count stood at forty thousand. Most bird conservation work during this period was done at the state-run Bird Protection Stations and focused on the economic aspects of birdlife.

In Britain, by contrast, membership in the RSPB reached one hundred thousand by 1970. The focus of the work, however, had not changed since the postwar years, as the annual report for the year stated. "The main priorities have remained the same; acquiring reserves of international or national ornithological importance in Britain; fostering by every means the widest possible interest in birds; fighting pollution" (Royal Society for the Protection of Birds 1970/71, 2). The RSPB was not more successful than the league because it was more progressive. It is true that the German organization had used the decades after the war to continue the work it had engaged in during the decades before the war. Yet the British organization had done just the same. The reason that the RSPB membership exploded while the league's membership stayed stagnant and even declined was not a change in direction but the fact that bird watching became increasingly popular. Economic ornithology, on the other hand, lost importance, and there was no field-ornithological tradition in Germany that could fill the gap produced by the diminution of practical bird conservation for economic ends.

SILENT SPRING AND THE FALL OF ECONOMIC ORNITHOLOGY

Despite the long-term nature of these organizational developments, it is the use of DDT and its effects that has frequently been credited as causing an environmental revolution. The publication of Rachel Carson's *Silent Spring* in 1962 is often taken as the event that gave birth to a new kind of environmental awareness (Carson 1962b). The book detailed the adverse effects of the misuse of synthetic chemical pesticides and insecticides. The story of this publication and its impact in the United States has been masterfully chronicled (Kinkela 2011; Lear 1997; Lytle 2007; Murphy 2005). *Silent Spring* was first published in serialized form in the *New Yorker* in 1962 and as a book in the same year. The book sold half a million copies in hard cover and stayed on the *New York Times* bestseller list for thirty-one weeks. It was translated into many languages, including German (Carson 1962a).

Carson's main argument was that pesticides have detrimental effects on the environment. They are more properly termed "biocides," she argues, because their effects are rarely limited to the target species. Pointing out that pesticides such as DDT had been developed during World War II, she used the imagery of nuclear fallout—known since the bombings of Nagasaki and Hiroshima—to illustrate their potential danger. The imagined "silent spring" was the outcome of a silent war. "No witchcraft, no enemy action had silenced the rebirth of new life in this stricken world. The people had done it themselves" (Carson 1962b, 3). Carson addressed the effects of pesticides on all the better-known aspects of the country's wildlife and then turned to their possible harms to humans, including poisoning, cancer, and other illnesses attributed to pesticides. It was the two chapters dealing with birds, however, that provided the title for the book.

At first sight this account seems to resonate with the environmental risks described by Ulrich Beck as evidence for his argument on the emergence of a risk society in the second half of the twentieth century. Although he does not explicitly reference Carson and her work, chemical contamination and its health effects for human beings are among his premier examples of the environmental risks produced by industrial society. Beck describes a risk society as a society in which the center of societal conflict has shifted from the question of the production and distribution of "goods" to the question of the production and distribution of "bads." "The argument is that, while in classical-industrial society the 'logic' of wealth production dominates the 'logic' of risk production, in the risk society the relationship is reversed. . . . The gain in power from technoeconomic 'progress' is being increasingly overshadowed by the production of risks" (Beck 1992, 12–13). Beck uses the framework of modernization theory to formulate his argument. While the industrial society that produces new risks qualifies as first modernity, the emerging risk society describes a new stage in history that he refers to as a second modernity. The environmental movement activism of the postwar period is accordingly interpreted as a reaction to this transformation of society. Yet Beck does not look at any specific social movement organization to make his argument. The existence of nature conservation organizations such as the Royal Society for the Protection of Birds and the League for Bird Protection that were founded in the late nineteenth century is not taken into account. What is more significant than the mere historical omission, however, is the misplaced conclusion to which it gave rise. The contrast between first and second modernity that Beck postulates does not on closer inspection describe a succession of stages in history, but merely gaps in his empirical data on history. While it is true that synthetically produced chemicals such as DDT were a recent development, it is likewise true that the perception of the risks posed by these substances and the reaction to them played out within the context of already established environmental movement organizations.

Carson herself, for instance, never claimed to be the first to voice concern about pesticides or to have started a new movement; on the contrary, *Silent*

Spring not only referenced the two decades' worth of research on the issue in the United States and Britain, but also named environmental movement organizations that had facilitated this research. Among the first publications was an article by Vincent B. Wigglesworth on "DDT and the Balance of Nature" in the *Atlantic Monthly* in 1945 (Wigglesworth 1945). The entire argument of the book was based on evidence drawn from the scientific literature on the topic available at the time.

Rachel Carson's own connection to the topic was shaped by the same tradition of natural history collecting that was so central to nature conservation in Britain. She grew up on a farm and spent her childhood recording the animals and plants around her, under the guidance of her mother. By the time she published *Silent Spring*, she was a recognized natural history writer, with a focus on marine biology. Her books made Carson famous as a naturalist and science writer, and she resigned from her government position at the US Fish and Wildlife Service in 1952 to devote herself to writing.

Her scientific background notwithstanding, among her most severe critics were those who argued that the success of the book derived from its rhetoric rather than the facts it presented. John Maddox, the editor of the journal *Nature*, was among Carson's most prominent critics in Britain. In *The Doomsday Syndrome*, published in 1972, Maddox aimed to discredit *Silent Spring* and other writings on environmental dangers. "Why was the influence of *Silent Spring* so great?" he asked. Going through the famous opening lines of the book, he concluded, "Much depended on Miss Carson's technique of calculated overdramatisation" (Maddox 1972, 15).

The statement is a rather charming display of a belief in the power of rhetoric by a man who could not understand how a point of view that he considered to be evidently wrongheaded could win the day. Yet it was hardly Carson's rhetoric alone that accounted for the success of *Silent Spring*. As Priscilla Murphy (Murphy 2005) has pointed out, the opening passage of the book is an almost verbatim copy of a passage on the same topic published by John Terres in the *New Republic* in 1946. If it was rhetoric alone that could get that much public attention, then one has to ask why it was only Carson's but not Terres's work that produced this effect? Whatever Carson's influence on nature conservation and environmentalism, it did not derive from any unique power—rhetorical or other—in her work. Rather, Carson and her book had resonance because they were part of a larger transformation of practices and institutions. The book was an outcome of these changes rather than their cause.

In Britain, it was one of the population density surveys conducted by the British Trust for Ornithology that instigated a large-scale monitoring program for populations of common birds and facilitated the debate on pesticides. In 1960, Derek Ratcliffe conducted a survey on peregrine falcons breeding on the Isles. It was commissioned by the Nature Conservancy not out of concern for the falcons but on behalf of pigeon fanciers who saw peregrines as a threat to their birds. The main purpose of the enquiry was to gather accurate data that would allow the conservancy to make recommendations on whether the peregrine should remain a legally protected bird. Because many complaints had mentioned a considerable increase in peregrines, the first need was a census of the resident population, for comparison with the figures from a peregrine survey conducted from 1947 to 1950. Data also existed for the years 1930 through 1939, when the War Office had attempted to reduce the number of peregrines to protect carrier pigeons, which were used for communication during the war (Bijleveld 1974; Mellanby 1967; Moore 1965, 1987; Sheail 1985).

It was more or less by accident that the data on the peregrine falcon and its eggs turned out to be the first to provide direct evidence on the impact of toxic chemicals on birdlife. The issue had been debated for more than a decade, yet up to that point all evidence had been circumstantial, that is, no causal connection had been proven. The story that has been told in virtually all accounts of the development of nature conservation and environmentalism in Britain after the war begins with the observation of increased mortality in wild birds in the mid-1950s. At the time it was not altogether clear whether this should be seen as a cause for concern. By far the largest decline was in the population of wood pigeons, not a species considered to be the farmers' friend. Some were rather happy to get rid of them.

Declines among other species caused more concern. Dead partridges and pheasants worried landowners and sportsmen. The causes of death were not immediately obvious, but the recent increase in pesticide use in agriculture was the prime suspect. The term *pesticide* covers all chemicals used to kill some form of plant or animal life, including herbicides, fungicides, insecticides, acaricides, and rodenticides. Among the most lethal substances for wildlife were aldrin, dieldrin, and heptachlor. For agriculture these substances had been a blessing: beginning with World War II, Britain put emphasis on increased agricultural production at home, and the effort continued after the war. The Agriculture Act, passed in 1947, underlined the need to increase agricultural yield, and pesticides were thus promoted with this end in view.

Efforts to control insect pests with poisonous substances had been around for more than a century, beginning with simple salt or arsenic. Their success, however, had been modest. Chemicals such as arsenic did not discriminate sufficiently between beneficial and harmful organisms. At the same time, the chemicals intended for use in agriculture had to be cheap and easy to use, which arsenic was not. The new pesticides developed during the war fulfilled all these requirements. Their use rose steadily in Britain during the late 1950s and early 1960s, and with it the number of reports of birds and other wild animals found dead in the countryside.

In the late 1950s the issue was put on the agenda in the House of Lords, and a species advisory committee was created in the same year. The BTO and the RSPB started to publish reports on the impact of pesticides on birdlife in 1960, based on the postmortem analysis of birds (Cramp and Conder 1961; Cramp, Conder, and Ash 1962; Cramp, Conder, and Ash 1963). The joint reports by RSPB and BTO and the debates in the House of Lords were among the evidence referenced in Carson's *Silent Spring*. The publication of the book did thus not start the debate over pesticide use in Britain, although it had an impact on it. In a next step, the Nature Conservancy was commissioned to inquire into the matter. It had set up a department in the mid-1950s, as part of the Monks Wood Experimental Station, to research a possible connection between the increasing use of pesticides and the increase in bird mortality. The problem for the Conservancy was that the mere potential of pesticides to kill birds did not prove that the birds found dead in the countryside had actually been killed by these substances.

Ratcliffe's results from the peregrine survey provided the first direct proof of the causal relation between pesticide use and bird mortality and became a milestone in environmental science. The survey found that the number of peregrines was in decline, and that in some parts of Britain they had ceased to breed. The causes of the decline were persistent pesticides, most notably DDT. Wild birds were not killed immediately by the pesticides they consumed but by an accumulation of these substances in their bodies. An additional effect was the thinning of eggshells, which led to a decreased rate of reproduction. The conclusions from the data of the peregrine survey, along with a prediction of the future prospects of the species, were first published in an article in the BTO journal *Bird Study* in 1963 (see also Ratcliffe 1958, 1962, 1969, 1980): "The present state of the Peregrine population thus matches the varying use of pesticides in different parts of Great Britain. . . . Future prospects for the species are extremely bleak. . . . At the present rate of decline, numbers and breeding success will soon be reduced to an extremely low level, and final extinction cannot be dismissed as a possibility" (Ratcliffe 1963, 77).

Ratcliffe transferred to the Nature Conservancy's Monks Wood Experimental Station in 1963, where he continued his work on peregrines and other raptors (Ratcliffe 2000). He published a classic paper on eggshell thinning in the journal *Nature* and a more detailed paper in the *Journal of Applied Ecology* (Ratcliffe 1967; Ratcliffe 1970). These articles turned into two of the most frequently cited publications in scientific ornithology.

A substantial amount of the research in Britain on the effects of pesticides on wildlife was carried out at Monks Wood, both before and after *Silent Spring* was published. Ian Newton was the head of the Pollution Research Unit at Monks Wood, later a part of the Centre for Ecology and Hydrology (CEH), from 1979 to 1984. Some of his most important publications dealt with the population dynamics of peregrine falcons and other birds of prey, documenting their gradual recovery after the banning of DDT in Britain (Newton 1979, 1986, 1989, 1998). Newton's interest in birds of prey developed in childhood:

> I used to be particularly interested in sparrow hawks. It was really when I began to climb trees. I suppose age 11 or 12 or something like that, because sparrow hawks nested in all the woods around my home. I was interested in birds in general but I spent probably more time on sparrow hawks and finches than any other birds. . . . At that time I was really limited to within cycling distance of my home, because my parents didn't have a car and everywhere I wanted to go I had to go on a bike. . . . I knew the birds in my local area extremely well. (Newton, interview, July 2006)

Newton retained his interest in birds as he grew older and eventually decided to turn it into a professional career. He did a PhD with David Lack at the Edward Grey Institute of Field Ornithology (EGI) at Oxford. Ornithology had just been recognized as an independent scientific discipline at British universities, and the EGI was one of the leading institutions. It was established in 1937 by the same group of actors that had set up the BTO. Newton took his first job at the University of Edinburgh in 1967, just when the environmental problems introduced by DDT and other pesticides were a central focus of the work of bird conservation organizations and the work at research centers such as Monks Wood. Jobs in ornithology, however, were scarce.

> In the fifties or sixties you were a really eccentric pioneer if you were interested in these things, and people thought you are rather odd in some way. I mean, my parents were pleased that I was interested in birds as a boy, but they soon began to hope that I would eventually grow out of it and do something worthwhile. Because I think of my parents' generation nobody really, at least nobody that they knew, made a living out of doing anything with birds. You just didn't do that. You became a doctor or a lawyer or whatever. . . . Playing with birds

> was OK as a hobby, but you really got to do some proper work eventually. So it really was a complete amazement to them that I eventually got a job working in birds. (Ian Newton, interview, July 2006)

Newton subsequently became vice president of the British Ornithologists' Union from 1989 to 1993 and president from 1999 to 2003. He was in addition chairman of the RSPB conservation committee from 1997 to 2003 and chairman of the council from 2003 to 2008. His career exemplifies the close connection between bird watching, field ornithology, and nature conservation in Great Britain.

The fact that there was continuity over time in this triangular relationship is not to suggest that the pesticides experience did not have an impact on the valuation of birds. Through the consequences of the use of DDT a novel place was assigned to birds as indicators for environmental changes, which transformed the focus of field ornithology and bird conservation. The traditional collector's interest in rare species gave way to the gathering of data on common species in order to measure the effect of pesticides on birds and other wildlife. There was no way to tell by what proportion the abundance of birdlife of the British Isles had been reduced based on data on rare birds alone, and it was exactly this information on abundance that was needed to make a case for the prohibition of pesticides. The records on rarities were abundant in British ornithology, but data on common birds were almost completely absent. With the exception of Max Nicholson's heron survey inaugurated in 1928, no recording scheme for common birds comparable to those for rare birds had ever been set up. Now data were needed on average songbirds rather than on a species like the heron that, because of its size, was much less sensitive to accumulated toxic chemicals than other species.

The first large-scale population survey on common species was published by Wilfred B. Alexander and David Lack in *British Birds* in 1944. Based on county avifaunae and biographies of every bird species breeding in the British Isles, it aimed to show whether each species had decreased, increased, or remained stable during the previous hundred years (Alexander and Lack 1944). The next serious attempt to systematically study the distribution of birds was made by Tony Norris. In 1950 he organized a pilot project to map the distribution of one hundred species based on fourteen hundred questionnaires in the West Midlands. The project was followed in 1952 by a survey that mapped the distribution of thirty species in Britain and Ireland (Norris 1960). Since this survey investigated only distribution and not changes over time, it could not provide information about the effects of DDT on populations. Many proposals and tentative attempts notwithstanding, longitudinal population monitoring started only when pesticides came on the scene.

It was again Max Nicholson who initiated a survey that would produce the needed data. In 1961, under his leadership, the Nature Conservancy requested the British Trust for Ornithology to conduct a survey of common breeding birds in order to provide baseline data. A Common Birds Census (CBC) was organized by the BTO in 1962 with the aim of establishing basic information on the population fluctuations among common British birds. With these data, conservationists hoped to determine annual fluctuation in population levels, explore the probable causes, and discover whether there were definite trends in numbers. "The results could be significant in monitoring the effects of any widespread poisoning of the environment due to the use of persistent toxic seed-dressings, weed-killers and sprays" (Williamson and Homes 1964, 254). In retrospect, the census had a rather modest start. In the first year only forty-nine sample areas, amounting to 9,980 acres, were surveyed. The census was carried out by volunteers coordinated by the BTO, and the choice of the area of investigation was left to the ornithological field workers. The selection of survey plots thus mirrored the geographical distribution of BTO members. The far southwest and north of England and a good deal of Scotland and Wales were underrepresented (Taylor 1965). Because the CBC was intended not to map distribution but to produce data on population trends over time, this selection bias was not considered to be a major problem.

The CBC was the first survey of its kind. No other country had ever aspired to organize a wildlife survey involving so many volunteers recording so many birds over an indefinite period. It continues today (albeit using somewhat different methods) and became the model for monitoring schemes of common breeding birds in other countries. Data from the census went far beyond answering the specific question it was designed to tackle. In the long run, it benefited nature conservation in general and strengthened the position of field ornithology within the broader science of ornithology.

Even more substantial than the list of publications that were based on the data was a novel status assigned to birds. With the beginning of monitoring the effects of pesticides on common birds, the concept of indicator species was born. Norman Moore, as the head of the Nature Conservancy's research unit on pesticides, was among those who promoted the new concept in print (Moore 1966, 1973, 1975). An indicator species, according to Moore, is a species that points to larger changes in the environment, changes that might prove dangerous to humans as well: "While wildlife becomes increasingly valuable for economic, scientific and aesthetic reasons, it is also developing a new value;

technological change today is so extensive and rapid that we require pointers to show us what we are doing and to help us to predict. Increasingly, it is realized that wild plants and animals have yet another role to play in the modern world—that of biological indicators of unforeseen problems" (Moore 1973, 3).

Birds are so high up in the food chain and so widely distributed that fluctuations in their population levels can be taken as a measure of changes in the larger natural environment. Collecting population data was helped by the fact that birds were more popular among human observers than any other taxonomic group. Once the new concept of indicator species was coined, predecessors could be named. Moore pointed out that the concept had a precedent in the canaries kept in coal mines, whose distress or death gave warning of the presence of poisonous gases. Aldrin, dieldrin, heptachlor, and DDT have long gone from Britain—the last of these substances was banned in 1982. The Common Birds Census, the concept of indicator species, and the accompanying metaphor of the canary in the coal mine have remained.

The publication of *Silent Spring* in Britain increased the public awareness of the environmental hazards of pesticide use and fostered cooperation between experts. Things looked different in Germany. As in Britain and the United States, bird conservationists had researched and published on the effects of pesticides on birdlife, but in Germany it had been the conservationists themselves who promoted pesticides. What they knew about the topic derived from firsthand experience, because they had been among the users. Research on the effects of pesticides on birdlife had been produced by the state-run Bird Protection Stations. Herbert Bruns at the Bird Protection Station in Hamburg produced crucial publications on the topic. At the suggestion of Hermann Hähnle, then president of the League for Bird Protection, he conducted experiments with three hundred nest boxes treated with various pesticides in the early 1950s. The aim was to keep the boxes free of hornets and wasps and to get rid of bird parasites such as fleas. Bruns considered the experiments a success and recommended the use of pesticides: "A comparison of the untreated nest boxes with those treated with contact insecticides shows that the treated boxes are almost free of hornets and wasps. . . . One will now question if the birdlife is threatened by such a treatment with contact poison; the results dispel such concerns. The number of successful broods is even proportionally higher in the treated boxes than in the untreated ones" (Bruns 1953, 184–85).

Bruns was a council member of the League for Bird Protection and one of the most prolific authors and editors on bird conservation since the 1950s (Bruns 1975). He was the major promoter of the use of the term *applied biology* (*angewandte Biologie*). In Germany he was best known as the publisher of a series of short monographs on the topic, including the work of Sebastian Pfeiffer and other researchers at the Bird Protection Stations as well as his own (e.g., Bruns 1969). One of the more frequently referenced monographs in the series was a literature review by Wilfried Przygodda, published in 1955, titled *Pesticides and Birdlife under Consideration of Other Wild Animal Life* (*Pflanzenschutzmittel und Vogelwelt mit Berücksichtigung der übrigen freilebenden Tierwelt*). Przygodda cited substantial research on the effect of pesticides on birds in Germany. Echoing Bruns, he concluded that their use posed little or no harm to birds, arguing that "an adverse impact of DDT through direct contact could not be proven" (Przygodda 1955, 7).

In Germany the experiments on the benefits of pesticides in treating nest boxes came to an end in the early 1960s, and no research on the long-term effects of DDT and other pesticides on birds in the field replaced it. Rachel Carson's *Silent Spring* was translated into German in the same year that it was published in the United States, but, except for increasing awareness of the topic among ornithologists and bird conservationists, it did not alter the agenda of nature conservation and field ornithology. The League for Bird Protection conceded in its 1964 annual report that it lacked the financial resources and the scientific expertise to draw any conclusions from the study.

> At the national as well as at the state level, numerous parliamentarians of all parties have paid much more attention than before to the problem of nature conservation and biological plant protection. The publication of *Silent Spring* in German has without doubt contributed to this and caught the parliament's attention. . . . In my opinion, we have to closely affiliate with the United States, where they have completely different possibilities than we have, since we have no financial tools available for this issue. . . . The proof that pesticides do indeed have these disastrous unintended effects is much easier to provide in the United States, precisely because more money is available for their detection. (Bund für Vogelschutz 1964, 5–7)

By contrast with work in Britain as well as in the United States, Germany lacked even a national field-ornithological organization that could have organized a project on the scale of the Common Birds Census. When such an organization was eventually founded, it had little to do with research on pesticides. The Association of German Avifaunists (Dachverband Deutscher Avifaunisten, or DDA) was created in 1968, initially as a section on field ornithology within the German Ornithologists' Society (Deutsche Ornithologen-Gesellschaft, or

DO-G). The aim was to put the work of the DO-G's field ornithologists on a more institutional footing. The section turned into an independent organization in 1970, representing fifteen established field-ornithological organizations (i.e., bird watching clubs). There was, however, no monitoring scheme set up to allow for the assessment of population changes in common birds. Bird conservation and the provision of data needed for the task was initially not part of the agenda of the association.

Einhard Bezzel was one of the founding members of the DDA and indeed one of the few members of the DO-G who had an interest in field ornithology, or *Avifaunistik*, as it was called (Witt 2003). Publications on the topic in the *Journal of Ornithology* were rare at the time—he calculated that it accounted for only 5.2 percent of the total pages published between 1967 and 1969. Bezzel had started watching birds as a high school student when he was about fourteen years old, using a pair of old military binoculars that his grandfather had given him. He found his first bird book in his father's library: a print of a dissertation on the birds of Bavaria, the state where he grew up. He shortly thereafter joined the Ornithological Society of Bavaria (Ornithologische Gesellschaft in Bayern). Bezzel used the marginal status of field ornithology within the DO-G as the justification for organizing a session on the topic at the annual meeting in 1968.

> The story is simply that it was said that the society needed the avifaunists, that is, the amateur ornithologists. . . . The initial question was, How can field ornithologists gain a footing within the DO-G? Yet in 1969 the DO-G decided that it was more sensible to find a solution outside the society: the avifaunists could remain independent, but receive financial assistance from the DO-G, within modest limits. Here are the steps: it was first declared that we wanted to have the amateurs, by all means, but then one saw—goodness, that is a rather big thing after all—particularly because the resonance was rather high. (Bezzel, interview, March 2012)

At the time, Bezzel was the only DO-G member under the age of forty. In the early 1960s the DO-G was still led by Erwin Stresemann, who after the war had retained all three offices to which he had been appointed in the 1920s: president of the professional ornithological association, editor of the leading journal in the field, and curator of the country's largest ornithological collection. It was only in 1962, when he was in his seventies, that Stresemann ceded the editorship of the journal to Günther Niethammer, who also took over the presidency of the DO-G from Stresemann in 1968.

> The leadership of the DO-G was quarrelling back then, because Stresemann was becoming too old to take the journal entirely under his wing and consequently transferred it to Niethammer, . . . but then interfered tremendously all the time. In addition, there was the fact that Niethammer had a strongly burdened past—at Auschwitz. I even have the cover page of the journal somewhere, where he thanks Höß, the camp commander at Auschwitz, for allowing him to watch birds. There were DO-G members who, quite rightly, opposed Niethammer vehemently for these reasons: "It's not possible that such a Nazi should become president of the DO-G and edit the journal as well." (Bezzel, interview, March 2012).

Bezzel eventually became the editor of the *Journal of Ornithology* in 1971, as a consequence of a quarrel between Niethammer and Stresemann. He held this position until 1997 and became one of the most noted ornithologists in Germany, publishing not only a standard textbook (Bezzel 1977, 1990) and a handbook (1983–1984) but also one of the first books that used systematic data on bird population trends (1982b) and a book on environmental threats to birdlife intended for a popular audience (1973). He was also the editor of the popular field-ornithological journal, *The Hawk* (*Der Falke*), from 1996 to 2007. For most of his career Bezzel was the head of the Bird Protection Station at Garmisch in Bavaria. There he succeeded Otto Henze, the outspoken proponent of economic ornithology, who after the war continued to publish revised versions of his *Check-List for Nest Boxes in Agriculture and Forestry* (*Kontrollbuch für Vogelnistkästen in der Land- und Forstwirtschaft*), first published in 1936 (Henze 1936, 1949, 1958, 1983, 1991). In stark contrast to his predecessor, Bezzel was among those German ornithologists who abandoned the focus on economic ornithology in the work of the Bird Protection Stations.

> Back then bird conservation was characterized by idealistic bird conservation and by economic bird conservation, and economic conservation included everything that has to do with the promotion of insect-eating song birds. My predecessor, the forester Henze, a wicked guy, talked about the labor birds in the forest. He said straight away—that was his most famous example—songbirds in the forest are labor birds, they work against insect infestation; sparrowhawks slay songbirds, they live off songbirds, thus sparrowhawks have to be fought against. He propagated the control of the sparrowhawk to such an extent that people called him crazy. That went so far that I welcomed Henze at a conference by saying, Well, Henze, how many sparrowhawks did you have

for breakfast today? and everybody burst out laughing. Yet Henze was a stubborn Swabian [*schwäbischer Sturkopf*] who was flattered by being teased in such a way. (Bezzel, interview, March 2012).

Bezzel led the organization away from promoting checklists for nest boxes and toward publishing field guides comparable to those in Britain. The aim was to make field ornithology more popular. Growing up, Bezzel and many others in his generation used the Peterson, Mountfort, and Hollom field guide that was originally published in Britain. For his generation, he recalls, this volume was something like a bible, one that converted many young people to ornithology. At the time, field guides were in short supply in Germany. The country had not experienced anything akin to the British boom in bird watching, and publications facilitating this development were accordingly limited in number. One of the postwar classics of the genre was Heinrich Frieling's field guide *What's flying there?* (*Was fliegt denn da?*), published in 1950, which was a slightly updated version of a book originally published by Wilhelm Götz and Alois Kosch in 1936 (Frieling 1950; Götz and Kosch 1936). The innovations in British field guides were not matched by this best seller.

Almost all of the newer field guides available in Germany were in fact translations of British texts. A revised translation of Peterson, Mountfort, and Hollom was provided by Günther Niethammer in 1954 (Peterson, Mountfort and Hollom 1954b). The introduction pointed out that the focus of the presentation had shifted from British to German birdlife, although the publication remained the same otherwise. In 1972, Niethammer, together with Hans E. Wolters, also translated the Heinzel, Fitter, and Parslow field guide, which was at least partly homegrown given that the illustrator, Herrmann Heinzel, was German (Heinzel, Fitter and Parslow 1972a). A revised translation of James Ferguson-Lees and Ian Willis, *The Shell Guide to the Birds of Britain and Ireland*, originally published in 1983, was provided by Einhard Bezzel in 1987 (Ferguson-Lees and Willis 1987). Bezzel had also published his own popular field guide, *Watching Birds* (*Vögel Beobachten*) in 1982 that went through several editions until 2002 (Bezzel 1982a).

In 1974, Bezzel, together with Peter Berthold and Gehard Thielcke, published a professional book on field ornithology—the first of its kind produced in Germany after the war. *Practical Bird Study: A Guide for Field Ornithologists* (*Praktische Vogelkunde: Ein Leitfaden für Feldornithologen*) aimed to facilitate the survey work of the DDA (Berthold, Bezzel, and Thielcke 1974). One of the coeditors, Peter Berthold, became the president of the DDA in 1975 and recalls the circumstances that initially led to its establishment as well as the problems the association faced during its early years. It was founded primarily as a reaction to the resistance to research on field ornithology within Germany's professional ornithological organization: "Initially there was a great aversion among the field ornithologists to the DO-G, because the DO-G was a very academic society, whose council members, to which—thank God—I did not belong back then, said, that's all birdy-ology [*Piepmatzologie*], bird conservation included. We don't want to have that at the DO-G. That was a closed group of council members and advisory board members who only let a very few young people come up, and then there was a kind of palace revolution" (Berthold, interview, July 2007).

The association did initially not fully live up to expectations. Given that the practice of bird watching was not as established in Germany as in Britain, it was difficult to find enough volunteers to collect ornithological data in the field and to fulfill the purpose of the new organization: "For God's sake, it existed, one met from time to time in Hesse, and representatives came from the state organizations, but the data that they were supposed to provide has mostly been missing. They came with a piece of scrap paper with a few numbers and said they would deliver later, and that didn't happen. This association had a very modest existence for a long time" (Berthold, interview, July 2007).

The DDA in its early years was not involved with monitoring common bird species. It did, however, come to play a key role in the assessment of rare birds. The DDA became the major collaborator and coauthor of the Red List of endangered birds in Germany, first published in 1971 by the Federal Office for Nature Conservation (Bundesamt für Naturschutz), and it has been updated about every five years since then (Binot-Hafke et al. 1999; Blab et al. 2005; Blab and Nowak 1989). The association began to systematically monitor six protected and endangered bird species in 1977. In 1980, eight more species were added to the list and in 1982 another twelve. Even for this small number of species, the available data were considered insufficient, as the first report for the revised scheme revealed (Witt 1983).

Throughout this period, national lists of endangered birds and other taxonomic groups nevertheless became the major instrument of organized nature conservation in Germany. The idea was copied from the Red Data Book pioneered by the IUCN. Eugeniusz Nowak was a research scientist at the Federal Office for Nature Conservation and involved in publishing the cumulative Red List on animals and plants since the first edition in 1977.

> The red list of the IUCN was of course an inspiration. The one by Peter Scott, who was very active in Switzerland at the IUCN back then. That has

> influenced us as well, but at the beginning of the seventies there was an awakening (*Aufbruch*) in Germany in terms of nature conservation and environmentalism. That was the period when some of us were searching for something new, for new methods, for new possibilities. The bird conservation societies started with a critique of the conservation strategy that "what's good should be spread universally," without trying to find a focus, and slowly the idea developed that we have to deliberately focus nature conservation on those topics where a real threat exists. Then the idea slowly crystallized that comparable to the global Red List of the IUCN, we have to have a nationally limited Red List. (Nowak, interview, July 2007)

The first edition of the Red List in Germany did not use systematic criteria for degrees of endangerment. It used a common grading key from one (critically endangered) to four (least concern) for all taxonomic groups, but there was no shared set of criteria to determine how to apply this key. It was left to experts on the various species to make their own judgment. For some species these judgments were based on quantitative estimates of the total population level; for others they were not. The quality of the data used in these cumulative lists thus differed among species and taxonomic groups.

Birds are not only the first group for which a Red List was assembled; they are also the group with the best available data. The bird data currently rely on a monitoring program on protected and threatened bird species conducted in West Germany since 1977 and in East Germany since 1956, the longest-running monitoring program in the country that is administered by the DDA. Since a restructuring in 1995 the goal of the program has been to monitor every species with a population of fewer than a thousand breeding pairs (Boschert 2003). Comparable monitoring programs for other taxonomic groups came into existence much later. With this data the list has become an established practical tool for environmental management, as Nowak explains, "Even courts of law have lauded this. The judges used the red list as evidence when there were disputes about species conservation. It has pleased us tremendously that it was adopted. There was hardly any criticism. We have made an effort to find additional coworkers to expand the list by many groups of invertebrates, and now it is three times as long, and that was a scoop. That has strongly promoted species conservation targeted on endangered species and taxonomic groups" (Nowak, interview, July 2007).

The Red List played the same role in Germany as the data provided by the British Birds Rarities Committee in Britain. It was, however, not until 1977 that a comparable committee was instituted in Germany. The German Rarities

Commission (Deutsche Seltenheitenkommission) was created with the declared aim of compiling and coordinating all rarity records in the country, past and present, in one place in order to produce a German List. The first list was eventually issued in 1993 (Barthel 1993; see also Bauer and Berthold 1996). Data on rare and endangered birds entered the work of the League for Bird Protection in the late 1960s and early 1970s. Renamed the German League for Bird Protection (Deutscher Bund für Vogelschutz, or DBV) in 1966, it was no longer concerned exclusively with the population levels of common species considered economically beneficial.

Theories of rising demands, such as the one advanced by Ronald Inglehart, which locate the concern for the environment in the changing economic structure of industrial society, seem to account for this development. According to Inglehart, economic progress in postwar industrial societies produced an amount of wealth that allowed new cohorts to experience unprecedented levels of economic security during their formative period. It is this experience of security during the phase of socialization that causes people to rank postmaterialist values, such as quality of life and subjective well-being, above materialist ones, such as economic growth. Based on longitudinal survey data collected in forty-three societies, engagement in environmental movements and a preference for Green politics is shown to be associated with such intergenerational value change. "The rise of Posmaterialist values helps account for the spectacular rise in the salience of environmental issues which has taken place . . . as the postwar generation emerged into political relevance" (Inglehart 1997, 241). While the survey data support the claim, the causal mechanism that Inglehart proposes remains incomplete. He does not take changes in the environment into account in order to explain changing environmental values. Inglehart thus ignores that it is not merely the amount of affluence in industrial society that changes over time but also people's physical relation to the environment—that is, the way they experience nature in everyday practices. Industrial society does not merely produce wealth and economic security; it also structures the way people interact with the environment. The industrial mass production of pesticides after the war is a crucial aspect in this development. Once pesticides were introduced, birds were no longer part of the world of work. It was not the increasing level of wealth in society in general that changed people's attitudes to them, but the fact that birds lost their very specific economic significance in agriculture and forestry.

The crucial insight in this instance is that the turn away from economic ornithology—the thinking about nature in terms of utility and harm—took place at the very time that birds had effectively lost their economic value.

Ornithologists and bird conservationists alike now agreed that birds alone could not prevent insect infestations of trees and crops. Biological pest control was at best a supplement to chemical control. Birds thus became valued as an end in themselves when they had lost their instrumental value. This transition, which occurred during the 1960s and 1970s, can be traced throughout the five editions of Sebastian Pfeiffer's *Pocketbook on Bird Protection* from 1957 to 1980. The range of topics discussed by the various authors throughout the book exemplifies this change as much as the preface, introduction, and articles by the editor. Conservationists had initially used birds as instruments in pest control, but now the birds themselves had to be protected from the more efficient chemical instruments of control. It was not the attitude toward birds that had changed but the circumstances.

The protection of the peregrine falcon (*Wanderfalke*) in Germany illustrates this point. As in Britain and other parts of the world, the species was observed to be in decline in the 1960s. Bird conservationists in Germany took note of Ratcliffe's research findings in Britain but did not consider them to prove that the causes for decline were the same in Germany. Many argued that pesticides were a major cause of decline but not the only one: equally detrimental were hunters, egg collectors, and falconers who took birds from the wild. No inquiries comparable to those conducted by the Nature Conservancy, the RSPB, and the BTO in Britain had been carried out in Germany. There was hardly anything published on the topic except for a series of scattered observations. Peter Berthold, for instance, argued in an article titled "Forerunners of Silent Spring" ("Vorboten des Stummen Frühlings") that the number of bird species living in the country was in decline. Berthold assumed pesticides to have an impact on common birds, yet "in the Federal Republic of Germany, investigations and institutions that could provide such records are so far lacking" (Berthold 1974, 47).

All that was known with certainty was that birds of prey were in decline (Fischer 1967; Mebs 1966). Gerhard Thielcke, for instance, observed that in West Germany the population of peregrine falcons was reduced from 350 breeding pairs in 1950 to seventy-eight pairs in 1967. Yet the cause was uncertain. "At present it is not possible to say whether pesticides in the Federal Republic of Germany are threatening species of birds of prey in their existence in the way that has been proven to be the case in England and North America. This could well be the case, because so far no systematic inquiries have been conducted in the FRG, since it has no corresponding research institute" (Thielcke 1969, 25–26).

A Working Group for Peregrine Protection (Arbeitsgemeinschaft Wanderfalkenschutz) was founded in Baden-Wuerttemberg in 1965 (AGW-Vorstand und Mitarbeiter 2005; Rockenbauch 1970, 1998, 2002). It soon received the support of the minister for agriculture and food. One of its organizers was Claus König, an ornithologist working on the population dynamics of owls and raptors, who became president of the DBV in 1969, succeeding Sebastian Pfeiffer. He was also the director of the Bird Protection Station in Ludwigsburg from 1962 to 1971. Among his publications during this period was an ornithological field guide, *European Birds* (*Europäische Vögel*) (König 1966, 1967). Like other German ornithologists, König was also involved in the publication of revised translations of British field guides, in his case Bertel Bruun and Arthur Singer's *Hamlyn Guide to Birds of Britain and Europe* (Bruun 1970), published in Germany as *Der Kosmos-Vogelführer* (Bruun 1971).

König, together with Friedrich Schilling, initiated some of the first investigations into the effect of pesticides on birds of prey in Germany. His support for the protection of the peregrine falcon notwithstanding, König initially voiced doubts that pesticides were a major cause for their decline in Germany. The results of an investigation on pesticide residues in peregrine falcons and their eggs published in 1970 seemed not to confirm the observations made by Ratcliffe and others: "Apparently, in the area under investigation in Baden-Wuerttemberg, no such developments as they are described for the peregrine falcon populations in Finland, Great Britain, and the American East Coast have taken place until now. . . . Several domestic and foreign experts who have access to our data cast doubt on any kind of impact of pesticides on our peregrine populations. In our area other conditions apparently dominate than in the USA, and confirmations of serious pesticide poisonings of peregrines are lacking so far" (König and Schilling 1970, 174).

The authors revised this assessment in a publication ten years later, based on the data for the years 1966 to 1978, and argued that the recent population decline was due to eggshell thinning caused by pesticides (Schilling and König 1980). Subsequent studies by Schilling and Wegener confirmed this relation between pesticides and species decline, showing that eggshell thinning diminished again after the DDT ban in West Germany in 1972 (Schilling and Wegner 2001; Wegner et al. 2005).

While the studies by Schilling and König were among the most rigorous in the country, they were not the most influential. What captured media attention was a report by Bernd Conrad that largely replicated for Germany Ratcliffe's findings in Britain. Conrad's report examined the thinning of eggshells of birds of prey in connection with contamination by DDT. He compared 457 eggs of nineteen bird species—peregrine falcons included—for current years with 1,867 eggs of the same species deposited in museums between 1875 to

1930. Regional differences were taken into account. "The data currently available for Baden-Wuerttemberg show that peregrine falcons have contaminations that are comparable to British and North American results" (Conrad 1977, 57). Conrad recalls that the publication turned into a genuine media event:

> That caused a furor. The year it was published I traveled all over Germany speaking to all kinds of radio and TV stations. Requests came from various talk shows and specialized broadcasts. . . . For many people it was surprising. They had not been aware that it was something so dangerous. Many probably still remembered their own war and postwar experiences of DDT treatment at the camps, where DDT was praised as a universal remedy. . . . We had no investigations like those in England. Ratcliffe's investigations were published, to put it bluntly, in obscure journals and newsletters that normal people did not look at. (Conrad, interview, March 2012)

Conrad began his research at the suggestion of Gerhard Thielcke during a summer internship at the bird station in Radolfzell. The aim was to test the validity of the British results for Germany. With this research project to his name, Conrad eventually gained a permanent job at the Bird Protection Station in North Rhine–Westphalia. He was the direct successor of Przygodda, one of the most vocal supporters of DDT. Their opinions on pesticides were not the only difference between the two. Conrad belonged to the postwar generation of ornithologists who rejected economic arguments for bird conservation. Part of his day-to-day work at the Bird Protection Station was to convince people that the distinction between useful and harmful species has no significance for conservation.

> I don't want to count the hours and days that I have spent on the phone trying to sweet-talk people. The classification into good and evil still existed, and still exists today. Sure, publications are not like that anymore, but the general issue still exists. When you talk with people, normal garden owners, with nice tidy gardens where they put their bird feeders and then find pleasure in the birds that come, when they see how naturally the blackbird mother is feeding her young, for them that is nature, and woe betide the magpie or whatever that comes along and catches the brood. (Conrad, interview, March 2012).

The goals of bird conservation changed throughout the postwar period, even though the organizations that were implementing these goals stayed the same. Many of the research networks and programs established by the Bird

Protection Stations for the purpose of furthering economic ornithology were later redirected to other ends. The station in North Rhine-Westphalia had since the 1950s coordinated a network of three hundred to four hundred trusted volunteers (*Vertrauensleute für Vogelschutz*) who, equipped with an official passport that enabled them to access private property, monitored local bird populations. Conrad described this program:

> If I wanted to know something about a topic, I could call someone and say "Hey, look, I need information about this area," and most of the time they had already collected some information over the years that one could make use of and that was generally free of charge. They did not sell the data, but instead were indirect coworkers. That developed in North Rhine-Westphalia, interestingly out of the issue of economic bird conservation. Initially, they were supposed to be the people who took care of economic bird conservation, who, for instance, should catch sparrows. That was turned upside down within a few years. To put it this way, they turned from Saul to Paul, that is, they turned into genuine bird conservationists. (Conrad, interview, March 2012)

Nothing could describe this fundamental transformation of the ends of bird conservation throughout the 1960s and 1970s better than Conrad's colloquial formula of a shift from Saul to Paul.

YOUTH ORGANIZATIONS

While the ends of organized bird conservation changed throughout this period, the means stayed largely the same. Nest boxes and bird feeders were still considered the major tools of bird conservation, whether to protect useful species or endangered species. Peregrine falcon nests were protected around the clock against egg collectors and illegal animal traders. Bird conservation organizations such as the Working Group for Peregrine Protection that became affiliated with the DBV in the 1990s did not engage in any form of public protest against the use of pesticides in agriculture. The DBV itself was an apolitical organization in the sense that it engaged in no contentious political action and restricted its involvement with the state to the cooperation with the Bird Protection Stations and the Federal Office for Nature Conservation.

This was to change in the 1980s in a process of generational replacement. The DBV transformed into an environmental lobbying organization. The youth organization of the DBV that formed during this period of transition became the major motor of change. Previous to the establishment of an independent youth organization, children and teenagers were hardly what the generation then in office looked toward as the future of the DBV; the organization's attitude toward youths had been strictly paternalistic: at times, it appeared that the youths taking part in the work of the DBV were subsumed under the list of helpful tools. The 1960 annual report, for instance, described a situation in which 1,450 school students participated in planting trees and shrubs in a suburb of Heidenheim: "In assigning teenagers to their work stations, mixed groups of boys and girls are useful, because the boys are better suited for the hard work, while the girls attend better to the measuring, providing, the bringing of humus, peat, and water, and piling up retrieved stones. Where the ground wasn't too hard, girls have proved successful at digging holes as well. The deployment of, on average, 200 to 250 students per day, divided into planting groups, is quite manageable. An increase in numbers is possible if the site can be well observed" (Bund für Vogelschutz 1960, 6). Throughout the 1960s and 1970s, the young members of the organization were educated to become active bird conservationists in the future, but they were not encouraged to develop their own opinions and engage in any projects of their own.

Eventually a clash ensued between the teenaged and young adult members of the DBV, who so far had had no say in the work of the organization, and its officers, the majority of whom were in their fifties and sixties. The first step in this direction was the foundation of the DBV Youth (DBV Jugend) as an independent, self-governing youth organization. Jochen Flasbarth became the head of the youth organization in 1983. Like many other members of his cohort, he was introduced to the DBV by a member of the older generation. The story of his involvement in nature conservation and environmentalism begins with volunteer work at the bird station on Heligoland, where he learned about the contamination of bird eggs by pesticides.

> That awakened my youthful interest. And then I asked, "When I am back on the mainland, where can I get involved?" and they replied, "Well, go to the German League for Bird Protection—they do such things on the mainland as well." I then called at Wesel, which was the closest local group of the DBV, and said who I was and that I would like to participate. I found a patron there, the county representative [*Kreisvorsitzende*] of Wesel, Dr. Hermann Klingler. He picked me up for a lecture the day after I introduced myself, and after the lecture, when he brought me back home, he said, "Boy, out of you I'll make something big in nature conservation," and that's how it started. (Flasbarth, interview, August 2007)

FIGURE 5.3 Jochen Flasbarth at an information booth of the DBV Youth.

Flasbarth founded a youth organization as part of the local DBV group in Wesel and later became the head of the national youth organization.

The DBV Youth was focused on a broader set of environmental topics than the parent organization. With its declared aim of being progressive and forward-looking, it had also a more political emphasis, seeking to go beyond nest boxes and bird feeders in protecting birds. The DBV leadership, however, did not perceive the need for any such change and eyed the work of the youth group with suspicion.

> Around '83 or '84 the DBV Youth gave itself a basic program on environmental policy, copying a little bit from the political parties. While the basic program of the DBV was still dealing more with the question of, let's say—I caricature a little bit—the size of nesting tubes for little owls, we fought the NATO Double-Track Decision, the existing nuclear energy plants, and the much too dense road system. Let's just say that all environmentalist peace-policy-oriented demands that could ever be formulated found their way in there. In retrospect I can understand that this made the gentlemen up there a bit nervous. There was the point when that came to a head and it was questioned whether this was still

acceptable. "Can the DBV still support what the young people are doing?" (Flasbarth, interview, August 2007)

The conflict between the main organization and the youth organization escalated into an official expulsion proceeding against Flasbarth, with the result that the ranks behind him closed even more tightly. Flasbarth stayed in office, and the youth organization's causes began to dominate the work of the DBV when he became its chief executive in 1985.

The position of chief executive, a paid, full-time post, had been introduced in that year to put the work of the organization on a more stable footing. The presidency remained a volunteer post. The changes were made under the DBV president Hans Scholten, an undersecretary (*Ministerialdirigent*) at the Ministry of Finance in North Rhine-Westphalia by profession, whom Flasbarth had only shortly before helped into office. Scholten moved the DBV to Bonn, thereby effectively unseating the previously dominant power center in Baden-Wuerttemberg. He then asked Flasbarth to take over the position of chief executive, which had become available as a result of this strategic maneuver.

The account of the transformation of the DBV seems to fit a shared expectation of both Beck's theory of need defense and Inglehart's theory of rising demands. Both approaches have in common that they describe environmental movement activism as driven by a rejection of the core values of modernity centered around the parameters of progress and productivity. Industrialization, urbanization, and bureaucratization, as well as advances in sciences and technology are regarded as the central moving forces of modernity. Environmental values, by contrast, are described as anti-, counter-, or postmodern values that developed—as their name already suggests—either in opposition to or as an extension of these.

Yet such a conflict between modernity and the opposition to it that fueled the generational schism just described only emerged in the case of the German organization. No such revolution took place in Britain's bird conservation organization, where the process of change can be described as a gradual evolution. The youth organization of the RSPB was founded in 1943 with the aim of introducing teenagers to the joys of bird watching. It was called the Junior Bird Recorders Club (JBRC), and its activities focused on field outings and essay competitions rather than on planting trees and putting up nest boxes. From the very beginning, the RSPB's youth work was geared toward the study of natural history. As early as 1904, the RSPB had sponsored a school essay competition (which was discontinued in 1964 only because more successful schemes had taken over). The JBRC was renamed the Young Ornithologists' Club (YOC) in 1965, and an independent magazine for youth members, called *Bird Life*, was established in the same year. There was no generational conflict: membership in the YOC was open to all young people up to a certain age, and it was hoped that after ceasing to be YOC members these young adults would continue as members of the RSPB. The YOC membership was always counted as part of the overall RSPB membership. Throughout the 1960s and 1970s YOC members consistently accounted for more than a quarter of the total membership, with as many as 110,000 YOC members in 1980.

The YOC was a bird watching club for teenagers without any wider environmentalist or even political agenda. Although the retention rate of YOC members as adult members of the RSPB left much to be desired, there are numerous accounts of former YOC members becoming leading bird conservationists. For a long time, virtually all the presidents, directors, and council members of the organization were lifelong bird watchers many of whom had started their conservation careers in the YOC.

One of these former members was the biologist John Lawton, who served as the chairman of the RSPB council between 1993 and 1998. When interviewed in 2006, he was chairman of the Royal Commission on Environmental Pollution. Growing up in a small town north of Manchester, he was as a child a collector of dragonflies, frogs, skulls, and other natural history specimens, not least bird eggs.

> I was just a passionate childhood natural historian, it just filled my life. . . . My father was a very keen amateur botanist, so he encouraged me. . . . I joined what was then the Young Ornithologists' Club of the RSPB. . . . It was all very serious and very birdy. I still got the magazines, and we all sent our records in. . . . Right about that time, . . . I read James Fisher's *Watching Birds*, and then I read it again. You know, how could anybody possibly know so much about birds? . . . After that, I realized that birds were both a hobby, but also that there was some serious biology—biology was what I was going to do. Right about that stage in my life I decided I was going to be a biologist." (Lawton, interview, June 2006).

Lawton's turn from bird watcher to conservationist took place when several bird species that he had taken for granted disappeared from his local surroundings. Lawton recalls how the landscape was transformed in the 1950s and 1960s by agricultural intensification, and birds that he loved and cared about, such as the corncrake, disappeared. He joined the YOC because he had grown concerned about these changes in the countryside. Lawton remained a member of the youth organization until he went to university; he dropped out only because he could no longer afford the membership. As a student he was still engaged in conservation efforts, working as a warden for the RSPB during the summer. After completing his studies, he moved to York, joined the RSPB, and helped to create a local RSPB group, of which he became the leader. He describes the group as a club of people who enjoy birds—that is, a bird watching club—first and foremost. Lawton was eventually invited by the RSPB director Ian Prest to join the council of the organization as its scientific ornithologist. He was a council member for five years before he became its chairman.

> That just got me more and more involved in the politics of conservation, because as chairman of the RSPB you have, I mean, I had direct access to ministers. If I wanted to go and see the minister of agriculture he'd see me, because we had nearly a million members. We had more members than the Conservative and Labor Party. The then minister of agriculture was John Gummer, . . . [and] I used to go and see John about . . . the intensification of agriculture that I began to witness as a child, as a teenager. . . . I got to learn how to deal with senior politicians, not to be fearful of senior politicians. Actually, they often want help, because they are struggling with very difficult issues, and if you can help them you are more likely to make progress. They sometimes contacted me, as the chairman of the RSPB. You do a lot of work behind the scenes. (Lawton, interview, June 2006)

Lawton's career exemplifies the cooperation between the youth organization and the main organization of the RSPB in the postwar period. His career as a conservationist developed within the established organization and was in alignment with the main focus of its work. In Germany, on the other hand, the previous focus of bird conservation organizations on economic ornithology was perceived as part of the problem that the postwar generation of conservationists turned against. It was accordingly only in Germany, not in Britain, that the developments from the 1960s to the 1980s could be described as an environmental revolution. In Britain, the development was a gradual process better described as evolution. Both trajectories, however, took place within the context of the already established institutions of field ornithology and bird conservation, which continue to shape the work of the RSPB and the DBV today. Beginning with the late 1990s and early 2000s, the latter organization would moreover start to copy the strategies of the former.

CHAPTER 6

Bird Watching as Organizational Strategy

Peter Holden has his office at the headquarters of the Royal Society for the Protection of Birds (RSPB) at the Lodge in Sandy, Bedfordshire. Holden is the director of the Department for People Engagement and has been working for the organization since 1969. Before he started working for the RSPB, he was a member of its youth organization, the Young Ornithologists' Club (YOC): "I've been passionate about birds as long as I can remember. I had my membership bought for me when I was ten years old, but actually being a child they gave me Junior Bird Recorders' Club, which predates the YOC, and I absorbed that stuff, probably didn't read it very thoroughly, but felt quite privileged to be a member" (Holden, interview, February 2008).

Holden organized his first local YOC group when he was sixteen. When he started working for the RSPB, he was responsible for administering the RSPB's youth membership as a Junior Education Officer. At the time the YOC had a membership of about twenty thousand. The club was renamed Wildlife Explorers in 2000 and had 190,000 members in 2010, making it the largest organization of its kind in the country (RSPB Annual Review 2009–2010, 5).

As organizer of the YOC, Holden initiated the RSPB's most successful public relations campaign, the Big Garden Birdwatch.

> In 1978 we were sitting around saying, What can we do in January? We can't take journeys, it may be too cold, too cold to go to local parks and gardens—save that to March. What could they do looking out of their window? And so we invented the Big Garden Birdwatch. I said, What we will do—if everybody looks out at exactly the same time for exactly the same hour—we will have the

top ten garden birds of Britain. No thought of science in it at all—we were just going for the top ten garden birds. (Holden, interview, February 2008)

To his surprise, the scheme was an instant success. More than thirty-four thousand people took part during the first year, and fifteen hundred of them joined the organization as new members. The scheme has been used ever since to recruit members and to communicate the conservation message of the RSPB. In 2001, the activity was extended from children to the membership at large. In that year a record fifty-two thousand people participated, and the survey has been growing almost every year since. In 2010 nearly 530,000 people participated. The RSPB initially advertised the event to draw more participants, yet over time the Big Garden Birdwatch became so popular that the British news media report about it on their own.

> We invested in it. I spent forty-five thousand pounds on advertising. I've actually advertised in all the national press or the significant daily papers. . . . The last three years have just been phenomenal for the press coverage, and this year it's the biggest RSPB story, and virtually every newspaper will have covered it in one way or another. . . . So you get the *Times*, the *Telegraph*, *Daily Mail*, even the *Sun* and the *Mirror* will take this. . . . This year we all thought it is a breakthrough cause they said, "And it's Big Garden Birdwatch time again." It's not RSPB—it's a bit like, remember it's Sunday, it's like a date on the calendar you ought to remember. (Holden, interview, February 2008)

The Big Garden Birdwatch is an example of what has been called citizen science, meaning a large-scale data-gathering enterprise conducted by volunteers rather than by paid professionals. It takes place once a year on a weekend in January. Participants are asked to record as many birds as possible in their garden or in a comparable locality. Complete instructions are provided, along with an illustrated checklist of the most common garden birds to record how many have been observed. The results can be mailed to the RSPB or submitted online, and are published on the organization's website. Since its inception, survey participants have clocked more than three million hours of garden bird watching.

In Holden's view the campaign is a way to make people look at birds and develop an attachment to them. Without instruction, he argues, most people go through nature with tunnel vision.

> You can see somebody walk down the path in one direction and they just go for a walk, but if you got the right person with you when they come back and

> they are saying "Have you ever thought how beautiful that is?" or "Look, over there are these birds and look what they did" and suddenly your whole view of that walk changes. . . . One of the drives for me, whether it's children or adults, is actually helping people look on either side of the path. "Look through this telescope, look through there and you will be surprised." (Holden, interview, February 2008)

Holden has translated this insight into action by writing or coauthoring more than half a dozen ornithological field guides and handbooks, some of which are marketed under the RSPB's name, such as the *RSPB Guide to British Birds* (Holden and Sharrock 2002), the *RSPB Handbook of British Birds* (Holden and Cleeves 2002, 2006), and more recently the *RSPB Handbook of Garden Wildlife* (Holden and Abbott 2008). In Holden's view the practice of bird watching, encouraged by campaigns such as the Big Garden Birdwatch and guided by his publications, is a way to turn people into bird lovers and conservationists. The effectiveness of this strategy has earned him official recognition: in 2009, after a lifetime spent nurturing enthusiasm for birds, Holden received an MBE in the New Year's Honours list, recognizing him for inspiring more than one million children during his time at the RSPB.

SEEING, LOOKING, AND WATCHING

The game of bird watching, as in the case of the Big Garden Birdwatch, is a constitutive element of the concern for conservation and the work of voluntary bird conservation organizations. This chapter describes and explains what kind of valuations and commitments emerge from playing the game of bird watching and how it is employed as a strategy by the RSPB and the German League for Bird Protection (Deutsche Bund für Vogelschutz, or DBV) to increase the public appeal of birds and to recruit new members.

This recognition of the potential of practices of observation to transform individual preferences has much in common with the work of Michel Foucault on surveillance and scientific investigation. For Foucault knowledge is a form of power. In his view knowledge is not merely a value-neutral cognitive datum but actively shapes the human self and with it the way actors behave. This characteristic of knowledge to create the world, rather than just to reflect it, is described as power/knowledge. The Foucauldian perspective shares with pragmatism the central insight that the self cannot be separated from the knowledge that it possesses, or, to use Dewey's language, that one cannot separate thought from the thinker (Dewey and Bentley 1949). Knowledge is not just a neutral tool for action, as in the Baconian expression according to which knowledge is power, but it is motivated and constitutive of the ends of action. There are, however, also major differences between Foucault's perspective and a pragmatist position. For Foucault there is an unavoidable dark side to knowledge, which he traces in practices of scientific knowledge production: according to Foucault scientific knowledge is a technology of control or tactic of power that is exerted through hierarchical observation, normalizing judgment, and the examination (Foucault [1975] 1979).

Corrigan and Sayer have extended this perspective beyond scientific practices and turned it into an argument about disciplinary practices more generally, which they address as moral regulation (Corrigan and Sayer 1985). Scholars working from this perspective have included the regulation of leisure activities such as gambling, drinking, and prostitution in their analysis (Rojek 1992). The regulation of the self described by Foucault and by scholars influenced by him is a form of control that constrains the individual and makes his or her world smaller. While this argument does not contradict a pragmatist perspective, it also does not exhaust its possibilities. From a pragmatist perspective knowledge also has the potential to make the world bigger. The appropriate slogan for the ability of knowledge to open up a space of possibility would be the formula of love/knowledge, rather than power/knowledge. While the formula of power/knowledge captures the effect of surveillance in the world of work very well, as in the prisons and poorhouses described by Foucault, the formula of love/knowledge would be more appropriate for the world of play. This chapter shows that conservationists use the game of bird watching and ornithological field observation in order to make people fall in love with birds. It is a form of knowledge production and data collecting that does not try to discipline people, but one that aims to open a new space of possibilities for them.

This effect differs according to what extent the experience of birds is organized as a game. There is an essential difference between watching birds as a coordinated effort to spot, identify, and record them and simply seeing birds as a by-product of the routines of everyday life. Birds can be seen everywhere, but not everybody watches them. Between watching and seeing stands looking: the experience of actually paying attention and engaging with birds, yet not to the point where the activity is structured by the rules of a game. People look at the birds in their garden, but they do not necessarily watch them.

Seeing, looking, and watching are thus three crucially different ways of experiencing birds. To the extent that these experiences are engaged in a playful way, they describe different degrees of institutionalization of the experience of birds as a game. One of the main strategies of the RSPB for recruiting new members and gaining public support is to move people as far as possible from seeing birds to watching them, that is, to engage them in the game of bird watching. During the 1980s, the organization changed from a bird watchers' club to a mass-membership organization. To sustain and expand its membership, which now exceeds one million, the organization tries to link the aims and interests of the more committed bird watchers who run the organization to the interests of the general public, who experience birds in their own gardens but make no systematic effort to watch them. The organization has developed strategies to bridge this gap, such as designating certain garden bird species as "flagship species" and fostering interest in garden bird watching. The RSPB also employs the appeal of birds to raise awareness for larger environmental issues such as climate change, which affects the habitats of birds and is considered a major cause of species extinction. It is concern for birds that gives credibility and urgency to environmentalism, not a general concern for the environment that makes birds a relevant topic. Although the RSPB is engaged in environmental issues, it carefully avoids aligning its public image with that of environmental organizations such as Greenpeace or Friends of the Earth.

In Germany the organizational strategy works in the other direction. When the DBV merged with its East German counterpart in 1990 after the fall of the Berlin Wall, it used this merger as an opportunity to significantly change its public image. An exclusive focus on birds was perceived as limiting the organization's growth. The organization dropped birds as its central focus and renamed itself the Naturschutzbund Deutschland (German League for Nature Protection, or NABU) with a broader focus on nature conservation and environmentalism. The rebranding was followed by a rapid growth in membership.

Bird conservation has become one issue among others in the work of the NABU. Bird conservation as advanced with nest boxes and bird feeders is now considered old fashioned and ineffective. Whereas the RSPB produces and sells these devices to support its strategy of promoting garden bird watching, the NABU temporarily stopped promoting them altogether, because they were regarded as diverting resources from more important environmental issues. Despite an overall focus on nature conservation, the NABU now strongly resembles environmental organizations such as Greenpeace and the German affiliate of Friends of the Earth, the League for Environmental and Nature Protection Germany (Bund für Umwelt und Naturschutz Deutschland, or BUND). The strategies of organized bird conservation thus moved in opposite directions in the two countries in the 1980s and 1990s. In more recent years, however, the NABU, recognizing the RSPB's success in recruiting members and raising public awareness through garden bird watching, has gradually begun once again to support the practices of feeding and watching garden birds.

GETTING HOOKED

In Great Britain bird watching and bird conservation are no longer the preoccupation of a minority. From the late 1970s to the early 1990s, bird watching gained in popular appeal and public attention. Its growth was marked by the expansion of bird watching journals, magazines, clubs, and online services; bird watching fairs and rallies are regularly organized and covered on prime-time television and in daily newspapers (Cocker 2001; Wallace 2004).

Stephen Moss is among the people who have contributed to the trend. He is the author of more than twenty books on birds, including an ornithological field guide (Moss 2006b), a manual on feeding garden birds (2006a), a popular ornithological handbook (2007), and a history of bird watching (2004). He also writes a monthly bird watching column for the *Guardian* (2006c). Moss has been a bird watcher since his childhood days and he still remembers how it all started:

> My mother took me to feed the ducks. . . . I was about three, and I asked her what the funny black ducks were and she didn't know, and she took me home and she found someone had given us the *Observer's Book of Birds*. . . . I looked them up and they were coots and that was it. I got hooked by them. The reason I kept bird watching was down to the RSPB. I was a member of the YOC—Young Ornithologists' Club was the thing everyone joined. I became a member of that when I was about eight. . . . Peter Holden was running it. . . . The thing about British conservation and bird watching is that there are these seminal figures like Peter who must have influenced literally hundreds of thousands of people, and without them you wonder whether it would have happened. (Moss, interview, June 2006)

Moss's autobiographical account exemplifies the most common narrative of becoming a bird watcher put forward by people of his and previous generations. Interest in birds begins between the ages of six and ten, when children are first able and allowed to roam and ramble on their own in the countryside. Most grew up in a family that was supportive of the hobby. Playing the game has to be learned—it does not come about by simply seeing birds. Many different people are commonly reported to have introduced children to the game, including relatives, neighbors, and friends. Next, fledgling bird watchers meet fellow bird watchers in the field. The more committed join local natural history societies or conservation organizations. Many also report a deep admiration for the author of their first bird watching guide and other ornithological publications. They frequently subscribe or even contribute to bird watching and natural history magazines.

The experience of birds in the practice of bird watching is made meaningful by the institutional context of ornithological handbooks, field guides, and lists that allow for the classification of birds. It is the practice of observation, identification, and recording that produces the meaning of birds. Moss points at a chaffinch in his garden to explain:

> When I look at that I think several things. . . . One was, it's one of the commonest birds in Britain; another was when I first ever saw one, which was at Hampton Court, down the road here, when I was about eight, . . . and another one was an anecdote about Gilbert White who two hundred years ago wrote about the chaffinch and explained that Linnaeus gave it the Latin name *Fringilla coelebs*. . . . So that's a cultural and historical thing about the chaffinch and a distributional thing about it and my personal memory. That, to me, seeing that bird there triggers off various random and fairly unconnected thoughts about it, which enhances the way I look at things. (Moss, interview, June 2006)

There is more than one way to play the game of bird watching and to produce the kind of meaning attributed to birds that Moss describes. British bird enthusiasts, Moss included, distinguish between the effort to collect field notes on as many different bird species as possible and the practice of observing one's local "patch" over a long period. Observations are usually made with binoculars, and records are collected on special checklists or in notebooks—or simply by "ticking" off the species one has seen in one's field guide. The former practice, which bird watchers call "twitching" or "tick hunting," can be considered a game of skill, and the latter, referred to as "birding" or simply "bird watching," a game of chance. In both cases, birds are observed in the built as well as the natural environment: not only in rural fields and forests, urban parks and suburban gardens, but also in landfills and wastelands. There are also more specialized forms of the game, such as photographing, filming, sound recording, and radar tracking.

The practice of collecting is of course not restricted to birds or other objects of nature. Many committed bird conservationists interviewed reported having competing hobbies throughout childhood. Many continue to make a connection between their interest in birds and other collecting pursuits, such as train spotting and stamp collecting. These comparisons place bird watching in the context of other hobbies and pastimes, not in the context of a general concern for the environment.

The comparison of bird watching to these other forms of collecting is intriguing for both their similarities and their differences. All three hobbies are about collecting, yet what stamp collecting and train spotting miss, in the eyes of bird watchers, is the high level of equilibrium of forces and unpredictability of outcome characteristic of bird watching: "My take is that bird watching is different from train spotting, because it's unpredictable. Most bird watchers are driven by unpredictability. You don't know what you are going to see. . . . The difference is that you can fail. If you are train spotting you have a time schedule—you know what's going to come. In the end you could see everything. With birding you could never see everything" (Moss, interview, June 2006).

In differentiating collecting hobbies, bird watchers did not draw the distinction that birds are part of nature, whereas other collectibles are not. It is the activity of play that aligns these three pastimes. Birds do not carry an intrinsic and fixed meaning that motivates the activity of bird watching. It is the activity itself that gives meaning to birds. Conservation was accordingly not compared to other moral or political causes, such as child protection, cancer research, or consumer safety. Bird conservation does not begin with other moral, political, or religious causes: it begins with birds—birds as collectibles.

Personal accounts of this process of meaning production frequently highlight transformative experiences, mostly during childhood, that produce deeply ingrained commitments. Bird conservationists talk about "becoming passionate," "getting hooked," and "falling in love." Moss even makes an analogy to religious conversion and the experience of being "born again."

> I suppose I always had a strong inclination, mission almost, to popularize bird watching. I think it's a good thing for lots of people to get into wildlife or birds. . . . It's a bit like—I mean it's a very strange analogy, but to me it is a little bit like a religious conversion. Born-again Christians want to convert other people to Christianity, because they can't understand why people wouldn't want what they have out of it. And I feel the same about birds. I am not making an exact correlation, but for me being into wildlife, being into your local patch, be it your garden or your little local area or anything or going abroad, to many people is very fulfilling. It's fulfilling on a personal level, the fun you get from watching birds. It gives you a challenge. (Moss, interview, June 2006).

Moss's account of how he got involved in bird watching is mirrored in the reasons he provides for bird conservation. He describes what he gets out of bird watching—the place it has in his life—and uses this descriptive account as an argument for engaging in conservation. The analogy to art looms large in his account. Objects of nature are like works of art, unique and irreplaceable. If we destroy one, it is gone, and nothing can bring it back.

> I often think an interesting analogy is with art. Why would it matter if all Raphael's paintings were destroyed? Suppose they have all gathered at one exhibition, and there was a fire and they're all destroyed, why would it matter? . . . I think we would all feel that it was a tragedy for human culture. And I think it's like that when birds become extinct. . . . There ought to be a word that explains the fulfillment you get from doing something, be it supporting a football team or playing music or listening to music, you know, satisfaction, pleasure, fulfillment. . . . You can get fulfillment from your work, but that's a different kind of fulfillment from the fulfillment you get from listening to a Beethoven sonata or seeing a golden eagle. (Moss, interview, June 2006).

There is nothing mysterious about the phenomenon Moss tries to find a word for: the word for it is play. It is the very experience that scholars working on this form of practice have described—the complete absorption in the moment that is experienced as gratifying. Play is an activity that is not about achieving an extrinsic reward but experienced as rewarding in itself. It is what Csikszentmihalyi ([1975] 2000) has described as an autotelic experience. What makes the analogy to art so apt is that works of art (such as a Beethoven sonata) are usually also engaged with in situations that are free from formal constraints and thus allow for autotelic experiences. The activity of play, whether engaged in the realm of art or elsewhere, is transformative of the actor. Linking bird watching to art in the way Moss does is an attempt to make his own experience intelligible to people who have never experienced birds as he does, yet who might have experienced art in a similar way. Making the analogy is accordingly not an effort to create a symbolic boundary and to mark golden eagles as part of high culture but an attempt to cross a boundary and point out what he and other people who are not bird watchers nevertheless have in common. This commonality derives from a shared experience: the enjoyment of being absorbed in the moment that is created by play as an autotelic activity.

GARDEN BIRDS AS FLAGSHIP SPECIES

Most bird conservationists in Britain narrate comparable life histories and provide arguments for conservation similar to that of Moss. The majority of the actors who run the RSPB developed their commitment to birds as bird watchers. Most started birding early in childhood. The exact status of birding has changed over time, yet it remains a central element in their motivation to engage in conservation and to engage others in it. Bird watching is the RSPB's main tool for attracting new members, and the Big Garden Birdwatch continues to be the organization's most successful campaign. Richard Bashford works in the RSPB Department for People Engagement and organizes the Big Garden Birdwatch. During his childhood, when Peter Holden was still the organizer, Bashford was himself a participant.

> An activity for our members is a good thing, it helps keep them warm, it helps to retain them as members, they are less likely to leave the organization. We are speaking to a lot of new people. They are more likely to join having done an activity with us; they are more likely to buy a membership with us. . . . They have already given some of their time, and we say, "Thanks for helping us, and we are a charity, and we need your support. Would you like to join us?" So we do mailings to these people afterwards. . . . The conversion rate—the rate of return on investment—is extremely high in marketing terms, in these terms it's very good. (Bashford, interview, July 2006)

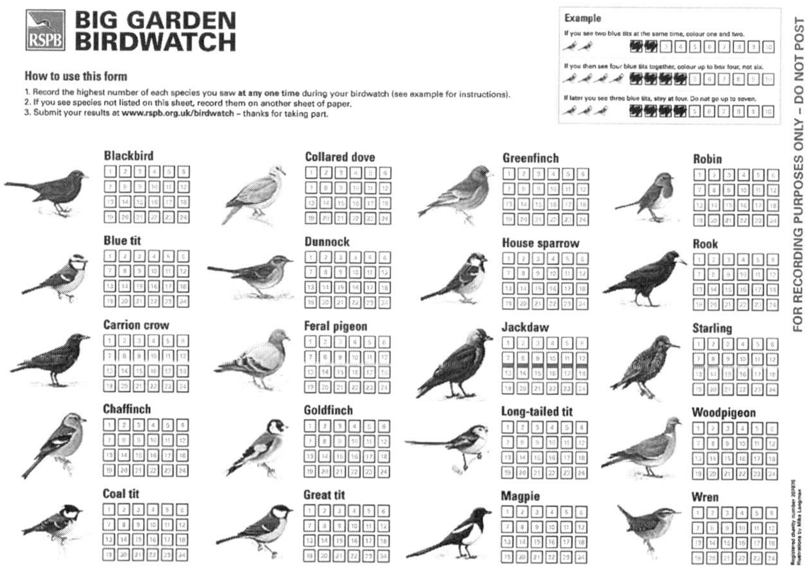

FIGURE 6.1 RSPB Big Garden Birdwatch Checklist. Illustrations © Mike Langman (rspb-images.com).

FIGURE 6.2 RSPB Big Garden Birdwatch leaflet. Greenfinch photograph on twig with berries © Mark Hamblin (rspb-images.com).

The Big Garden Birdwatch is set up as a recording game. The RSPB distributes a chart with drawings of the most common garden birds and space for noting the numbers observed. The chart serves as a collecting list, encouraging people to experience the same pleasures of collecting that have been formative for most of the already committed conservationists.

The RSPB does not try to teach abstract lessons about conservation with this campaign but instead aims at manufacturing an experience. The task is to move people as far as possible from just "seeing" birds to actively "watching" them by making them "look." In 2006, for instance, the flyer announcing the event was in fact titled "Look!"

British newspapers and radio stations announce the event and report the results. The announcement of the rise or decline of the observed numbers of different bird species does not merely carry a potential conservation message but also serves as a game score, enabling people to compare their individual results to larger national trends. The true value of the scheme for the RSPB, however, is not the data collected but the public participation in the activity.

> We have found out that most people take part in this because they believe that their information is helping bird conservation. Now, that strictly is true, but not in the way they might imagine. It helps conservation because we get lots of publicity, we get membership, we raise awareness, we get more people interested, therefore we are better equipped to do our work. So you could say this activity does help conservation, but indirectly. Their view—that they are counting birds and that the bird information is the key—is vitally important. . . . To a small extent the scientific information is useful, but it's superfluous in many ways. We already have this information from other surveys. (Bashford, interview, July 2006)

In addition to the Big Garden Birdwatch, the RSPB arranges local events, such as an observation point for watching peregrine falcons nesting on the chimney of the Tate Modern in London, a webcam that documents the breeding success of the osprey at Loch Garten, and guided tours on its reserves, such as those at Minsmere starring the avocet, the species in the RSPB logo. The organization also sells bird tables and nest boxes in its gift shops at the reserves, through the members' magazine, and online or by mail order. The sale of these devices is intended not only to increase revenue but also to encourage people to watch birds in their gardens, thus drawing them into the game of bird watching.

Yet not everybody who watches birds and develops an emotional attachment to them automatically develops an interest in conservation. The RSPB therefore strives to attach a conservation message to people's experience of birdlife. Birds that are part of everyday experience, such as most garden birds, are employed as so-called flagship species to communicate the importance of conservation. Sarah Niemann is in charge of coordinating the work of the RSPB's Conservation Department with the Marketing Department. The major challenge of her work is to make a link between birds that are known and popular, and hence have a marketing appeal, and birds that are the actual target of the RSPB's conservation work.

> A problem that I find we have continuously is how to help my marketing colleagues to engage people's attention, the general public's attention on the kinds of things we need to talk to them about. What we tend to do is to talk to them about things that we know are their particular interests. And that quite often is birds in the garden. . . . It's huge in this country, it really is huge. The British

> seem to like the birds in their garden even though they don't probably know very much about them, really. (Niemann, interview, July 2006)

The crux of the matter is that most of the bird species that are popular and well known, such as the blue tit, are common species with high population levels, while conservation work focuses on rare and endangered species.

> We have to try and get people to understand our work for these birds which are not necessarily so familiar. If you say to the general public, What is a redpoll? they'd have no idea at all, because it's a bird they don't see. . . . But if you said to them, a blue tit, a robin, they'd know. . . . That's what we call a flagship bird. We can say to people, House sparrows, you like house sparrows, you are concerned about house sparrows, and the house sparrow represents the wider conservation work. . . . We can use these birds to represent a whole group of other birds that are in similar problems, which is a good way of getting people to take an interest in some of these birds. (Niemann, interview, July 2006)

These flagship species are those that appear most often in the RSPB's brochures, advertisements, and products. They also figure prominently in the membership magazine, *Birds*. For many members, unless they visit reserves or belong to a local group, the magazine is the main point of contact with the RSPB. Unlike the field ornithologists and bird watchers that run the organization, most ordinary members do not have any expert knowledge about birds or the environmental dangers the various species are facing. According to Rob Hume, the editor of *Birds* from 1989 to 2009, to confront them with this knowledge in the magazine would not be the best way to gain new members or to retain old ones. The success of the RSPB as a mass-membership and lobby organization hinges not on ornithological experts but on people who want the birds in their garden to thrive.

> All members get *Birds* magazine. . . . It has to keep members interested enough to retain their membership so they renew their subscription. . . . They are interested in reading what they already know to a certain extent. If they know that a blue tit is in a nest box and it rears so many young and it feeds on caterpillars, then you write and tell them that blue tits nest in nest boxes and rear ten young and they feed them on caterpillars, and they all say, Yeah, that is just what it does in my garden! If you say, an antshrike in a South American rain forest feeds on so-and-so, they say, I have never heard of that, and they turn over. (Hume, interview, July 2006)

Presenting information that readers already know from their own experience leads them to regard the RSPB as an organization that cares about the things they value. A closely related insight is that a focus on environmental dangers does not produce the same effect as a focus on the bright things in nature. Good news, in Hume's view, works better than bad: "That's the biggest worry, that if you tell them all the problems in the world, climate change, rain forest destruction, the loss of farmland birds in Europe, the habitat destruction in Eastern Europe, the loss of wetlands in Poland, anything you like, everybody just—it makes people just so depressed. . . . You have to be much more positive and optimistic and give them good news. There is a lot of good news as well" (Hume, interview, July 2006).

The membership magazine thus represents what the majority of members of the organization like to hear, not the focus of the RSPB's conservation work. The role of familiar garden birds is to get people hooked and to induce them to join the organization. Their membership fees then provide the funds that are needed to engage in larger conservation issues that do not take place in people's back yards.

FROM NEST BOXES TO POLITICS

Garden birds have a very different standing in the work of the German NABU than in the work of the RSPB. The practice of feeding garden birds and providing shelter for them, initially something of an afterthought to conservationists in Britain, was in the late nineteenth and early twentieth centuries at the very heart of bird conservation in Germany. In recent decades, however, this emphasis has reversed. It is now the RSPB that promotes bird feeding in its publications and sells bird tables and food, while the NABU has moved in the opposite direction. In the 1970s and 1980s, such measures were increasingly referred to as "grandfather's bird conservation" by the young generation of German conservationists. They shifted the organization's focus toward not only the protection of other animal and plant species but also larger environmental issues, from river pollution and dying forests to radioactive waste. A first decisive step in that direction was accomplished in 1981 through a change of the organization's bylaws. Article 2, paragraph 1, was revised to state, "The purpose of the German League for Bird Protection is the conservation and care of nature, with special consideration for wild birds" (Ertel 1981, 34). In the 1980s the NABU, then stilled called DBV, evolved into a full-blown environmental lobbying organization.

This transformation was addressed in a lead article published in the May 1988 issue of the organization's magazine titled "From Nest Boxes to Politics."

FIGURE 6.3 "From Nest Boxes to Politics," cover story of the NABU membership magazine, May 1988, photograph by Armin Maywald.

It was accompanied by a cover image displaying a songbird sitting on a flagpole and defecating on the German flag.

The article was written by Klaus Ruge, the editor of the magazine and former head of the DBV's youth organization (Ruge 1988). It was a triumphant yet carefully drafted farewell to the agenda of practical bird conservation that had dominated in previous decades. The goal was to transform the DBV into an organization focused on environmental policy issues. Bird conservation was presented as an integral part of the protection of entire ecosystems. This change of focus was argued to also have a political dimension, since human beings are dependent on the environment and functioning ecosystems for their survival. To protect nature is to protect human society as well.

> In the face of global environmental problems from the depletion of the ozone layer to the nuclear scandals, bird conservationists are often smiled at as small minded and far removed from reality. Those who only put up nest boxes will not save nature. Nowadays bird conservation can only mean the protection of habitats, the creation of a network of ecological zones. . . . Bird conservation is not a game. Bird conservation is necessary if we want to protect habitats for humans and animals. Bird conservation is also very political. After all, being political means feeling responsible for the well-being of the community. (Ruge 1988, 4, 9)

At the time, Ruge was employed as the director of the Bird Protection Station Baden-Wuerttemberg in Karlsruhe. His work for the DBV was done on a voluntary basis, although he was at times able to use the facilities of the station for this purpose. Ruge also published guidebooks on bird conservation in which, again, bird conservation was presented as part of a wider environmental agenda. In 1989 he published *Bird Protection: A Practical Handbook* (*Vogelschutz: Ein Praktisches Handbuch*), which, despite its traditional-sounding title, had a broad environmental focus that represented a significant departure from previous publications on the same topic (Ruge 1989). Ruge looked in particular to the younger generation to push his agenda.

> Up until the seventies things were very conservative, and the old gentlemen—to whom I now belong myself—were still around, but they were from a completely different generation, raised completely differently, where hierarchy played a much bigger role. . . . The youth work at the local groups proceeded in the way that the youngsters were allowed to accompany them, that they were allowed to clean and carry the nest boxes, and things like that, but they were not entitled to make their own programs. When I was the youth organizer we tried to run seminars, courses, and campaigns, where we pointed out what can be done in youth work. (Ruge, interview, August 2012).

The period in which Ruge was active with the DBV Youth brought about a reevaluation of the practice of bird feeding. Providing food in winter to ensure birds' survival had been at the center of the organization's work since its foundation. Recommendations on how best to feed birds were grounded in scientific research and disseminated in instructional materials such as the *Pocketbook on Bird Protection* (*Taschenbuch für Vogelschutz*) (Pfeiffer 1980). By the time the fifth and last edition of this guide was published in 1980, however, bird conservationists within the DBV were beginning to point out the downsides to this longstanding practice. An article by Rainer Ertel published in the DBV magazine *We and the Birds* (*Wir und die Vögel*) in 1976, for instance, argued that feeding birds favors those species that are already common and not in need of protection. Moreover, artificially supporting the population of resident winter birds results in increased competition over food and nesting sites for rare migratory birds on their return in the spring. The strongest argument in favor of bird feeding, the potential to foster children's interest in birds, was not given much weight. "Everybody should realize that although feeding birds can be a lot of fun, it is of quite doubtful utility for our birdlife. Those who care about the protection of nature should in the first place support the conservation of valuable habitats and the protection of endangered bird species" (Ertel 1976, 15).

Opponents of bird feeding further argued that it diverts financial resources from more important conservation issues. In a 1986 article for the magazine, titled "Winter Donation Initiative," Jochen Flasbarth argued that feeding birds helps the purveyors of bird food but not the birds themselves. "The DBV therefore asks all its members to pursue bird feeding in moderation this winter. We call on everybody to purchase one less packet of bird food and instead donate 10 mark to the DBV" (Flasbarth 1986). The donations could be used to buy and manage nature reserves and thus protect entire habitats and ecosystems. The issue was presented as a zero-sum game in which all funds spent on bird food were funds lost to more important issues.

When he wrote this article, Jochen Flasbarth was the CEO (*Geschäftsführer*) of the DBV and a major force in transforming it into an environmental lobby organization. He was one of the leading figures to argue that to focus solely on birds was to ignore the world's environmental problems. While Flasbarth is well aware of the personal fulfillment and joy that watching birds can bring, he does not consider this experience to be either the sole or the most important reason to argue for bird conservation. He accordingly does not consider ethical arguments that postulate the existence of an intrinsic value of nature to be a persuasive justification for conservation.

> I believe that nature conservation today . . . can definitely not do with that alone. Nature conservation is important through the protection of ecosystems, through the function it has for the entire process of life on earth. . . . Even if one considers

> intrinsic value to be nonsense, if one does not say, That is simply something nice, and I want our kids to be able to find that nice as well, or at least be able to decide for themselves if they like it or not. Even if one does not see any of that, there are still tangible reasons, in the larger sense economic reasons, to say, A comprehensive conservation of nature is important. (Flasbarth, interview, August 2007).

Flasbarth is very outspoken on the fact that an attachment to birds was among the reasons that got him into conservation. As such his personal experience and biographical trajectory is the same as that of many of the bird conservationists in Britain. Yet there is a difference in the way Flasbarth translates his experience into discourse compared to the translation offered, for instance, by Moss. In the case of Moss, the discourse he chooses to communicate to others why birds are important matches his personal motivation. He uses an analogy to art to express the enjoyment that he derives from watching birds. Flasbarth, by contrast, does not engage in such a discourse. He seems not to have a language available that matches his own experience. The justification that he chooses to communicate to others why conservation is important is somewhat remote from the reasons that turned him into a person to whom birds matter. Flasbarth presents the reasons for conservation as a choice between intrinsic (ethical) and instrumental (economic) values, that is, birds are valued either as pure means or as pure ends. Since he assumes ethical arguments to have little resonance, he chooses economic ones. The possibility that there could be a third option is not considered and no analogy to art is discussed, whether to affirm or to deny it. Both Flasbarth and Moss describe a human benefit as the justification for conservation, but the benefits they describe are fundamentally different: in Moss's case this benefit is a subjectively experienced pleasure, while in the case of Flasbarth it is an objectively valid economic fact. The discrepancy between the arguments reflects the difference in the way bird conservation in the two countries is institutionalized: in the former case it is located in the world of play, in the latter in the world of work.

Flasbarth points out, however, that in his efforts to transform the DBV, he did not lose sight of the great appeal birds have for many members of the organization. He considered it key to the success of the organization that birds should stay on the agenda and that conservationists interested primarily in ornithological matters should continue to see the DBV as the organizational representative of their concerns (Flasbarth 2001). He chose the representation of voter preferences in the German party system as an analogy.

> I have always said—the way the CSU [Christian Social Union] has always said that there should be no space for anything next to them at the right fringe—I have

always said there should be no space for something birdish (*vogeliges*) next to us. That would be the end, since there has of course always been the worry that a new bird conservation organization would be founded that would then absorb the membership potential. That was not supposed to happen. For that reason we have always continued to support bird conservation, interest in nature study, and traditional habitat conservation. (Flasbarth, interview, August 2007).

The issue was perceived as particularly important because the DBV, unlike the RSPB, is not centrally organized but consists of approximately fifteen hundred local groups (*Ortsgruppen*) that are administered by a network of fifteen state associations (*Landesverbände*) plus its Bavarian partner, the State Association for Bird Protection (Landesbund für Vogelschutz). At the local level, the work of the organization is focused on practical issues, mainly bird conservation and the management of nature reserves, rather than broad environmental problems and their political solutions. Diverting the work of the national organization away from the focus of the local branches thus risks alienating the members. For Flasbarth the experience of the National Audubon Society in the United States served as a negative example of the consequences of neglecting the concerns of the established membership base.

I had observed the development of the associations in America, in particular of the Audubon Society, which had very similar roots to the DBV but then shifted completely towards sustainability earlier in the eighties. . . . My impression was that they told their members in this tradition-rich association, Everything we have done in the past was nonsense, and now we will do something new and better. They had to register an incredible decline in membership and then had to reverse course, as I observed. And that, I said, ought not to happen to the DBV or NABU. (Flasbarth, interview, August 2007)

The compromise was to continue practical conservation work at the local level while carrying out environmental lobbying at the headquarters in Bonn, which was then the capital of the country. Flasbarth describes his work in those days as an effort to navigate between two audiences—that is, between the organization's members and politicians. Flasbarth mentions his past meetings with a high-ranking politician of the Christian Democrats as an example.

I met a couple of times with Wolfgang Schäuble, then the parliamentary party leader, and those were always very informative, interesting, and inspiring conversations. . . . But I didn't get those appointments with him because of bird

> conservation or classical nature conservation: I got them because we were one of the leaders in the debate on ecological taxes. And then, of course, I also said, By the way, we have this EU Birds Directive, something has to happen with that. I never neglected those issues, I never left them out, but I said we would only have a societal impact if we discussed the issues that are considered important within the political elite as well. (Flasbarth, interview, August 2007)

The change in the organization's direction was reflected by the change in name from DBV (German League for Bird Protection) to NABU (German League for Nature Protection). What was noteworthy about this change was the fact that the term *bird* was replaced by *nature*. The change was intended not only to better reflect the widening agenda of the organization's work but also to attract additional members. The name of the organization's magazine also changed from *We and the Birds* (*Wir und die Vögel*) to *Nature Protection Today* (*Naturschutz heute*). The initial name had already become somewhat anachronistic, since the magazine displayed not only birds but also other species on its cover.

The change in name was followed by a professionalization of the organizational structure of the NABU, also initiated by Flasbarth: "In 1992 I became president. I got myself somebody from the corporate sector as a treasurer, because I said that cannot be done with a little bit of dilettantism anymore—we need somebody who really knows about these things. A very successful businessman from Cologne then worked with me, turning the entire association inside out. . . . That is, the association not only got an environmental-policy-oriented profile, but it also organized itself differently. We hired new personnel" (Flasbarth, interview, August 2007).

After serving as the president of the NABU from 1992 to 2003, Flasbarth moved into politics. He became the head of Department for Nature Conservation and Sustainability at the Federal Ministry for the Environment, Nature Conservation, Building and Nuclear Safety (Bundesministerium für Umwelt, Naturschutz, Bau und Reaktorsicherheit) in 2003 and president of the Federal Environment Agency (Umweltbundesamt) in 2009. The changes that Flasbarth introduced during his time at the NABU were a true success if assessed by the membership numbers. Beginning in the 1990s, the membership of the NABU increased steeply, in a trajectory comparable to the rise experienced by the RSPB two decades earlier. In 1990, when the organization changed its name, the membership stood at 138,900; by 2010 it counted 445,000 members and forty thousand supporters (NABU Jahresbericht 2010, 27).

Today the organization continues to follow the strategy of addressing a wide range of nature conservation and environmental policy issues. Birgit

Fahrenholz, head of the Membership Department of the NABU since 2007, observes that the distinction between these two broad objectives is not an important one to most members or to the public. The public presentation of these issues operates accordingly more on a catchall basis.

> Ask a passer-by in the street, I would almost say they don't even know the difference. I don't mean that in a bad way—that's simply how it is: I go somewhere and want something good to be done for nature. Now, where does nature stop? Nature, is that animals, or only plants, or do I have to call that the environment? Instead, I simply say, well, there is somebody doing something good for nature and the environment, and that's all right. . . . We cover a lot of topics. We say, No matter what your interests are, we are surely taking care of that. (Fahrenholz, interview, May 2011)

In Fahrenholz's view, being able to claim more than half a million members would make a significant difference to NABU's political influence, and passing this threshold is a major goal for the organization.

One of the NABU's major vehicles for gaining new members and sustaining old ones is the Bird of the Year (Vogel des Jahres) campaign, which has been running since 1971 and receives national media coverage. In this campaign, one bird species that carries a conservation message is selected by NABU leaders every year and promoted by the organization. The first bird of the year was the peregrine falcon, which at that time was in steep decline. Helmut Opitz is a member of the council and one of the vice presidents of the organization. He is in charge of organizing the campaign—his first contribution was as an expert on the Eurasian curlew, the 1982 bird of the year. Opitz describes the importance of the campaign:

> The Bird of the Year is the biggest marketing campaign run by the NABU. In terms of publicity and name recognition it is probably the most effective one. . . . The day after it is announced it is in every newspaper in Germany—it happens only very rarely that a small paper doesn't cover it. In up to ninety percent the NABU is also mentioned, and that is of course very important in the first place for marketing reasons. . . . That's why there is material: there are brochures, there are stickers—it is something the members are really interested in, and it has the most online traffic. (Opitz, interview, March 2012)

The campaign's success has led to the adoption of the model by other organizations. According to Opitz, there are around twenty different "species of the year" campaigns, including mammal of the year, reptile of the year, butterfly of the year, insect of the year, tree of the year, shrub of the year, and so forth. The bird of the year designated by the NABU, however, continues to receive the most attention.

The criteria for selecting the bird of the year have changed since the campaign's inception. Initially, the focus was on bird species that were endangered or in decline, such as the peregrine falcon. Yet since endangered species are by definition rarely seen and narrow in distribution, they often lack broad popular appeal. Not only have few people seen these species but even fewer can actively participate in their protection. For this reason, a different set of criteria is currently in use, Opitz explains:

> First of all, general degree of familiarity and sympathy. Second, wide distribution—that is, it's not about extremely rare species, which are, for example, only found in Mecklenburg-Western Pomerania or in Bavaria. Such a species has no chance anymore, because then the other groups can't do anything with it. The third criterion is activities for groups . . . , activities that one can do by oneself, also in order to make it accessible to the general public. A species that is living somewhere hidden in the twilight or in the reeds and is never to be seen would not be well suited. . . . The fourth point is the political message that's in the background. (Opitz, interview, March 2012)

Thus birds are now more often chosen for their suitability as flagship species (*Leitarten*) that can raise awareness about larger environmental issues. In 1980, for instance, when the NABU was running a campaign on wetlands, it chose the black grouse; in 1981, when highlighting forests, it chose the black woodpecker; and in 1982, when the focus was grassland, it selected the Eurasian curlew. In this the strategy of the NABU resembles that of the RSPB as described by Sarah Niemann: the choice of popular and easily seen flagship species to raise awareness about less known conservation issues. The list of birds of the year over the last decades also clearly indicates that the NABU has moved away from its initial focus on birds useful to agriculture. The first noxious species on the list—that is, one traditionally identified as harmful—was the rook in 1986 (Opitz 1997).

In 2002 the NABU went one step further and chose the house sparrow as bird of the year, the species whose population bird conservationists had tried to reduce in previous decades due to its detrimental effects on agriculture. The NABU did not conceal this history. The brochure on the bird of the year contained a review of the now discarded distinction between useful and harmful birds and pictures of sparrow-proof bird feeders that the organization had offered for sale in the first half of the twentieth century. By 2002, sparrow population numbers were in decline across Europe. Opitz considers awareness of this decline to be an effect of the campaign: "Choosing the house sparrow was very controversial. It only came out afterwards that it was actually threatened. Through the Bird of the Year campaign many people took a look at their house sparrows, which was previously not the case, and have noticed, Wait, they're not here anymore, they're not there anymore, here it has declined, there it has declined. . . . The house sparrow was a very important species. . . . For many people that made it click: Even the sparrow, even such a common bird, is threatened" (Opitz, interview, March 2012).

In addition to running the Bird of the Year campaign, the NABU borrowed the concept of the Big Garden Birdwatch from the RSPB. In Germany, the monitoring scheme was first introduced in several state associations and established as a national scheme in 2005 under the title of Garden Bird Hour (Stunde der Gartenvögel). About forty-five thousand people participated during the first year. Taking into account that the RSPB scheme started out with no more than fifty-two thousand participants when it became a national event, the turnout is quite substantial, albeit much lower than in Britain during that year. Bernd Pieper was the head of the Department of Communications of the NABU at the time and involved in organizing the event: "England was the role model—that was a very clear-cut story. . . . It was surely our largest campaign at the national level. Compared to England, where millions are reached with comparable campaigns, it is not as successful, but for the NABU it's a successful campaign. . . . It's an attempt to make the topic of bird conservation, bird watching, more popular, to show people, Here in your immediate surroundings there are species that are interesting—it's fun to watch them, that's a nice leisure activity to engage in, especially with others" (Pieper, interview, August 2007).

Pieper, who for several years was the editor in chief of the NABU membership magazine *Nature Protection Today*, noted, like the editor of *Birds*, Rob Hume, that good news sells better than bad. In the late 1980s and early 1990s, the NABU had some of its biggest successes with campaigns that drew attention to environmental threats and disasters. Yet this success, Pieper argues, was short-lived.

> In the last couple of years the nature conservation associations have learned that it is not effective to always only lament the state of the world. Instead, one should also sometimes say, We do many nice things, and it's worth it. It's much easier to get people interested in nature conservation if you offer them

> something pleasant, something to participate in. That is a learning process that I think has been completed three or four years ago—that it is much more effective than always saying, No, that's bad, the world is going to pieces. That was eighties-style nature conservation. (Pieper, interview, August 2007)

Building on its success with the Bird of the Year and the Garden Bird Hour, the NABU has tried to broaden the range of species used for campaigning in recent years. Birds still have the greatest popular appeal, but the organization is also trying to attract potential members with interests in other species. Heike Finke, a NABU council member engaged in an effort to move the organization beyond birds, administered a project on sharks at the time of the interview.

> There are people after all who are interested in something other than birds and whom we should maybe attract to the NABU as well, if we want to have more support, a larger clientele, or more members. . . . We are doing tests. We have a shark project running, Shark Tracker—it's online. It's not on the NABU website. We are running it separately for good reasons. . . . It's actually not that far removed from the NABU, insofar as we have also attached transmitters to birds in order to track bird migration. Now the point is to track the migration of sharks across the world's oceans. That's actually very similar. (Finke, interview, August 2007)

The project is explicitly geared toward recruiting young people to the NABU. Finke describes it as a trial balloon. Not everyone is optimistic that the project will succeed. The Marketing Department, she reports, was particularly skeptical about a promotional image of a white shark with its mouth wide open—reminiscent of the movie *Jaws*—and advised her to choose a less ferocious example. Yet Finke thinks this new campaign has the potential to reach out to people who find garden birds too tame.

Even with a wider environmental agenda for the national organization, much of the work at the local level continues to focus on nature conservation, particularly bird conservation. Markus Nipkow is the bird conservation officer at the NABU and reports that not all members were happy with the changes since the 1980s and 1990s.

> Those were difficult times for the active members. It was all happening very fast, and suddenly bird conservation was not that important anymore. That at least is how it looked to some people. Suddenly it was about a lot of other things, where the expertise was not always in existence. I mean, today, but even

> more back then, our active members had a background in species conservation and above all in bird conservation, and for that reason one can imagine that some of them were not too pleased that now other things were suddenly equally important or even more important. (Nipkow, interview, May 2011)

Although today it is only one of many fields of the NABU, the fact that bird conservation is administered by a specialized officer like Nipkow indicates its continued importance on the organization's agenda. As bird conservation officer, Nipkow, together with the help of others, formulates the organization's priorities for bird protection. In 2010 he coauthored the NABU's position paper on bird protection (*Grundsatzprogramm Vogelschutz*) (NABU 2010). The document outlines conservation goals that differ substantially from the organization's original objectives. The distinction between useful and harmful bird species is nowhere to be found in this publication, and the issue of endangerment is instead emphasized as the main criterion for conservation. Bird feeders and nest boxes, once the primary tools of economic ornithology and practical bird conservation are now given secondary importance. Their main function is not to actually increase population levels but to foster people's attachment to birds. According to the document, feeding birds has no overall effect on population dynamics, which are held to be largely dependent on the availability of suitable habitats. Habitat protection is thus considered the primary method for conservation. Yet more recently debate on the topic has reopened, Nipkow reports.

> This discussion actually came up again in the last couple of years because there was suddenly a protagonist of year-round bird feeding—this professor, Berthold, who has visited talk shows and advertised disseminating bird food by the hundredweight in the garden and preferably from helicopters. He is serious about all that. . . . One can have that discussion, but what I found annoying was that Berthold has misrepresented some facts, basically describing the NABU in a way it may have partially looked twenty to twenty-five years ago. There was actually a time—that was during this period of transition in the eighties—when putting up nest boxes and feeding birds acquired a bad aftertaste. Some people have simply stopped doing that—it was partially frowned upon. (Nipkow, interview, May 2011)

The NABU's more recent position on bird feeding can to a certain extent be seen as a reaction to a book on the topic published jointly by Peter Berthold and his wife, Gabriele Mohr, in 2006, which is the point of reference for

Nipkow's statement. At the time Berthold was the director of the Bird Station Radolfzell, Max Planck Institute for Ornithology, in Bavaria. In *Feeding Birds Properly* (*Vögel Füttern aber Richtig*), which gained considerable media attention, the authors explicitly attacked the position of the NABU on bird feeding (Berthold and Mohr 2006). They presented British data that provide evidence for the effectiveness of bird feeding in raising population numbers and cited the RSPB as a positive example for encouraging the practice.

This, of course, was also the earlier position of the NABU, or rather its predecessor, the DBV, although it was a position that the organization had been at pains to abandon when it was restructured in the late 1980s and early 1990s. By the time Berthold and Mohr wrote their book, the roles had been completely reversed. Now it was the British organization, once a latecomer in this field that had copied the practice from Germany, that was heralded as an example to be emulated. Why did this reversal take place? Berthold proposes an answer:

> There was this development of the bird conservation societies and closely related societies, which were suddenly dependent on using a substantial amount of the money that was initially intended for bird conservation purposes for all their centers and such like, to keep their machinery running, their administration. . . . That has cost an unbelievable amount of money, of course, and then one has tried to devalue peripheral areas of bird conservation that were considered less important, by saying, Bird feeding is no good, putting up nest boxes is entirely unimportant, the birds have enough nesting sites, etcetera. Only acquiring and securing large nature reserves counted for something. This strange development never existed in England, where these things have actually always had their place. (Berthold, interview, July 2007)

Berthold holds that there is no empirical evidence for the arguments made against bird feeding since the 1980s. In his view the scientific expertise that is available, mostly from Britain, points in the opposite direction. With common bird species evidently in decline, Berthold asserts that there is no good reason to reject bird feeding on the grounds that it benefits only these species. He claims that the NABU advanced its position in disregard of such data and mostly with the financial needs of its administrative structure in view.

> They need the donations. If some granny has ten euros and she gives the entire ten euros to the NABU, then it is of course cumulatively possible for a lot of NABU people to live off that. If every granny spends five euros on bird food, then the NABU is missing out on them. Because if that many people had

no qualms about saying, Don't feed the birds anymore—it's better to give the money to us, and that's when we said, to put it in a nutshell, that they had started to steal the birds' food. . . . These arguments [against bird feeding] are constructs of some white collar criminals [*Schreibtischtäter*], who are simply not able to develop any reasonable ideas. One has to say that, in all clarity, that is such an idiocy that it almost hurts, if one notices that people who are working in ornithology and in conservation come up with such moronic arguments. (Berthold, interview, July 2007)

The book itself was somewhat less outspoken on the topic, but it left no doubt as to the authors' opinion that bird conservation in Germany had gone in the wrong direction in recent decades. It received some very critical reviews from field ornithologists and bird conservationists (e.g., Flade 2006; Opitz 2006). Yet the overall effect of this critique and counter-critique was, in Berthold's view, to bring the NABU closer to the model advanced by the RSPB.

> We have supported year-round bird feeding, which the English have already had for some time. That was completely off-limits in Germany—it was looked upon as almost perverse—and now we have a large movement in Germany supporting year-round feeding. The large bird food companies have just gotten together and started to develop new products and to establish marketing. Well, it is not possible to stop it anymore . . . and the bird conservation societies, which have opposed it for a long time, now have to backpedal, of course. That is, they have to argue more and more carefully. (Berthold, interview, July 2007)

At the NABU, on the other hand, people suggest that Berthold might in fact be the one driven by financial considerations, presumably by a contract with the German manufacturers of bird food. At around the time the book was published, the British market for bird food was several times bigger than the German one: while around €150–225 million were spent on such products in Great Britain annually, the German market amounted only to about €20 million (Schäffer and Schäffer 2006). Promoting unlimited bird feeding would certainly be beneficial for the German manufacturers. The NABU, however, continues to argue against such measures. It currently supports bird feeding in winter but not year round. This position, stated for instance in the NABU position paper on bird conservation authored by Nipkow, can be considered a middle ground between the position formulated in the 1980s by people such as Flasbarth and that of Berthold and Mohr.

Among those advocating this moderate position is Norbert Schäffer, a German-born ornithologist who was active in his youth in the State Association for Bird Protection (Landesbund für Vogelschutz), the Bavarian partner organization of the NABU. He wrote his dissertation at the Max Planck Institute for Ornithology while Berthold was a director there and knows him well. By the time of the interview, he had been working at the RSPB headquarters for twelve years. Schäffer is thus familiar with the debate in both Germany and Great Britain.

> Berthold has missed an important opportunity, since his book had a lot of resonance, but it was completely pulled to pieces by the experts. . . . He had the opportunity to address this dogmatic German treatment of bird feeding that states, Feeding birds, for heaven's sake, only at temperatures under zero Celsius, when there is snow, otherwise the birds get a salmonella infection, which has not been proven. [Or] if one feeds birds in summer, they will feed their young with the wrong food and they will then die. There is no evidence for that, nothing. He should have addressed that and taken it apart scientifically. But he has—apparently in order to provoke the bird conservationists—gone to the other extreme. (Schäffer, interview, February 2008)

In 2006 Schäffer, with his wife, Anita Schäffer, wrote his own guidebook on feeding garden birds (Schäffer and Schäffer 2006). Unlike Berthold's book, this publication gained the support of the NABU. Markus Nipkow was among the contributors, and the introduction was written by Olaf Tschimpke, the president of the NABU. The book was intended as a guide for the Garden Bird Hour in Germany. Schäffer addresses many of the points addressed by Berthold's book but comes to different conclusions. He notes that although conservationists display strikingly different attitudes toward the practice in the two countries, there is a lack of empirical evidence for both the potential negative and the potential positive effects of bird feeding. The currently available data do not answer the question of whether the increase in bird numbers in gardens where feeding takes place is the result of higher survival rates for resident birds or of attracting more birds from adjoining areas.

One of the benefits of feeding birds that Schäffer is certain about is its potential to attract people to birdlife. In his view, the NABU has failed to take advantage of the potential of the practice to attract people to nature conservation. Schäffer argues that bird watching is an outdoor sport like any other, and it is misleading to attribute its popular appeal to a presumably eccentric character of those individuals who practice it.

> What's wrong with saying, On Saturday morning I went outside, and it was a nice Saturday morning because I watched a grey heron during sunrise searching for food? A lot of people do that around here. In Germany that always has a strange connotation. At the same time it is entirely normal to go to a tennis match in order to watch a felt-made ball being hit across a net by two adults. One can actually ridicule every kind of leisure activity, and the German bird and nature conservation scene still doesn't dare to stand by what they do. (Schäffer, interview, February 2008)

In Schäffer's experience, German conservationists tend to distinguish between "soft" and "hard" topics, the former encompassing the sentimental appeals of an intact nature, the latter the looming threats of environmental disaster. Bird conservation is in this scheme a quintessentially soft topic; climate change and its potential dangers are examples of quintessentially hard topics. Soft topics work with a positive appeal, hard topics with a negative threat. The NABU had intentionally moved its agenda from appeal to threat (i.e., from soft to hard topics) in the 1980s, but Schäffer sees this strategy as a mistake. He argues that the strength of the RSPB derives precisely from the fact that it does not privilege hard topics over soft ones but instead uses the popularity of birds to raise awareness of larger environmental issues.

> If one starts with climate change, people immediately lose interest because they simply don't want to hear that anymore. But if I make it accessible to people—and in that the English are world-class—by saying, Look at what's happening in front of your door, you can observe that, then the problem suddenly becomes concrete, it becomes relevant. Then I want to do something about it, and that is the foundation of the nature conservation movement in England. It starts with enthusiasm and not with concern. In Germany the nature conservation societies always gained members when there was a catastrophe: the dying forests, the pollution of the river Rhine, Chernobyl—then there was an increase, but that isn't something long-term, because nobody wants that forever. I don't want to hear how bad the world is every time I open my membership magazine. (Schäffer, interview, February 2008)

Schäffer holds that the positive appeal of bird watching is a more effective strategy for increasing support for conservation. For the same reason he is not of the opinion that economic arguments for bird and nature conservation are particularly convincing. There is no doubt for him that the conservation of species and habitats may produce economic benefits, yet in his view economic justifications do not reflect the actual reason why people such as himself want to protect these species and habits. Here he departs from the point of view of many of his colleagues in Germany, who put forward an economic argument for conservation that highlights the importance of functioning ecosystems for human survival. Schäffer, by contrast, advances the analogy to art, which is more frequently encountered in Britain, although he couples it with a notion of intrinsic value rather than a reference to people's enjoyment of nature.

> For me a wood grouse in the Bavarian Forest has a right to exist, whether I can see it or not. Whether that has an economic justification is a completely different discussion. Why engage in nature conservation? The answer to that is—well, we have to protect the rain forest for one thing because of the ecotourists . . . and then of course there's the possibility of finding a cure for cancer there, and for that reason we protect the rain forest. Those might both apply, but I think that the rain forest as such has a right to existence simply because it has been around for ten million years, just like the wood grouse in the Bavarian Forest. . . . I think one would not say, Let's tear down the Acropolis because it makes more economic sense to build a parking lot there—nobody would accept that. Just in the same way I cannot say, Let's cut down the Bavarian Forest because we can make money that way, or let's protect it because we can make money that way. . . . For me personally, taking that as the only justification for the conservation of species or habitats is not enough, and I also consider it to be dangerous, because what is going to happen once the cure for cancer is found? Will we then chop down the rain forest? (Schäffer, interview, February 2008)

Economic arguments for conservation can of course be powerful discursive strategies. They provide a way to make birds count in the eyes of people who have no emotional attachment to them. Yet the opportunity to form such attachments to birds in people through engaging them in the game of bird watching is lost to this strategy. This neglect of play is not new. The practice of bird watching was belittled at the beginning of the twentieth century when conservationists in Germany were eager to point out that nest boxes are not toys and bird feeding not a game. Now as then many conservationists in the country try to actively distance themselves from this presumably less serious engagement with the natural world and instead highlight the either purely economic or purely ethical reasons for conservation.

The large-scale transformation of the DBV into the NABU in the 1980s and 1990s did not imply that economic arguments were entirely abandoned.

The major difference from the practical bird conservation of previous decades was that this time not a single species but the environment at large was represented as important to human survival and material well-being. In Schäffer's view this effort to highlight economic arguments indicated not that German conservationists derived no enjoyment from nature but that they failed to translate this enjoyment into an explicit discourse to guide organizational strategies. Experience and discourse simply did not match each other.

Only in more recent years has the argument for the joys of bird watching migrated from Britain to Germany. Schäffer is a champion of the effort to supplant the German tradition of bird conservation rooted in the world of work with the British one rooted in the world of play. The introduction of the Big Garden Birdwatch in Germany and the accompanying guide that he coauthored are a central part of this effort.

FROM LOCAL BIRDS TO GLOBAL WARMING

The RSPB not only serves as a model for conservation organizations in other countries such as the NABU but also offers support in the form of both expertise and funding. Ken Smith is part of the organization's International Department, which investigates which organizational strategies work in which countries. The RSPB has attained its current success by encouraging bird watching, but this approach does not work in every country, Smith points out:

> Birds are a minority interest. In lots of countries there is a small number of people who are very keen, very enthusiastic bird watchers. And if you say to the public, Come and be like them, and the public looks at them and says, We have loads of other things to worry about. . . . Even here it only works to a degree, and I think there has been quite a major realization here in the last, well, quite recently, in the last few years. . . . Broadly speaking, in the early stages our message was that birds are fantastic, and the places where birds live are fantastic; and we love them; and we think they are great; and you, the great general public, we think you should come and be like us and come and enjoy these birds and enjoy these places like we do. (Smith, interview, July 2006)

Some at the RSPB argue that the effectiveness of this strategy has reached its limit. Historically, the RSPB has largely been a club for bird watchers run by bird watchers. At the present, however, this automatic identification of all the members of the organization as bird watchers no longer works. Stephen

Noble in the Marketing Department, for instance, argues that if the organization wants to increase its membership further, it needs to tap into additional social groups. He sees a potential problem for membership recruitment with campaigns that focus exclusively on bird watching:

> People that work here, certainly the people who are trustees, tend to come here because they love bird watching. They love the hobby and the pursuit of bird watching, and secondary to that they are trying to save the planet as well. And so they see nothing wrong with either deliberately or implicitly producing communications for the wider public that present the RSPB as a bird watching organization. And that's a problem for us, because we know that there is very limited potential for getting people to say, Yeah, I'll give you money, I'll be a member, if we continue to pursue that. (Noble, interview, June 2006)

This insight has led to a discussion of possible changes in the public presentation of the RSPB. One suggestion was to change the organization's name by changing the meaning of the *B* in RSPB from Birds to Biodiversity. This suggestion was rejected on the grounds that biodiversity does not have the same resonance as birds, despite its broader meaning, as Graham Wynne, CEO of the RSPB at the time, points out:

> The biodiversity word is not a good word. It is dry, it is policy. A lot of people still don't know what it means. So it does not have mass or popular appeal in our country. . . . It's not very neat, and I don't think we are quite ready for it yet, but I think part of the idea would be that we stop saying Royal Society for the Protection of Birds, we say RSPB, and the RSPB just has a life of its own. People to some extent stop bothering that the *B* is for birds, and you change the imagery around RSPB. You show other wildlife as well as birds, and people just absorb the idea that RSPB is broader than birds. That's certainly where I want to go for the next five years or so. (Wynne, interview, February 2008)

In the end, the society opted to keep the name that it has had since 1904, when the Society for the Protection of Birds received the royal charter. What has changed over time is the organization's motto, which has altered from "Saving the birds, saving the future" (pre 1998) to "For birds, for people, for ever" (1998) to "A million voices for nature" (2008), to "Giving nature a home" (2013) (Danielle Meyer, RSPB, pers. comm., July 2014). The replacement of the word *birds* by *nature* in the motto constitutes a compromise between those who wanted to drop birds from the name of the organization altogether and those who wanted to keep the name as it is. These changes are marginal compared to the almost complete change in agenda and public presentation that accompanied the transformation of the DBV into the NABU in Germany. The name change resulted in a public presentation of the NABU similar to that of the BUND, the German affiliate of Friends of the Earth International.

The RSPB, on the other hand, carefully tries to avoid aligning the organization's image with that of environmental organizations such as Greenpeace or Friends of the Earth, even though nature conservation and environmentalism are part of its agenda. Changing the society's image is assumed to decrease membership, not to increase it. The RSPB's key to success remains its focus on birds, not nature or the environment in general. Its new environmental agenda is advertised only to the degree that it connects to bird conservation. Birds continue to be used to gain public support, while many of the organization's broader environmental initiatives are conducted behind closed doors.

John Lawton, the chairman of the RSPB council from 1993 to 1998, was among those who initiated the new environmental agenda of the organization. He was a passionate bird watcher during childhood and continues to engage in outdoor record collecting, both for his professional research and as a leisure activity. Lawton keeps a world list, a UK list, and a garden list. He is a collector, or a "lister," as he calls himself, and this passion finds a reflection in the arguments he gives for conservation.

> I think there are all kind of reasons for conserving wild nature. . . . I am not religious in any sense, I am an atheist, but I actually do believe there is a moral and ethical imperative on us to try and hand on to future generations a world that is no less interesting than the actual one we inherited. I happen to believe that—maybe you don't, but I happen to believe that. . . . I mean there is a straight argument. There is an awful lot of nature that is simply wonderful, and we wouldn't hesitate to use that argument for Monet paintings, or Mozart concertos, or medieval cathedrals. Nobody would dream of asking, How many Mozart concertos do you need? . . . How many cathedrals do you need? It's just a no-brainer. I have huge amounts of pleasure, as I have from listening to music, from watching birds. (Lawton, interview, June 2006)

A view of objects of nature as unique items akin to works of art or historical treasures does not imply that Lawton is unaware of larger ecological relationships: he is in fact a professional ecologist. Lawton was the president of the British Ecological Society from 2005 to 2007 and was knighted in

2005 for his contributions to ecological science. He served as the chief executive of the Natural Environment Research Council from 1999 to 2005 and as the chairman of the Royal Commission on Environmental Pollution from 2005 to 2011.

One of the major policy issues that Lawton dealt with at the commission was climate change. He was thus working on an environmental issue that goes far beyond the conservation of any single species. Climate change was already on his agenda when he was still the chairman of the RSPB. He recalls that, back in the 1990s, many people at the organization thought that the issue had nothing to do with the RSPB. The way to make the argument, Lawton explains, was to tie the issue of global climate change to local British birds.

> It was the realization that if we continue just to buy nature reserves, look after species and habits, we would win some battles, but we would lose the war, because there are much bigger forces at work. Of course common agricultural policy was one; drivers of massive habitat destruction, the fisheries policy, and global climate change—and then actually the RSPB had to engage in those big issues. It had to carry the membership with them. I used to say: "If you can't get from Dotterels in the top of the Scottish Highlands to global climate change in three sentences, don't do it." You know, "Pull the kettle on—carbon dioxide—carbon dioxide destroys Dotterel habitat in the Highlands, because of climate change." That is a legitimate thing for the society to be involved in. (Lawton, interview, June 2006)

Tying concern for birds to wider environmental issues has been one of the major challenges at the RSPB since the mid-1990s. The link is far from self-evident. People who are passionate about birds do not necessarily care about the environment in general. It is the other way around: the wider environment and the concern for its protection derives much of its appeal and sense of urgency from the implications of environmental degradation for birdlife.

Lawton was also involved in the effort to professionalize the RSPB's work, comparable to the effort made by Flasbarth in Germany around the same time. One of the major insights at the time was that an organization with more than one million members could no longer be run by ornithologists alone: it needed a CEO with administrative expertise. Lawton was instrumental in selecting Barbara Young for the position in 1991. As he recalls, her appointment was controversial because she was not an ornithologist or conservation biologist: she was "just a bloody good manager, which is what we needed at the time" (Lawton, interview, June 2006).

Until her hiring by the RSPB, Young had pursued a career in public relations and health services management, as she explains:

> It was a big wrench to me to leave healthcare, but I had done twenty years in it and I wanted to do something different. I got into the environment by chance, but having gone into it, I am now completely hooked. . . . I mean, when I went to the RSPB I knew nothing about birds. I didn't know the difference between a robin and a ready-wrapped turkey. When I first went there, I went because they needed somebody to manage the organization and to campaign, and that's the two things I do, I campaign and I manage, and I got the bug in a big way. I became completely fixated within about six months. (Young, interview, September 2007)

Young was soon influenced by the genuine passion for birds among the people at the RSPB. She now has her own life list—that is, a list of all bird species she has ever seen—and, even more important, she has a life list of environmental goals. Her major strategy for achieving these goals is to win other people over to birds in just the same way that she was won over. One reason why birds are effective in winning people over to environmental issues is that almost everybody can experience them firsthand. Young later applied this insight in a new context. She moved into politics, working from 2000 to 2008 as the CEO of the government agency English Nature. She was made a life peer in 1997, and as Baroness Young of Old Scone is now a member of the House of Lords. She now takes the political elite of the country bird watching. In this way, she is applying the same strategy that Peter Holden used to appeal to children and teenagers in the country: to get people outdoors, give them binoculars and a field guide, and make them look at birds.

> The good thing about birds is they are very visible to ordinary people. They're there, there are a lot of them, so you can spot them. It's not like mammals, who you have a hell of a job seeing. Birds are publicly available. And I've not met anybody yet and I've taken umpteen ministers out who couldn't care less about birds—if you take them out with a good bird book and a set of binoculars on a nice day to a really good place with somebody who knows their birds inside out and can tell them all about them—I have not yet met a minister who comes back from that untouched. . . . And it's the same with the public, showing people birds pays off, because all of a sudden they get excited. It's good stuff. (Young, interview, September 2007)

It is indeed "good stuff" and not "bad news" that makes for the success of the RSPB and increasingly also of the NABU. The task is not to shock people with horror stories about environmental hazards—the trick is to make them see and love what it is these hazards could destroy. This appeal of birds experienced in one's own surroundings is also driving nature conservation in the international arena.

CHAPTER 7

Data Power and Geographical Reference Frames

On a weekend morning in fall 1984, a car carrying members of the German Committee against Bird Slaughter (Komitee gegen den Vogelmord) drove up to the Belgian border with Germany. The border is not policed: it is a "green" border, and only the contrast in color between two meadows and the occasional ditch or cattle fence indicates where one country ends and the other begins. The arrangement has a certain resemblance to the demarcations on a soccer field, and on this autumn morning the game was indeed on. The members of the committee crossed the border and removed the traps and nets set by Belgian bird catchers for their traditional autumn hunt for migrating birds. The German team brought the referees: on the German side of the border, a van with officers of the Federal Border Protection Force (Bundesgrenzschutz) kept the Belgian bird catchers from entering German territory to reclaim their traps. The event continued until the Belgian police arrived to prevent the Germans from entering Belgian territory. Ten such weekend trips were made that fall by the members of the committee, regular enough to be dubbed "bird wars" by television and newspaper media.

One of the players was Alexander Heyd, who joined the organization set up by members of the Naturschutzbund (NABU) who were disappointed with their organization's lax stance toward bird hunting and its reluctance to engage in any form of contentious action. Heyd was approached by a member of the organization when he was in his late teens and in search of a position for his civilian service, the alternative to mandatory military service in Germany.

FIGURE 7.1 Belgian bird catcher confronts members of the Committee against Bird Slaughter with a pitchfork at a fence demarcating the German-Belgian border. Photograph by Alexander Heyd, taken at Kreis Euskirchen, October 1984.

> He knew that I always did such *sponti* things [the term *sponti* is shorthand for a contentious form of spontaneous political activism]. I was a member of the Antifa [the antifascist movement] running after Republicans [a German right-wing political party] and throwing stones at demonstrations. He knew that and told me, there is a bird conservation society that's doing the same as what you are doing with the Republicans, just in Belgium, with the Flemish Bloc [a Belgian right-wing party]. They are all Nazis, and they are in the European parliament. It's a party of bird catchers, and they are right-wing—that's a good combination. That's where you should go. And that is how I joined the committee. (Heyd, interview, August 2007)

The event at the border, like many other initiatives by the organization, is a prime example of contentious politics. In his years with the committee, Heyd reports having been hit on the behind by an angry bird catcher with a shotgun, having his nose broken twice, and spending several short periods in prison.

Today, Heyd is the chief executive of the organization and looks back at the episode at the Belgian border as a battle that has been won. Bird catching with traps was banned in Europe by the Birds Directive in 1979. In Belgium, this

European directive had not been translated into national legislation when the committee started its initiative. The organization thus acted in the semilegal realm by obstructing a practice that was officially illegal under international law but had not been outlawed under national law. Stealing the traps thus remained illegal, and the committee was sued several times on that account. Yet like the police officers of the Federal Border Protection Force, the German judges were on the side of the committee and never followed the calls for action by their Belgian counterparts. In 1988 the European Court of Justice (now called the Court of Justice of the European Union) sentenced Belgium for transgression of the Birds Directive, and in 1993 bird catching was banned under Belgian law. The end of the bird wars was a victory for the committee.

The Committee against Bird Slaughter was founded in 1975. In 2010 it had about 450 members and 12,500 supporters. About forty local NABU groups support the organization, and many of the leading activists are members of both organizations, although they are not formally affiliated. The hunting of migratory birds outside Germany was among the key concerns of the organization, but the committee took action within Germany as well, for instance in a campaign against the annual spring hunt in the Wadden Sea on the northwest German coast. The EU Birds Directive has been key to all of the committee's actions. "The EU Birds Directive, that is our directive, that's our most important tool. That's what we pray to every morning. It's our daily bread, the EU Birds Directive and the Habitats Directive" (Heyd, interview, August 2007).

Members of the committee were in fact involved in the origination of the directive. A petition to the European Commission against bird slaughter in Italy, with more than seven million signatures, was one of the more noteworthy lobbying efforts in Brussels. The committee was involved in action abroad well before the directive was issued. One of the first campaigns in the 1970s was called "Kein Urlaubsort wo Vogelmord" ("No vacation chilling where bird killing," if one gives precedence to a translation that preserves the ring of the original to a more literal one) and urged German tourists in Italy to boycott locations where birds were shot. Fearing the loss of income from tourism, several Italian villages at the Adriatic Sea began to advertise themselves as free of bird slaughter.

Bird catching in Italy has remained the focus of the committee by the time of the interview, long after the enactment of the Birds Directive. Its major projects are annual camps in Brescia, a province in Lombardy with a particularly high number of bird catchers. Every year during the autumn migration, German members of the committee, together with their Italian colleagues and friends, gather to collect illegal traps and nets. The camps have turned into extended events that last up to three weeks and attract more than a hundred participants.

In the style of hunters who count up their bag, the committee has compiled statistics recording its success. Between 1985 and 2005 over 340,000 traps, 170,000 snares, and 1,800 nets were collected in Italy (Heyd 2006).

Brescia is a suitable spot for these activities not only because of the high number of bird catchers but also because of the cooperation of local police. The committee usually does not remove the traps but calls the police to the sites to wait for the bird catchers. The bird catchers in turn have become more careful, watching for members of the committee and at times even threatening them with guns. Heyd recounts the excitement of the chase: "I look at it from a sportsman's point of view. That's the way things are in Italy—it's like building up arms, it's an arms race with the bird catchers. They get better and better at hiding the traps, and we get better and better at finding them" (Heyd, interview, August 2007). Over the years, this arms race has been won by the committee.

A clear indicator of this victory is the fact that, until 2001, the numbers of traps collected rose in conjunction with the number of days committee members spent in the field. Since then, the rate of return on days spent in the field has been declining to the point where it would make sense to divert some of the effort to other places in Europe, where bird catching is still more common and more traps could be collected. Lately there have been discussions about moving the group's focus to France. Yet many members of the committee are reluctant to move on, not least because they have developed close friendships with bird conservationists in Italy. "We all like Italy, and we all speak Italian, and we have lots of Italian friends. Most of our active members are Italians, and there's of course no question that we will continue to work in Italy for the time being" (Heyd, interview, August 2007).

BORDERS AND FRAMES

Organized bird conservation in Europe has not always transcended national borders. To a large extent, it has been concerned with the protection of rare and endangered species on local nature reserves. What qualifies as *common* and what as *rare* depends on the geographical frame of reference. For most of the period covered in this study, this frame of reference has coincided with the political boundaries of the nation-state. Starting in the second half of the twentieth century, however, the national focus has increasingly been complemented by a European framework for conservation.

In this chapter I argue that the development of European nature conservation policies was established in those areas where conclusive environmental data had been collected. Field-ornithological data was the starting point for these policies. The Birds Directive, the first European conservation law, was enacted in 1979, and a Habitats Directive followed in 1992. Together these two directives created the network of protected areas called Natura 2000.

This development in the field of bird conservation is part of the larger downward trend in the significance of the nation state in contemporary societies. In the 1970s Herminio Martins (1974) influentially coined the term methodological nationalism to describe, on a critical note, the tendency of the social sciences to predefine social realities as nationally bounded (Chernilo 2006, 2007; Wimmer and Schiller 2002, 2003). Among critics of methodological nationalism there is widespread agreement that the boundaries of the nation state lose relevance in the present. Scholars have argued that processes of globalization have reconfigured the nation state. Urry (2000) and Papastergiadis (2000), for instance, have described the development of new technologies that facilitate global market exchanges, while Beck (1999, 2000) has highlighted the environmental side effects of technologies that create a world risk society, which transcends national boundaries.

The analysis in this chapter of bird conservation in the European Union shows that very little is won by replacing methodological nationalism with a methodological internationalism that simply transfers taken-for-granted territorial assumptions about social processes from a national to a European or global level. As Saskia Sassen and others have shown, globalization is a process that does not capture the entire globe at once and to the same degree but is taking place at particular locations and to different degrees instead (Sassen 1991, 2001). The way globalization is enacted at these places is moreover far from uniform. Karin Knorr-Cetina and coauthors (Knorr-Cetina and Bruegger 2002; Knorr-Cetina and Grimpe 2008), using the example of trading rooms, have shown that representations of the global economy employ different scoping mechanisms that can be either inclusive or exclusive; that is, globalization can include local and national idiosyncrasies, or it can exclude them. These different scoping mechanisms do not merely represent the world but actively produce it, a process that the actor network theorist Michel Callon has influentially described as performativity (Callon 1998).

The conservation of birds under EU legislation is an example of a process of globalization that does not eradicate national differences. The high appeal of bird watching and field-ornithological data collecting in Britain, for instance, did not become less of a national peculiarity with the establishment of the two directives—the institutional arena in which this peculiarity played out simply widened. British bird watching became the model for European nature conservation. As the boundaries and political influence of the European

Union expanded, a small group of national bird watchers seized the opportunity to translate their concerns into international policies. The two directives were the legal equivalent of a Trojan horse: the actors who drafted them had foreseen consequences that were not equally obvious to the national representatives who signed them. But the directives had effects that surprised even their authors. The geographical frame of reference established by the Natura 2000 network not only reflected and reinforced the valuations of that time but also produced and created valuations of its own along the way—valuations that were not foreseen and sometimes not desired. Changes in geographical reference frames change whether a species is considered endangered: many bird species that were target species for national conservation programs were assigned this status because they were represented by populations living at the edge of their species range: they were rare and endangered on a national level but abundant on a European or global level. The other way around, many species that are abundant and common in one country but whose total population is confined to this limited area are considered rare from a European or global point of view.

The introduction of European legislation thus implied that actors had to adjust their conservation priorities to new geographical reference frames. Emotional attachments to specific bird species, however, do not simply disappear with the introduction of new legislation. This chapter looks at how bird conservationists managed to create and implement European conservation laws and how they reacted largely successfully to their unintended consequences. Rather than abandoning national conservation frames in favor of European ones, conservationists began to use both geographical frames in parallel, with some significant differences in the way these parallel uses are carried out in Britain and in Germany.

EUROPEAN LEGISLATION AND THE NETWORK OF PROTECTED AREAS

The perils of migrating birds inspired the development of European legislation on bird conservation. Bird hunting in Italy, France, and Belgium was of concern to the members of the Committee against Bird Slaughter in Germany and to conservationists in many other parts of Europe. The major bird conservation organization in Italy, the Lega Italiana Protezione Uccelli (LIPU), for instance, was founded in 1965, and one of its concerns was the shooting of migratory birds. The Italian law dealing with the practice of bird catching had been reformed in 1967, and further discussions on the game laws were in process when the committee started its first campaigns. The restrictions on bird hunting that resulted were, however, rather modest compared to the legislation already in place in other European countries.

British, German, and Dutch bird conservationists were particularly concerned about bird hunting in southern Europe. One of the central institutional loci for the articulation of this concern was the International Council for Bird Preservation (ICBP), founded in 1922 as the International Committee for Bird Protection. It was initially the plumage trade that gave rise to the call for international coordination, but by the time it had been legally restricted in the fashion centers of Paris, London, and New York in the late 1920s and early 1930s, other topics, such as oil pollution at sea and migratory bird deaths caused by lighthouses, had already joined the agenda. International coordination ceased during the Second World War, and, although it was revived in the 1950s, the ICBP was eclipsed by other organizations founded after the war, like the International Union for the Conservation of Nature (IUCN) and the World Wildlife Fund (WWF, later called the World Wide Fund for Nature) (Boardman 1981, 2006; Hayden 1942; Holdgate 1999; McCormick 1989).

There is scant evidence that the ICBP achieved much in the 1950s and 1960s beyond the organization of international conferences on bird conservation and the publication of accompanying reports. Its most notable postwar achievement was the redrafting a 1902 International Law for the Protection of Birds, but the new version, the International Convention for the Protection of Birds, drafted in 1950, was anything but a success. Only twelve countries ever signed the convention, and twenty years later it was in effect in no more than eight countries.

When the ICBP in the early 1970s again took up the issue of shooting migratory birds, it was more successful. It owed this success to the institutional framework of the European Community (EC), the forerunner of the European Union (EU). Nature conservation had not been an established policy domain, much less a priority, during the first decades of the European Community, whose policies focused on the Common Market. Although the EC established some environmental policies, largely concerned with the purity of air and water and the control of chemicals, it did not concern itself specifically with wildlife and nature conservation. All of these regulations had moreover the aim of coordinating national policies, not of creating new ones (Caldwell 1996; McCormick 2001).

With the Birds Directive and later the Habitats Directive, however, the EC went far beyond the coordination of existing measures and created what remain the major legal instruments for nature conservation in both the EU and its member states. The first steps toward European nature conservation were taken by actors inside the European institutions, some of whom were also members of national bird-conservation organizations and the ICBP.

Bird conservation was first mentioned in 1973 in the EC's first Environmental Action Program (Cramp 1977, 1978). The program proposed to harmonize national regulations on the protection of animal species, migratory birds in particular. A section titled The Protection of Birds and Other Animal Species foregrounded the economic arguments for conservation: "Hundreds of millions of migratory birds and songbirds are captured and killed in Europe every year. . . . This massive destruction provokes a serious threat to the ecological balance in Europe, because it causes plant parasites to proliferate" (Council of the European Communities 1973, 40). The document went on to argue that the protection of the environment should therefore include measures to prevent the large-scale destruction of birds, particularly migratory songbirds. The possible extinction of rare birds and other species lacking obvious economic relevance was mentioned only at the margins.

The European Commission (the executive body of the EC) was a driving force in the establishment of the Birds Directive. Claus Stuffmann worked for the commission in the 1970s and was involved in the creation of the new legislation. He points out that the economic argument was not the driving force behind the project but merely a justification put forward to accommodate bird conservation within the EC legal framework.

> That was a new subject the commission was now suddenly supposed to work on. We of course asked what could be put forward as justification for an initiative on this issue. For normal economic goods—the harmonization of the size of screws and similar facilitations of the Common Market—that was relatively easy. When establishing the legal basis of any kind of initiative, one has to connect it to the functioning of the Common Market. That is where the idea came up that birds are not just popular with the people but also an economic factor. Continuing this line of thought, we came to the conclusion that birds have an impact on agriculture through their control of insect infestations. Afterwards, it sometimes looked as if bird conservation was indeed enacted for the one and only reason of sustaining the free movement of goods and its impact on the functioning of the Common Market. (Stuffmann, interview, July 2012)

Economic ornithology had of course a long tradition in Germany and many other countries in continental Europe. Yet by the time the Birds Directive was under consideration this economic rationale for conservation had lost much, if not all, of its convincing force. The postwar generation of conservationists had moved bird conservation away from economic goals and justifications. British conservationists, who were a driving force in the negotiation process, never had a strong tradition in this field to begin with. The economic justification was used not because it reflected the motivation of the involved bird conservationists but because it fit the requirements of European law for intervention into the sovereignty of its member states. Only policies that facilitated the common European market could be enacted. Attributing economic importance to migrating birds thus fit the bill perfectly.

This creative way to make bird conservation fit the agenda of the European Union was facilitated by the cooperation of a wide range of experts representing different ways of approaching the matter. This diversity of approaches turned out to be crucial. David Stark and coauthors have argued that it is the employment of different standards of representation and valuation that generates innovation (Beunza and Stark 2004; Vaan, Stark, and Vedres 2015; Stark 2009; Vedres and Stark 2010). Following Dewey's pragmatism, Stark proposes a sociology of worth that recognizes a heterarchy of multiple overlapping principles of valuation. Formulated in the context of the sociology of organizations, heterarchy contrasts to the notion of hierarchy in the account of bureaucratic organizations informed by a structural-functionalist notion of rule following and norm-guided behavior. In contrast to the image that hierarchical organization is the most rational way to achieve organizational goals, Stark shows that it is precisely the intersection of multiple logics of valuation, or, what Boltanski and Thévenot (2006) have called economies of worth, that makes organizations innovative and successful. The organizational setting of the European Commission provided such a heterarchy in which different orders of worth and different areas of expertise represented by legal scholars, economists, ornithologists, and conservationists could interact and produce innovative outcomes.

Stuffmann himself, for instance, was an economist by training who came to the topic of nature conservation through his work for the commission. Once he started work on the European environmental program, he began to take a personal interest in pushing this work forward. In the end, he counted himself among those actors for whom the issue was "a matter of the heart" (*Herzensangelegenheit*), as he calls it. In his view, the legislation would not have come into being without the many like-minded actors who collaborated behind the scenes, traversing the administrative boundaries between European institutions, national parliaments, and nongovernmental organizations.

The success of the legislation depended on backing it with scientific expertise. Much of this came from a study ordered by the commission and administered by Bernhard Grzimek of the Zoological Society of Frankfurt (Zoologische Gesellschaft Frankfurt). The resulting report was written by Bernd Conrad and Wolfgang Poltz, two doctoral students of Gerhard Thielcke at the Max Planck Institute for Ornithology at Radolfzell, who was the head of the German section of the ICBP. Submitted to the European Commission in draft form in 1974, it was published in Germany two years later as *Bird Protection in Europe* (*Vogelschutz in Europa*). The report compiled the available data on the population dynamics of bird species throughout Europe and their threat status. It concluded that the population levels of 125 bird species in Europe were increasing, while 221 were declining, and sixty-two had stayed the same. Only two European species were reported to have become extinct in the previous centuries: the great auk and the northern bald ibis. The list of national extinctions was decidedly longer: Italy topped the list with thirty-two extinct bird species (Conrad and Poltz 1976).

How the authors produced these numbers on population trends is not evident from the report. The data presented in the more detailed accounts for single species were sketchy at best. Most data were based on regional developments in Germany, and to a lesser extent also in Britain, over varying periods of time. The authors argued that this geographical bias did not affect the validity of the results, as they considered the trend in Germany representative of all of continental Europe. The analysis of the causes for species decline was somewhat biased as well. The authors focused on bird shooting and catching in three countries—Italy, Belgium, and France—not on habitat destruction across the continent.

> Well, it was the issue that was of course the easiest to pin down—something that we perceived as most likely to get resolved. . . . The task was to try to calculate the economic value of hunting. It was a part of the goal to be able to say, well, there are the captured birds, but that has no particular value for the nation or for the people. It's a fun activity, but it's of course not contributing anything if compared to the potentially increasing employment of pesticides, because no birds are left to devour insect pests. . . . It was of course a very difficult and idealized endeavor. We have tried to the best of our knowledge and belief to find out what can be found out within half a year. (Conrad, interview, March 2012)

The authors went to considerable lengths to estimate the numbers of birds killed by hunters, particularly in Italy. They produced five calculations based on the statistics of various Italian hunting associations and arrived at estimates ranging from 25 million to 440 million birds killed by Italian hunters each year. They also cited the estimate of the Committee against Bird Slaughter, of 250 million birds.

Once the report was completed and submitted to the commission, the authors were not involved in the way it was used in the legislative process. In December 1974 the commission presented the conclusions of the study and recommended that all member states accede to the 1950 International Convention for the Protection of Birds. In the meantime bird conservation organizations had presented a petition (no. 8/74) on "the need to save the migratory birds" to the European Parliament, Commission, and Council. This led the parliament to adopt a resolution in February 1975 urging the commission and the council "to propose and adopt in the near future practical measures for the protection of migratory birds." The commission's proposal to the council for a directive on bird conservation was, after amendments, adopted on April 2, 1979, as Council Directive 79/409/EEC on the conservation of wild birds (Lang 1982; Wils 1994).

By the time the council directive was passed in 1979, the economic argument for conservation had completely disappeared from the agenda. At issue were exclusively endangerment and population decline. Article 2 of the directive established that the protection of wild birds was conceived as serving ecological and scientific as well as cultural requirements, and that these criteria weighed more heavily than economic and recreational considerations. The directive has two groups of provisions: articles 3 and 4 deal with the habitat requirements of wild birds and thus with indirect impacts on birdlife, while article 5 to 9 are concerned with the direct effects on wild birds, first of all hunting. Annex I of the directive lists the threatened and vulnerable species for which, together with migratory species, member states are required to undertake special conservation measures, including the classification of Special Protection Areas (SPAs). Annex I initially listed 179 species to be protected in all member states. In an amendment in 1985, the number almost doubled. Annex II lists bird species that may be hunted under the legislation of the member states. The directive bans certain nonselective methods of hunting and defines the limits within which member states can set hunting seasons.

Lobbying by bird conservation organizations had a major impact on the European Commission, although this influence was not recognized as such at the time (Engels 2006; Meyer 2010a, 2010b, 2011). Alistair Gammell is one of the lobbyists who was involved. He had been a keen bird watcher since childhood and a member of the Young Ornithologists' Club, the RSPB's youth organization, in his teens. He got his first job at the International Department of the RSPB when he was nineteen. By the time of the interview he had been with the organization for thirty-eight years. He joined the work on what would eventually become the Birds Directive in 1974, when a first draft had just been circulated. According to Gammell, the commission was very responsive to the input by nature conservation organizations. He clearly sees a difference between that time and recent years: "It was so open. The commission was only quite a small organization. Europe was hardly even thought about very much, so if you were interested in something you just went to the commission and said, We are interested in helping, and they said, Oh, that's very nice, come in, sit around the table. It really was that easy. Once I was even given a commission ID, because I went so often. That doesn't exist nowadays. Today there are many lobby groups. That's how the world has changed" (Gammell, interview, February 2008).

The proliferation of lobby groups over the years has diluted their influence. Nature conservation and environmental organizations are, moreover, not the only international lobby organizations active in Europe (Hey and Brendle 1994; Roose 2003). One of the most influential forces opposing bird conservation is the European hunting lobby. The European Federation of Associations for Hunting and Conservation (FACE), founded in 1977 in Brussels in response to the negotiations on the Birds Directive, represented seven million hunters in Europe in 2010 and is one of the most powerful lobby organizations in the EU (FACE 2012).

Yet hunting is not the only issue the Birds Directive addresses. In the long run it turned out to be the habitat requirement, and not the exemption of certain species from hunting, that made for the truly decisive impact of the directive. Article 4(1) obliges member states to protect the habitats of species listed in Annex I. At the time of its enactment, not all signatories had been aware of all the implications of this requirement—that is, the designation of Special Protection Areas (SPAs) in the territory of the member states. Gammell describes the implications of the provision:

> Everybody thought it's about hunting of migratory birds, but it had built into it also habitat requirements. But nobody really thought a lot about the habitat requirements. Most governments, the British and the German included, thought this is about stopping Italians from killing birds. They thought it's no problem at all, so all the controversy that is going on now about the Birds Directive is that it protects habitats. It was not foreseen, except by those who wrote the article. It was done by an Irish guy, John Temple Lang, an Irish lawyer working for the European Commission, and he was very interested in birds. He worked

on competition policy, so it was not his professional work, but he was very interested in an amateur way in birds. (Gammell, interview, February 2008)

Article 4(3) of the directive requires member states to send the commission all relevant information so that it can ensure community-wide coordination and coherence. The Birds Directive had initially given members states a window of two years to take measures for reporting appropriate sites for SPAs. None of the member states complied, mostly claiming that they lacked sufficient data. In 1981 the European Commission requested the ICBP to compile the necessary data and to designate areas for potential SPAs, so-called Important Bird Areas (IBAs) (Parslow and Everett 1981). The first IBA inventory was produced by the organization in the same year (Osieck and Bruyns 1981), and a second inventory of IBAs, including countries not yet part of the European Community, was published in 1989, covering as many as 2,444 sites in thirty-nine countries (Grimmett and Jones 1989). A third inventory, published in 2000, covered 3,619 areas (Heath and Evans 2000). These Important Bird Areas are recognized by the Court of Justice of the European Union as the foundation for the classification of Special Protection Areas.

The organization's lobbying power rests on the field-ornithological data that it administers. BirdLife International, the successor organization of the International Council for Bird Preservation, does not simply make claims and formulate demands—it furnishes scientific data to support those claims and demands. In a *Science* article published in 2007, employees of BirdLife and the RSPB were able to demonstrate the statistically significant effects of the Birds Directive on bird populations in Europe (Donald et al. 2007). In Gammell's view, the IBA concept, with its basis in scientific evidence, is one of BirdLife's greatest assets: "The IBA books are used in the European Court. They are extremely good scientific work that is not open to much criticism—it is broadly accepted as [the] last word on the subject. So they are a bit frightened of us in Brussels, and they are surprised because we have a brain and a heart, and normally you don't find that in NGOs. We are rather well respected" (Gammell, interview, February 2008).

By 2004, 44 percent of all IBAs were designated as SPAs in the EU's then fifteen member states. The rate differed substantially between countries, ranging from 100 percent reported in Luxembourg to 22 percent in France. Britain and Germany fell in the middle of the spectrum, with 48 percent and 52 percent respectively (Papazoglou et al. 2004). Because of the expansion of the European Union, the Birds Directive has had a continuously growing impact (Boye Vischer-Leopold, Paulsch, Ssymank, and Beulshausen 2010). When the legislation was signed in 1979, the EC had nine member states. In 2010 the EU had twenty-seven member states. Every country that joins the EU has to comply with the laws set up by the initial signatories, including the Birds Directive.

The Birds Directive provides protection for birds and their habitats only. In 1992 a Habitats Directive (92/43/EEC) was created to provide protection for other animals, plants, and ecosystems as well (Diaz 2001). The ICBP and the RSPB were again involved in formulating this legislation. The first version of the Habitats Directive was drafted in the 1980s by Stanley Johnston, who worked for the commission, with the help of Alistair Gammell and Simon Lister. As with the Birds Directive and the involvement of John Temple Lang, the role of a conservationist within the commission has been central to the success of the directive. Johnston's draft was sent to nature conservation and environmental NGOs across Europe in the name of the commission, although without any official leverage.

> The question was, How do we take a draft and make it an official document of the commission? We sent the text to every NGO in Europe we could think of and said, The commission is thinking about producing this document, would you like to send us your thoughts? That way we got around the commission. Afterwards they couldn't say that they don't want to produce it, because thousands of citizens said it is wonderful, but it was a one-person initiative. I personally was shouted at by people—by ministers and other people. One British minister told me I had completely broken all rules, but I am not a government employee. I haven't broken any rules. All we were doing was to consult. (Gammell, interview, February 2008)

The ensuing negotiation process lasted almost a decade. The Habitats Directive, adopted in 1992, requires Sites of Community Importance (SCIs) to be designated for the conservation of specific types of habitats and for species other than birds. The SCIs, together with the SPAs designated by the Birds Directive, make up the Natura 2000 network of protected areas. The Birds Directive was incorporated, unaltered, into the Habitats Directive as the overarching legal framework by Article 3(1). In all, the directives protect over a thousand animals and plant species and over two hundred habitat types. In May 2010, the network of SPAs established by the Birds Directive was reported to consist of 5,315 sites in the EU's then twenty-seven member states. It covered a terrestrial area of nearly 490,824 square kilometers, that is, 11.4 percent of the total land area of the EU, including 6.2 percent in Britain and 12.2 percent in Germany. The Habitats Directive reported comparable numbers, although some areas are designated according to both directives, and it is therefore not possible to simply add the numbers. SCIs numbered 22,529 in the twenty-seven member states and covered a terrestrial area of 719,015 square kilometers: this constitutes 13.7 percent of the total land area in the EU, including 6.8 percent in Britain and 9.7 percent in Germany (European Environment Agency 2014).

BIRDLIFE INTERNATIONAL AND ITS PARTNER ORGANIZATIONS

In the hands of the established bird conservation organizations, the Birds Directive of 1979 turned out to be more than just a neutral legal instrument. Through its implementation, it transformed the organizations that were involved in setting it up. In 1994 the International Council for Bird Preservation (ICBP) was transformed from a rather ineffectual meeting platform for national bird conservationists into an international lobby organization with an independent policy agenda (Schäffer 2007). The restructured organization, now called BirdLife International, has its global headquarters in Cambridge, England, and coordinates the work of more than four thousand staff members and 2.5 million worldwide supporters. The organization makes use of the field-ornithological monitoring data that are collected in fulfillment of the habitat requirement of the Birds Directive. It was the availability of this data, first of all in Britain, that gave rise to the idea of restructuring the organization, as Gammell recalls: "Bird watchers really enjoyed doing this work. It was rewarding. People could see how it relates to protection, since government had to do it. That's also why NGOs started collaborating, and then BirdLife grew out of that collaboration" (Gammell, interview, February 2008).

The Swiss biologist Christoph Imboden was the head of the ICBP when it was transformed into BirdLife. At the time, he recalls, other international NGOs focusing on nature conservation had outstripped the once-pioneering organization.

> ICBP already had a certain reputation at the time, but the big problem was that it was operated on a volunteer basis. In the seventies it was noticed that it was falling further and further behind, because in the meantime WWF and IUCN had become very active and started to operate very professionally. The ICBP was getting rather less relevant. . . . The question was, What can be done? Some said this organization is no longer needed, considering that WWF and IUCN are doing this work so professionally now, and others said, No, we have

> to professionalize the ICBP as well. That is when I got recruited as the first executive director. That was 1980; that's when I started. That was in a small office in the British Natural History Museum, and then we moved to Cambridge very quickly, within two or three months. . . . We were about ten people for a long time, and from there on it was just steadily growing. (Imboden, interview, March 2012)

BirdLife moved to Cambridge primarily to facilitate collaboration with the World Conservation Monitoring Centre (WCMC) located there. This organization, created by the IUCN, is in charge of assembling the population data for the Red List of Threatened Species published by the IUCN. The data on bird populations for the list is provided by BirdLife, which receives this data from its national partner organizations. The influence of BirdLife thus rests on its national partners in the global network, including the NABU in Germany and the RSPB in Britain. The large and well-funded RSPB is considered the lead partner in the global network.

One of the main reasons for restructuring the ICBP into BirdLife was to increase the financial contributions of the national partner organizations: "The individual organizations contributed, but those were small, laughable amounts, on the order of magnitude of a couple of hundred pounds by the largest organizations, really ridiculously small. The RSPB, I don't recall exactly, they paid maybe fifteen hundred pounds, but with that one of course cannot finance an international organization. . . . When we realized that we really have to change the structure, we thought about the name as well and decided that we should actually find a new identity" (Imboden, interview, March 2012).

The British bird conservationists' experience with collecting population data was a crucial contribution to the formulation of the IBA program. The British Trust for Ornithology (BTO) had collected such data since 1962 through the Common Birds Census, which had originally been instituted to monitor the effects of pesticides such as DDT on birdlife. Other European countries, including Germany, lacked such large-scale monitoring schemes: although surveys were conducted on rare and endangered species, there were no comparable surveys for common species. Imboden recalls how the motivation for the creation of the IBA concept was an effort to find a survey instrument that did not focus exclusively on endangered species, but included the common ones.

> That was my idea. I had interesting coworkers with whom I did a lot of brainstorming, and the RSPB was heavily involved as well—the RSPB had already previously done such surveys in England. In such discussions the idea came

> up: Well, why don't we do that globally? That is, at first in Europe, and then it was so successful that we said our goal should actually be the entire world. That is what happened, and it happened on a scale that we would not have considered possible back then. I am still surprised today by the rigor with which that has been carried on after I left. It's still running, but the origins surely emerged in my office. (Imboden, interview, March 2012)

The work of the RSPB does not only provide a model for BirdLife, as Mike Rands, who succeeded Imboden as chief executive of BirdLife from 1996 to 2009, points out. He acknowledges that larger partner organizations, such as the RSPB, may have a disproportionate influence on BirdLife governance.

> We try very hard, I mean every partner in the BirdLife partnership has one vote, so to speak, so we try very hard, whether it's a small thirty-five-member organization in Ecuador or a million-member organization in the UK, to treat them all as equals, and have them all contributing something and receiving from being partner of the BirdLife International partnership, but that is honestly very difficult to do when you've got such a diversity of organizations and histories and skills and resources and everything else. I am sure therefore it is too dominated by the larger partners, and I think it is inevitable, although they are very sensitive to that. (Rands, interview, February 2008)

The British RSPB, with its more than one million members, accounts for slightly more than two fifths of the global membership of BirdLife, the German NABU for slightly less than one fifth. Yet while the RSPB has a disproportionate impact on shaping the priorities of the work of BirdLife, all national partners, including the less influential ones, benefit from the cooperation, as the former chief executive of the RSPB Barbara Young, who was on the council of BirdLife from 1991 to 1998, points out. By the time of the interview she was a vice president of the latter organization. In her assessment, the success of the RSPB and of BirdLife is mutually reinforcing. Uniting more than one hundred national bird-conservation organizations under one logo, in her view, conceals the fact that it remains a conglomerate of mostly rather small national organizations:

> It's almost a kind of a trick, you know. This is maybe a very tiny organization in its own country, but it's got this big global organization standing very close behind it helping it, so the local knowledge and the local political sensitivity and the local political links are helped by this big global organization and its

> funding and its expertise. It works really well. But the trick is that actually the big global organization isn't very big at all. It punches well above its weight because it now has a hundred and—I don't know—fourteen partners, and because it's got a good science base and therefore is very credible. It actually gives the impression of being much bigger than it is. It gets more support globally as a result. In a way the whole thing interacts very neatly, and both parts of the partnership gain from each other. (Young, interview, September 2007)

BirdLife works with this impression and uses an image of the more than one hundred logos of its national member organizations for advertising.

There were, however, some losers when the organization was restructured and renamed. Initially each country had an independent ICBP section, although the council members of the British section were largely identical with those of the RSPB, whereas the leadership of the German section drew on members from other organizations as well. When the ICBP was restructured into BirdLife, it adhered to its previous policy of having only one partner organization in every member country, but it affiliated with the major bird-conservation organization in each country, not necessarily with the former ICBP section. In Britain the RSPB took over this role, and the British section of the ICBP was dissolved. In Germany, on the other hand, the NABU became the BirdLife partner, and the German section of the ICBP was restructured into an independent national organization, the German Council for Bird Protection (Deutscher Rat für Vogelschutz). The council had sixteen partner organizations in 2010, including the NABU, the Interstate Working Group of Bird Protection Stations (Arbeitsgemeinschaft der staatlichen Vogelschutzwarten), and the German Ornithologists' Society (Deutsche Ornithologen-Gesellschaft). Among the noteworthy changes in recent years were the inclusion of the Committee against Bird Slaughter in 2005.

Hans-Günther Bauer works at the Max Planck Institute for Ornithology at Radolfzell and was the director of the German Council for Bird Protection from 1993 to 2008.

> I joined as a representative of the Institutes for Ornithological Research in 1988. During that time the ICBP had started to think about how to deal with its partners, trying to find out if it would be better to search for partners that are financially well endowed, in order for them to be able to better advance their strategies, and of course to be able to spend a little bit more money. Around the early nineties the NABU, one of our member organizations, was about to say, We will become the future partner, but that continued until '92. We were still

FIGURE 7.2 BirdLife International and the logos of its more than one hundred national partner organizations.

> an associate member at the time. It was obvious that the council was affected to the point that we said we will restructure entirely. The former head of the ICBP was Christoph Imboden from Switzerland. He visited us and discussed it with us, so it wasn't as if we were passed over. (Bauer, interview, July 2007)

Although not an official BirdLife partner, the German Council for Bird Protection has nevertheless continued to play a vital role in the organization's work. The provision of the German data for the IBA lists, for instance, continues to be administered by the council. Bauer considers the overall impact of the work of the NABU on BirdLife to be marginal.

> The NABU is only peripherally involved, even though they are the second largest financial sponsor globally. . . . The RSPB is the largest and the NABU the second largest, but as I said, they don't have the people who are able to do that work or who want to do it in any kind of way. Bird conservation as such is not important enough to them. That's why there are ambitions to create a NABU International. I don't know if that is still up to date, but those are the snooty ideas at the executive level at the moment. One has to learn to understand that—it's not entirely easy for us to comprehend what is driving them. Now, there is this international organization, and one could contribute to that at any time and advance one's goals there, but there is apparently more at stake. It's about one's own image. (Bauer, interview, July 2007)

In Bauer's view, the NABU's lack of independent scientific expertise and its concern with nature conservation and environmentalism in general, rather than a focus on bird conservation, are the reasons why it does not exercise a leading role in the BirdLife partnership.

The leaders of the NABU see the matter differently. Thomas Tennhardt has been a member of the council since 1995 and a vice president of the organization since 2003. He grew up in the former German Democratic Republic and was a founding member of the East German environmental network Green League (Grüne Liga). Since 2010 he has been the head of NABU International, the foundation mentioned by Bauer. At the time of the interview, NABU International supported fifteen projects in three regions in the world—the Caucasus, Asia, and Africa. These projects are administered in partnership with local conservation groups, some of which were created through the work on these projects and adopted the name of the German organization. There is, for instance, a NABU Kyrgyzstan.

The organization tries to focus on projects and regions that are not already covered by other conservation organizations. Birds are in consequence not a specific focus. In fact, none of the fifteen projects, which include the protection of snow leopards in Kyrgyzstan, European bison in the Caucasus, and marshland in Russia, concentrates on birds.

Although the NABU remains the official German partner of BirdLife, the decision to create its own International Department and a foundation to finance it was to a certain extent motivated by the aim to break away from BirdLife's exclusive focus on birds. This focus was perceived as restrictive because the NABU's work covers a wider range of issues in the field of nature conservation and environmentalism.

> That is one of our points of contention with them sometimes. That once escalated at a CITES [Convention on International Trade in Endangered Species] conference. . . . We wanted to know what their point of view is on this and that species, how BirdLife looks at that. They said, "We don't care, that's not a bird." That almost knocked our representative Ms. Finke from her chair. That is an attitude we really have problems with. . . . The worst thing of all was that they wanted to prohibit us from contributing to the other topics, because they thought that's not BirdLife policy, and we are BirdLife Germany, after all. (Tennhardt, interview, February 2012)

The creation of NABU International does not mean that the organization no longer contributes to the work of BirdLife. Yet this work is not part of the NABU's new International Department. The work on the Birds and Habitats Directives within the NABU, for instance, is done by Claus Mayr, who joined the organization in 1992 as a so-called IBA officer. His work has a strong national focus. In Mayr's view the European Union is less a place for the NABU to influence conservation policies in other countries than a place to push forward German national policies.

> By now, the stimuli for progressive environmental policies derive, in my perception, mostly from directives and regulations in Brussels. The nation-states admittedly stand behind that, because they have negotiated and adopted them, but they rarely work as progressively at the national level. That's because the collusion of lobby groups is stronger at the national level, so that one has to play it indirectly—we try to negotiate the highest possible standards in Brussels, and then of course we try to implement them nationally. That's essentially what we are trying to achieve with the Habitats Directive and the Birds Directive. (Mayr, interview, August 2007)

As an IBA officer, Mayr is involved in bringing complaints to the EU in Brussels about Germany's failure to implement the Birds and Habitats Directives. To address such complaints from member states, the commission can consult the Court of Justice of the European Union (Diaz 2001). What Mayr likes about his job is the fact that issuing complaints actually produces results.

> A real up-to-date case is the decision of the commission in June to call the European Court to rule against Germany because Germany has established too few bird protection areas. This case goes back to a complaint by the NABU. We established this network of Important Bird Areas in the context of BirdLife, and we simply compared what we had suggested as IBAs based on professional criteria and what the German states have reported as official European Special Protection Areas based on Article 4 of the Birds Directive. There is a discrepancy of up to 50 percent of the area in some of the states. I have been concerned with this case since . . . I started as a so-called IBA officer in 1992. I just found a letter from the commission dated July 16 [2007]: "Regarding your complaint from 2001, we are now going to court." That's something I rejoice over. (Mayr, interview, August 2007)

The EU Birds Directive has thus proven to be a powerful legal instrument for national nature conservation organizations. The Court of Justice of the European Union has been central in translating the complaints by NGOs from various countries into national policies. The lawsuit Mayr refers to resulted in the designation of twelve thousand square kilometers of SPA sites by October 2009, which almost doubled the area assigned since the inception of the directive in 1979. Previous judgments of the court condemned the Netherlands, Finland, France, and Italy for failure to classify a sufficient number of areas of SPAs in their territory (Boye, Vischer-Leopold, Paulsch, Ssymank, and Beulshausen 2010). The first of these court cases stated that the IBA inventories published by BirdLife, although not legally binding on the member states, can, by virtue of their acknowledged scientific value, be used as a basis for assessing the extent of member states' compliance with their obligation to create SPAs (Papazoglou et al. 2004).

The NABU and the RSPB both keep an eye on the implementation of the directive and make use of the commission's formal complaints process. Norbert Schäffer has for many years worked for the State Association for Bird Protection (Landesbund für Vogelschutz in Bayern, or LBV), the Bavarian partner organization of the NABU, and is now at the International Department of the RSPB. He says of the Birds Directive, "The EU Birds Directive is probably the best—at least the best transnational—conservation law worldwide. An incredible amount has been achieved with it. For me personally the Birds Directive stands on the same level with the abolition of capital punishment in Europe. It has an impact that is unbelievable. Fifteen to twenty percent of the land area of Europe is protected through it. . . . The European Union has really achieved something here, and the impact of the directive hinges on the availability of good data" (Schäffer, interview, February 2008).

Schäffer seems to derive a certain amount of satisfaction from the fact that the field notes collected by bird watchers are now regarded as legally binding scientific expertise. In the legal framework of the European Union, bird watching has turned from a hobby into politics.

> The people who were probably considered rather odd bird watchers twenty years ago have become people who are actually making politics. There are lawsuits regularly filed with the European Court against several countries, which these countries lost. Poland, for instance, tried to sneak out, and that's where we said, That's not possible. Suddenly the data of the bird watcher who is running around in March in his rubber boots at the Belarus border somewhere close to Poland to map wood grouse—those data suddenly become politics, because these areas have to be protected. If the governments had said, We have better data, the court would have accepted them, but the court said there are no better data: Dear governments, as long as there is nothing better available you have to protect what these bird conservationists have identified. (Schäffer, interview, February 2008)

The prerogative for providing these data was not given to BirdLife International, then still organized as ICBP, because of the organization's good intentions. The commission initially called on it because it was the only organization that had access to such data.

FIELD-ORNITHOLOGICAL MONITORING SCHEMES

The selection of IBAs and subsequently SPAs is based on field-ornithological data that the Birds Directive requires every EU member state to provide. The provision of these data depends on extensive area mapping and fieldwork. The request for such monitoring data would have been unrealistic if no such monitoring schemes had already been in existence and proven to provide reliable data. The most successful one was the BTO's Common Birds Census (CBC), established in 1962. Taken together, monitoring schemes in Britain exceed those in the rest of Europe for both the number of bird species monitored and the number of volunteers involved. The Breeding Bird Survey (BBS), the successor of the CBC, monitors more than one hundred common bird species through the annual observations of more than three thousand amateur bird watchers (Greenwood 1999; Greenwood 2000, 2005, 2007; Risely, Noble, and Baillie 2008). The Birds Directive resulted in the establishment of comparable bird monitoring schemes in all EU member states. In countries without such schemes or a tradition of amateur bird watching, such as Germany, some of this work has to be conducted by paid semiprofessional fieldworkers.

None of the data collected by the BTO or similar organizations had initially been collected with a view to the selection of nature reserves. Most of the data were collected to facilitate the science of field ornithology. The two decades that intervened between the establishment of the scheme during the DDT crisis and its utilization as an instrument for reserve selection were moreover hardly an intentionally planned test run. Jeremy Greenwood, chief executive of the BTO from 1988 until his retirement in 2007, emphasizes the unplanned character of the British monitoring scheme. "Of course different things work in different countries, and probably different things could work in the same country, but you stick with what history has given you. We've been very fortunate, I think, in Britain in that history has given us a BTO" (Greenwood, interview, August 2006). Greenwood has been a keen bird watcher since childhood and over the years not only involved with the BTO but also a council member of the British Ornithologists' Union, the RSPB, and the British section of the ICBP. He notes that it was data from the BTO's monitoring scheme that made it possible to formulate and support new policy goals. Yet the existence of these schemes remains a historical accident, no matter how useful and beneficial they turned out to be over time.

The simple truth is that the data were used because they were available. The means came first and found their ends later. This statement is more than an assertion of the fact that these data as means were enabling for a particular line of action. Means cannot only enable action, they can also motivate it. As Dewey has pointed out, means are not merely neutral instruments, they also serve as intermediate ends in a course of action and as such have a motivating force. "When you have a hammer, everything looks like a nail" is the proverbial expression of this pragmatist argument. Having a tool available opens up a space of possibilities of what could be done, possibilities that would not have come to mind otherwise. The BTO survey data were such a tool waiting for its application. They opened up a space of possibilities and imagination. Yet for most countries in the EU such an imagination was rather remote. A substantial number of member states did not have comparable survey data that could simply be put to use, not to mention a large number of amateur bird watchers and field ornithologists who were motivated to collect these data on a regular basis. The EU Birds Directive reflected the British experience—that is, a line of action became imaginable because the data were already available.

After these data had been successfully put to new uses, other countries could create similar monitoring schemes. When a monitoring scheme for common breeding birds was eventually established in Germany, it was explicitly modeled on the British example. Field ornithology had an established tradition in Germany as well, nurtured by the Association of German Avifaunists (Dachverband Deutscher Avifaunisten, or DDA), the German equivalent of the BTO. Yet no large-scale national effort at systematic annual monitoring of common breeding birds had been organized in Germany prior to the enactment of the Birds Directive.

Martin Flade and his friend Johannes Schwarz were involved in creating the new DDA monitoring scheme, which had as one of its main objectives the provision of data to implement the Birds Directive. They noticed that the organization was lagging behind not only Britain but most other EU member states as well, partly because of the lack of sufficient numbers of volunteer bird watchers willing to participate. The British Breeding Bird Survey served as an example for the DDA's Monitoring of Common Birds. The scheme, which has run continuously since its inception in 1989, monitors around 100 to 150 species, and the data collected serve as the basis for the selection of IBAs and SPAs in Germany.

Despite its success, the numbers of amateur participants in the German monitoring scheme do not compare to the British example. A review of all European schemes in 2007 showed that 2,810 watchers contributed annually to the data collection in Britain, compared to approximately a thousand in Germany. In proportion to the total population of each country, the German figure represents slightly more than a quarter of the British count. A review of the BTO's work in 2003 estimated that volunteers contributed approximately 1.5 million hours of labor to all bird monitoring schemes run by the organization each year (Greenwood and Carter 2003, 14). Data for the same year do not exist for Germany, but a survey conducted in 2009, after a noticeable rise in participation, estimated that volunteers contributed about 210,500 hours to all monitoring schemes administered by the DDA—that is, about 14 percent of the British total (Wahl and Sudfeldt 2010, 212). Lacking sufficient volunteer labor, some German states, for instance Brandenburg, rely on paid professionals to collect monitoring data.

Both German and British conservationists were involved in creating the Birds Directive, but it turned out to be the British experience that helped to craft the law and to shape the development of monitoring schemes in other

European countries, including Germany. What looks today like a unified European policy on protecting wild birds is in fact an outgrowth of the distinctively British interest in field ornithology and bird conservation. With the Birds Directive, the British model of nature conservation became the European model.

The aims and methods of the existing schemes were reformulated and refined once they were employed in a new institutional context (Dröschmeister 2000; Gibbons 2000; Sudfeldt et al. 2012). The British monitoring scheme itself did not remain unchanged by the new use to which it was put. After new ends had been formulated for the already existing data, the monitoring scheme was methodologically refined and adapted to the new purpose. At issue in this restructuring was the method of plot selection. Of the existing twenty-eight schemes in Europe, ten are based on a free selection of field sites by volunteer fieldworkers, rather than on a randomized selection. Most schemes created before monitoring was made obligatory by the European Union were free-choice based, most of those created afterward are based on a random or semirandom plot selection. In the British case, the difference between the two methods explains the replacement of the Common Birds Census (1962 to 2000) by the Breeding Bird Survey (since 1994). Yet the rules of the game cannot be changed at will. A change in the site selection method has to take account of the fact that the motivation of the participating volunteers to monitor birds relies on the equilibrium of forces and uncertainty of outcome that constitutes bird watching as a game. Being assigned to a plot by a formal statistical selection procedure could jeopardize this equilibrium and uncertainty. Jeremy Greenwood was involved in restructuring the BTO's monitoring scheme and is more than familiar with the problem: "In the past people were very free to choose their own areas. In the CBC people chose the areas that they would census, but in the BBS they're asked to go to particular places which are chosen at random, chosen by proper statistical design. . . . It raises a series of problems—sometimes people want to go somewhere close at hand and don't want to go fifteen kilometers away. . . . People certainly don't in general like to go to boring habitats where they are not going to see very many birds" (Greenwood, interview, August 2006).

The solution to the problem was to restructure the monitoring scheme in a way that again casts it as a game rather than as work. The situation is an ideal test case for the theory of games as institutions that both enable and encourage play. A game as an institution is constituted by rules and resources. In order to enable play, these rules and resources have to be arranged in a way that creates an equilibrium of forces between challenge and skill and thus ensures the openness of outcome of a course of action. Once this equilibrium is lost, and the outcome is entirely predictable, the given task or course of action becomes boring. Seeing hardly any birds of the species that one is supposed to monitor constitutes such a mismatch between challenge and skill that it causes the equilibrium to collapse. It is accordingly essential for the BTO to ensure that the survey is both methodologically rigorous and at the same time achieving an equilibrium of forces and openness of outcome. The challenge for the organization is a real one: if the activity ceases to be a game and to allow its participants to have what Csikszentmihalyi ([1975] 2000) calls a flow experience, volunteers might lose the motivation to participate. The largest, oldest, and most successful field-ornithological monitoring scheme in Europe is after all entirely based on volunteers who provide their service without pay. The BTO thus has to make sure that the activity of field monitoring remains a game. "For example, we did a survey of corn buntings some years ago. Now, corn buntings have become very restricted in their distribution in Britain. So . . . the vast majority of people wouldn't survey for corn buntings, if they are not seeing anything. They are getting very fed up. We then added into it all buntings, so that people could record the yellowhammers and the reed buntings as well as the corn bunting. That gave people something to record—as well as providing useful background information. We have to think of tricks like that" (Greenwood, interview, August 2006). Thus the revamped monitoring scheme produces some data that do not serve the main objective of the survey: instead, they are collected because they turn the practice into a game and sustain the interest of volunteers.

The monitoring scheme of the DDA faced the same problem (Mitschke et al. 2005). When it was created in 1989, it was based on voluntary plot selection. In 2004, again following the British example, it switched to random plot selection in order to provide more statistically reliable data. Martin Flade and Johannes Schwarz were involved in the restructuring. Schwarz has been a passionate bird watcher since childhood; he is an ornithologist by training, and works for the Berlin Senate Department for Urban Development and the Environment (Senatsverwaltung für Stadtentwicklung und Umwelt). Schwarz is interested not only in birds but also in computers. He wrote a program for the recording of the monitoring data and runs most of the analysis. The initial fear was that changes to the scheme might undermine people's motivation to participate.

> We now try to improve the monitoring; that is, we are first of all trying to attract more people to participate and to guide the data collection in a way that will gain statistically reliable results. That was the problem of the old monitoring

> scheme, that is was entirely at the discretion of the participant, meaning everybody could say, I do it at the front door, or, I do it at the next nature reserve, or wherever happens to be fun, because it promises more interesting results. With the new monitoring scheme the areas are designated. The areas have been selected by the Federal Statistical Office (Statistisches Bundesamt) based on statistical criteria. That is a one-time sample that remains fixed. (Schwarz, interview, June 2007)

One of the ways to make sure that people continue to participate is to publish the results of the survey as fast as possible and to circulate them among contributors. It ensures that people know their game score, which is an attempt to motivate them to continue playing the game. Yet organizing field-ornithological monitoring as a game is overall less successful in Germany than in Britain. A part of the data for the scheme is collected by paid professionals, rather than by volunteers. Among the contributors to the DDA monitoring scheme that are financially supported by the state is the German Interstate Working Group of Bird Protection Stations. Bird monitoring has become part of the work of these observatories, in stark contrast to their purpose at the time of their foundation in the early twentieth century, when their main aim was to increase the bird species useful to agriculture and to exterminate those considered harmful. Steckby, the model Bird Protection Station, first accredited by the state in 1908 and temporarily owned by the predecessor of the NABU, is the emblematic example. In 2002, the field ornithologists involved in the DDA monitoring program met at Steckby to formulate a resolution to guide their future work. The so-called Steckby Guidelines (Steckbyer Grundsätze) set out the principles of the Common Breeding Birds Survey (Gedeon, Fischer, and Sudfeldt 2003).

GLOBAL, EUROPEAN, AND NATIONAL CONSERVATION CRITERIA

The monitoring data produced by the various national field-ornithological organizations in Europe are collected by the European Bird Census Council (EBCC) and compiled into a continent-wide data set. This organization was established in 1992, merging the former International Bird Census Council (IBCC) and the European Ornithological Atlas Committee (EOAC). Like the ICBP before its restructuring into BirdLife International, this organization was originally nothing more than an institutional framework that enabled national organizations to meet and exchange their work. Martin Flade, who was on the executive committee of the EBCC for nine years beginning in 1995 recalls that the organization took the BTO's monitoring scheme as its model. In a sense, the EBCC is something like a European BTO. One of the major tasks that led to the organization's eventual restructuring was the compilation of the EBCC *Atlas of European Breeding Birds*, released in 1997.

> The EBCC is now practically the uncontested umbrella organization at the European level. It became heavily professionalized after the atlas was published. The atlas helped substantially to distinguish the EBCC. Until this spring [2007] it was essentially a phantom. The EBCC had no legal status. It was not a registered association or foundation. That was a problem, because if we wanted to hire people, we had to do that by means of third parties. The EBCC has admittedly functioned marvelously, but on paper it was nothing, it was simply a coalition of national organizations and institutes that wanted to do atlas work and population monitoring in Europe. A complete joke, actually. (Flade, interview, July 2007)

Through the publication of the *Atlas of European Breeding Birds*, the EBCC became a professional organization with independent funding (Hagemeijer and Blair 1997). Since 2002 it has administered the Pan-European Common Bird Monitoring Scheme (PECBMS). Gathering long-term monitoring data that capture changes in breeding populations across Europe, the survey uses common birds as indicators of the general state of the environment.

When analyzed in a new international framework, these data had some unforeseen and, to a certain extent, undesired consequences for the classification of threatened species. Many bird species that are target species for national conservation programs were assigned this status because they were represented by populations living at the edge of their species range: they are rare and endangered on a national level but abundant on a European or global level. Conversely, many species that are abundant and common in one country but whose total population is confined to a limited area are considered rare from a European or global point of view. For a handful of species, this change in the geographical frame of reference substantially alters their status of threat (Gregory, Greenwood, and Hagemeijer 1998; Gregory and Vorisek 2003; Hagemeijer and Blair 1997; Heredia, Rose, and Painter 1996; PECBMS 2006, 2007).

BirdLife International began early in the 1990s to systematically take national boundaries into account. On the basis of the aggregated data provided by the national field-ornithological organizations for the selection of IBAs, which provided a total population count of each bird species in each country in Europe, BirdLife introduced a new category, Species of European Conservation Concern (SPEC) in addition to the already existing national and global categories for status of threat. This category was included in the organization's two publications titled *Birds in Europe* in 1994 and 2004 (Burfield and van Bommel 2004; Tucker and Heath 1994). Christoph Imboden, then director of BirdLife, was involved in this innovation. He envisioned the SPEC category as something akin to a Green List, as he initially called it—that is, a label for common species that would supplement the existing Red List of rare species.

The establishment of the SPEC category was an indication of a change less in thinking about conservation than in the availability of new and more reliable data on European bird populations. The 1976 study *Bird Protection in Europe*, by Conrad and Poltz, testifies to the previous lack of comprehensive and systematic data and the rather sketchy information on population developments that could be derived from them. Less than a quarter of a century later, finer-grained assessments of the conservation status of birds in Europe came into existence. In the 2004 edition of *Birds in Europe*, out of 514 regularly occurring bird species, 278 were listed as being Species of European Conservation Concern. Twenty-four of these SPECs were considered to be of Global Conservation Concern.

The SPEC category is used to allocate EU funding for the national implementation of the Birds and Habitats Directives. Such funding has been made available by the LIFE program, the financial instrument for the environment, which is divided evenly between nature conservation projects under the two directives and more general environmental projects. In total, between its inauguration in 1992 and 2006, the LIFE program spent €1.36 billion on 2750 projects covering forty countries and territories. The subsequent phase of the program, LIFE+ ran from 2007 through 2013 and had a budget of €2.143 billion. A substantial amount of this total went to maintaining and restoring endangered bird populations and their habitats throughout the EU (European Commission 2004, 2017).

The legal obligations imposed on EU member states by the Birds and Habitats Directives and the financial incentives of the LIFE program shape the way national conservation priorities are determined. In addition, they lead to a certain amount of maneuvering in order to fit conservation projects under this regulatory and financial umbrella. The LIFE priority lists, for example, can be used as means to other conservation ends. Martin Flade describes an application for LIFE funding in the Schorfheide Chorin nature reserve, which he administers (Flade 1994; Flade et al. 2003; Flade et al. 2006).

> Most EU-financed nature conservation projects are based on some bird species as a major objective. It's like this: we had a LIFE project for the conservation of the bittern here in the Biosphere Reserve Schorfheide Chorin, but that was in fact about reedbed habitat, the reed belt of lakes, and that was running under the title of bittern conservation, but it was not about the bitterns. And at the moment there is a German-Polish LIFE project for the protection of the aquatic warbler—that is about fens, of course, but it is frequently tied to such priority bird species. (Flade, interview, July 2007)

Despite the availability of population data for all of Europe through the EBCC and the introduction of the SPEC category by BirdLife, most national conservation organizations continue to produce national Red Lists and to determine their conservation priorities based on these lists. What makes this lack of integration of the various global, European, and national lists particularly striking is the fact that the data used in regard to all three of these geographical reference frames are actually the same. There are, however, national differences in the way the European and global categories are taken into account. Following the publication of the EBCC atlas, Flade addressed the lack of an international perspective in German bird conservation in an article published in the popular ornithological journal *The Hawk* (*Der Falke*) (Flade 1998). He recalls his assertion that international monitoring data should be taken into account in setting national conservation priorities, because it changes the status of threat of some species:

> That is why I singled out the hoopoe—it has a relatively large range—it's fairly common in the entire Mediterranean and in Asia Minor and elsewhere—and the German population is at the edge of this range. The species has a high standing in national nature conservation, it is very popular, but I would say that it doesn't matter a damn shit for the hoopoe if there are a hundred pairs more or less breeding in Germany. But another species, the middle spotted woodpecker, for instance, that species has a quarter of the world population in Germany, and I said that is important. (Flade, interview, July 2007)

The article caused a stir and was reprinted eight times in slightly revised versions in various journals in the field. It led some of the German states to produce priority lists for their conservation work based on international rather than national or regional criteria. An additional priority list, comparable to Flade's, has since been published by the bird conservation expert of the NABU (Nipkow 2006) and is taken into account in the *Basic Program on*

Bird Protection (*Grundsatzprogramm Vogelschutz*) of the organization (NABU 2010).

Yet, despite these efforts, as of 2010 the German national Red List still failed to reflect international conservation priorities. The fourth edition of the Red List of breeding birds in Germany, published in 2007, remained exclusively based on national population counts and trends. Although threat status considered from a European and a global level is reported, this assessment is not a criterion for the determination of national-level threat status. The three geographical reference frames simply coexist in parallel (Südbeck et al. 2007).

The contrast between Britain and Germany in this respect is again striking. In Britain, the first national Red List, published in 1990, included in its criteria for the assessment of endangerment the international significance of British bird populations, whereby *international* meant northwestern Europe (Batten et al. 1990). The data available for other regions in the world were considered to be insufficient at the time. The Red List published in 1996 added the global IUCN assessment and the European SPEC assessment to the determinants for national status of threat (Gibbons et al. 1996). This practice has been maintained in the 2002 and 2009 editions of the list (Eaton et al. 2009; Gregory et al. 2002). International developments are thus a part of the national calculation. Yet even this assessment is limited in perspective because only population decrease and an existing status of endangerment are taken into account. The reverse possibility—that a species that is nationally rare or in decline might be abundant or on the rise from a global or European perspective—is not part of the picture.

This strong national focus of the German and to a somewhat lesser extent also the British Red List is one of the most clear-cut examples of the fact that conservation did not move from a national to an international level in the second half of the twentieth century as has often been claimed. Instead, a European and a global frame of reference were simply added to the national frame. Given the scarcity of funds for conservation, the various national and international NGOs can use these different frames selectively to advance their particular objectives. Once an institutional framework for conservation is established, it is likely to persist. A species invested with conservation value in any given framework is likely to retain this status even if a newly established framework attributes a different status to it. Geographical reference frames are not replacing each other, strictly speaking, but they are adding up.

The establishment of an international legal framework for conservation in the European Union is only one of several ways in which a geographical framework of reference can change. Over the past two centuries, the more frequent cause for such changes has been the alteration of the actual boundaries of nation-states, including the fall of the Berlin Wall and the reunification of Germany in 1990. The common crane is an example of a bird species whose conservation status was affected by this change. It is one of fifteen species in the crane family (*Gruidae*) worldwide. Eleven of these fifteen species are threatened to various degrees, according to the 2009 IUCN Red List. The common crane is one of the remaining four species that are classified as of least concern to conservation, and its global population estimate of 360,000 to 370,000 individuals is the second largest among the species in the crane family. Its total range is estimated at slightly above fifteen million square kilometers, stretching over most of Northern Asia, Eastern Europe, and Scandinavia. Germany is at the very southwestern edge of this range. Efforts at crane conservation have been in full swing in the country since the mid-1970s, after a period of massive population decline. In East Germany there were 370 breeding pairs remaining in 1969, while the count in West Germany stood as low as twelve breeding pairs in 1974 (Wilkening 2004, 13).

Carl-Albrecht von Treuenfels is one of the most influential crane conservationists in the country. He was born in 1938 on a family-owned farm in Mecklenburg-Western Pomerania. He grew up on a farm just a few miles away but across the postwar border: in the Duchy of Lauenburg in Schleswig-Holstein, the northernmost state in the former West Germany. In this rural and agricultural area, he spent much of his youth photographing birds and collecting their eggs: "Back then we stayed put in the countryside, and there weren't all the options people have today, no major traffic links, and so you had to keep yourself busy at home. One way to do that was to dig yourself in and become a bookworm. In my case, it was instead nature that always attracted me" (Treuenfels, interview, July 2007).

Treuenfels was briefly interested in hunting but soon abandoned it in favor of photography. For some years he contributed a regular nature column to the *Frankfurter Allgemeine Zeitung*, one of the leading national newspapers. He published several books, beginning with a guide on nature photography in 1968, several illustrated works on nature conservation, and two books on cranes (Treuenfels 1968, 1979, 1984, 1986, 1989, 2005). He also contributed photographs to an exhibition, The Magic of the Cranes, which opened at the Alexander Koenig Natural History Museum in Bonn and later traveled to museums and zoos throughout the country.

Treuenfels counts himself a member of the international community of crane lovers, sometimes referring to themselves as "craneniacs." His interest in cranes began during his youth in Schleswig-Holstein, where some of the last remaining exemplars of the species in Germany were breeding: "That here is an old administrator's lodge, and that's where two or three pairs of cranes had been breeding. They were the nucleus of the West German crane population. In this way I experienced cranes early on as a child and have always been fascinated by them, looked at them, and observed them" (Treuenfels, interview, July 2007).

In 1978 Treuenfels accepted the invitation of his former schoolmate Thomas Neumann to join the German branch of the World Wide Fund for Nature (WWF). He joined the council of the organization in 1980, became its chairman in 1989, and was the president of WWF Germany from 1997 to 2004. With Neumann, who had organized one of the first crane surveys in the country, Treuenfels put crane conservation on the agenda of the German WWF. In fact, the conservation of the crane population in Schleswig-Holstein was the first independent conservation project of the organization, which had so far spent its resources on conservation projects in other countries through the international framework of the WWF. In 1970 the German WWF bought thirteen hectares of the Dellstedter Marshland to prevent it from being drained. The organization invested continuously in wetland conservation and management to support crane conservation in the years and decades that followed (Groth 2003).

In the early 1970s, the number of breeding pairs of the species was at an all-time low. It was only through conservation efforts that the numbers eventually recovered. Yet the true rise in numbers of the German crane population must be attributed to the German reunification and the addition of the East German population to the total count for the country. When Treuenfels published his first illustrated book on cranes in 1989, he estimated that there were twelve hundred breeding pairs in East Germany and fifty-eight in West Germany (Treuenfels 1989). The numbers continued to rise for both parts of the country after reunification. Bird monitoring data for the *German Breeding Bird Atlas* showed that in 2004 there were 5,618 breeding pairs in the country. Of these, 2,954 pairs could be found in Mecklenburg–Western Pomerania alone. The population in neighboring Schleswig-Holstein was 156 pairs in 2002 (Engling 2003, 7; Wilkening 2004). The crane population in Schleswig-Holstein that was so crucial for awakening Thomas Neumann's and Carl-Albrecht von Treuenfels's concern for crane conservation turned out to be rather insignificant from a German, a European, and a global perspective.

Crane conservation continued after German reunification, although for different reasons. In spring 1990 crane conservationists from the former East and West first met in Schleswig Holstein at the invitation of the WWF to set up joint conservation projects. The so-called crane county, Lauenburg, where

Treuenfels lives, is still proud to have one of the largest West German breeding populations of the species, but the true crane hotspot in the country is in the northeast of Mecklenburg-Western Pomerania, close to the island of Rügen. The area is a stop on the migration route of the crane populations that fly from northern Europe and Russia to their winter quarters in southern Europe and Africa. Up to forty thousand birds stop in this area every year—on a good day, several thousand can be seen resting at a single spot. Much of this area is a nature reserve and is managed with special attention to migrating cranes.

In 1991 NABU and WWF organized the Working Group on Crane Conservation in Germany, and, since 1995, the two organizations have jointly administered a visitors center in Gross Mohrdorf, a town at the edge of the Western Pomerania Lagoon Area (Vorpommersche Boddenlandschaft) National Park. The center offers film screenings, lectures, and exhibits on cranes and provides access to a computer that tracks ringed cranes. The working group has put up blinds and organizes guided tours during autumn migration. Cranes now have the status of a flagship species: they are used to awaken enthusiasm for birds in people who have never before thought about them: "Many people who grew up in the city don't even have the opportunity for such thoughts to awaken in them, because they don't have a feeling for it, and they don't know it. That's why the crane is always such a good flagship species [*Leitart*] for nature conservation. That holds true even for people who are not that much interested in ornithology and nature—they always get very excited when they see hundreds or thousands of these huge birds arrive. That's always a great success" (Treuenfels, interview, July 2007).

For Carl-Albrecht von Treuenfels and his fellow crane conservationists the personal level of enthusiasm for these birds is in no danger of losing steam. The question now is whether their enthusiasm has found the proper object of concern. The breeding population of the common crane in West Germany has largely ceased to be such an object. As a flagship species, cranes make sense. As a target species, they are an example of the kind of misplaced priorities that Martin Flade has urged his fellow conservationists to reconsider. Treuenfels is aware of the issue: "Many nature conservationists have their favorites, and I don't want to exclude myself, but I always know in the back of my mind that it's only one specialty within a grand concert. Particular animals and plants have particular needs and requirements, but ideally that always evens out, since there are different people who focus on different things" (Treuenfels, interview, July 2007).

A call for diversity among biodiversity conservationists is certainly an elegant answer to the question. As a former president of the WWF Germany,

Treuenfels is among the actors who look at conservation from an international perspective. Yet his own attachment to cranes began at home, at the place where he grew up. It is the conservation of these birds at this locality that captures most of his passion and concern. From a global perspective, however, this very locality is among the least important places for conservation. Presenting the species as a *Leitart* or flagship species is a way out of this dilemma. Having such species available close to where people live allows conservation organizations such as NABU and WWF to raise awareness and gain new members. The fact that birds can be all three things—endangered species or target species, indicator species, and flagship species—leaves some room for maneuver. The categories provide a repertoire or toolkit for action, as Swidler (1986, 2001a) called it. Yet this strategic use does not imply that no emotional attachments or moral commitments exist in the case of the involved actors. To the contrary, it is the very purpose of the strategic use of this repertoire to protect these preexisting attachments and commitments. Conservation categories are repertoires that can be used as justifications for the allocation of resources—time and money—to the issues conservationists are most passionate about. In the case of Treuenfels this effort to find a justification to avert the unintended consequences of employing a global frame of reference seems to have been a success.

The case is not unique to Germany and the work of Treuenfels. The same holds true for Britain. Common cranes had not been breeding in Britain since the mid-seventeenth century. A pair was seen again on a farm in Norfolk, in the east of England, in autumn 1979. The farm's owner, an ornithologist, kept the discovery secret to avoid hordes of excited bird watchers. In spring 1982 the pair built a nest and started breeding. The species is now nesting at the RSPB's Lakenheath Fen nature reserve in Suffolk, a site that was a carrot field until the organization bought it in the 1990s and began its transformation into a square mile of marsh and fen. The site was initially created for a different bird species, and RSPB members came upon a nest of cranes by chance while carrying out a routine survey (Clare 2007). The organization launched a £1.5 million appeal to support crane reintroduction in Britain beginning in 2010. The Great Crane Project (GCP) is a partnership between the Wildfowl and Wetlands Trust (WWT), the RSPB, and other British conservation organizations and aims to reestablish a sustainable population of common cranes in Britain. The target of the project is to establish twenty breeding pairs by 2025 (Great Crane Project 2017).

As in the case of the common crane, bird conservation efforts in Europe have produced measurable effects. All European species listed in the Birds

Directive and protected by SPAs have done better than other species. A report by BirdLife on the population trends of birds in Europe published in 2004 showed that the directive was successful when implemented and backed by plans and resources, such as international Species Action Plans (SAPs). Despite overall unfavorable population trends for all 524 species assessed, the Annex I species in the EU of fifteen member states did better than non-Annex I species in the period from 1990 to 2000. They also did better than the same species in non-EU15 countries. All twenty-three species with an international Species Action Plan (SAP) did better than those without (Burfield and Bommel 2004; Papazoglou et al. 2004).

Yet only a small fraction of all known birds breed in Europe. The 524 species that have been assessed by BirdLife constitute only slightly more than 5 percent of the world's total of 10,027 described species. Most of the European species, those on the British Isles and in Germany included, can be found in other parts of the Western Palearctic. If all of Europe were to disappear, only very few bird species would be lost. As of 2010, only nine bird species are endemic to Europe, that is, they cannot be found in other parts of the world. Only three out of these nine are endangered (IUCN 2010; Sattersfield et al. 1998). For all the wealth and distinctiveness of its birdlife, the loss of the European habitats would matter comparatively little on a global scale. From a global perspective, there is a considerable imbalance between the distribution of concern for bird conservation and the distribution of birds of conservation concern. Roughly 80 percent of BirdLife's 2.5 million members live in Europe, but most of the world's 1,240 endangered bird species are found outside Europe (IUCN 2010).

In some of these places, however, the legislative and funding provisions of the European Union apply, because a number of European countries still hold overseas territories. Among these is a group of islands in the Pacific known as French Polynesia. Beginning around 2000, BirdLife started to produce an inventory of the birdlife in the Pacific. By 2010 the organization had identified thirty-two Important Bird Areas (IBA) in French Polynesia. More than three-quarters of the costs of this project, which totaled more than €1.6 million, were covered by the European Union. The project office is based in Suva, Fiji, and runs national programs in Fiji and in Palau, New Caledonia, and French Polynesia. The Société d'Ornithologie de Polynésie (MANU), an affiliate of BirdLife, oversees the project locally (MANU 2010).

From a global conservation perspective, the importance of the birdlife on these islands far exceeds that of the birdlife in Europe. According to the BirdLife database, thirty-two out of the overall ninety bird species in French Polynesia are globally threatened. Of the threatened species, eighteen are endemic species. Ten species that once lived on these islands are now considered extinct, among them the Society parakeet that was collected on Captain Cook's voyages and described and named by British and German ornithologists (BirdLife International 2014). The investment in conservation in other places in the world accounts for the most decisive overall impact of European bird conservation. Local concern in Europe has produced global conservation outcomes.

CONCLUSION

Studying Morality

In this conclusion I want to spell out the implication of this study for the sociology of morality. I want to provide a research guide, a how-to-do guide, of what to look for and focus attention on when studying morality. I will highlight insights derived from this study that not only show that the now defunct structural-functionalist view of morality described in the first chapter has been justifiably questioned by contemporary scholars working on the topic, but that the departure from its main theoretical assumptions has also methodological implications. Structural functionalists studied morality (i.e., cultural values) as a discourse—that is, as a cognitive script or actually written text that articulates moral values that motivate behavior and can be translated into normative constraints of action. This theory put discourse at center stage and viewed actual practices as an execution or enactment of these discourses. Moral philosophies, whether religious or secular, were accordingly seen as a privileged site to study morality.

One of the more striking insights of this study is that this form of discourse was never mentioned by any of the interviewees, nor was it discussed in the relevant publications on the topic. Environmental ethics as the endeavor to systematically formulate moral principles regarding the way human beings relate to the environment exist, yet they exist within the academic disciplines of philosophy and theology—they are not an integral part of what the actors engaged in bird conservation talk and write about. Academic debates on ethics were never mentioned in the interviews that I conducted. Asking actors explicitly about such debates turned out to be so disruptive and alien to what they wanted to talk about that I eventually gave it up. "My son read philosophy at Cambridge," a bird watching celebrity replied to one of my questions, "and he has no idea. I think you just got to enjoy them." When asked why birds matter to them the most committed conservationists told stories about their lives, that is, stories about how they became bird watchers. There was no difference between these stories about why birds matter to them and their individual life histories. The values attached to birds were virtually inseparable from whom they are as persons. For most of them the attachment to birds began with childhood experiences of bird watching, nest recording, and egg collecting, not with the exposure to moral lessons and abstract ethical principles.

Moral justifications entered the picture when conservationists were asked how they communicate to others why conservation matters, yet when asked why it matters to them personally, abstract moral principles had very little to do with it. The moral justifications that they evoked were not the causes of their action, but they were chosen post hoc. Moral discourse is accordingly not a privileged site when it comes to studying morality, and in some cases it is even the worst place to do so. As I want to highlight in this conclusion, focusing on discourse can be systematically misleading. Researchers have to focus on other aspects when they want to study the role of morality in social life. Three such aspects stand out: they concern, first, the relation of moral discourse to practice, second, the relation of practices to a wider institutional context, and, finally, the development of these practices and institutions over time. Taken together, the analysis of these three aspects gives considerable weight to the claim that morality is a central element of social life, yet it calls into doubt approaches that conceptualize morality as an analytically fixed category or operationalize it as a self-contained entity akin to a variable. The study shows that the boundaries of moral discourses and their relation to practices and institutions are contingent and should thus themselves be made the objects of empirical inquiry.

THE RELATION OF MORAL DISCOURSE TO PRACTICE

It is tempting to identify moral discourse with actual practices, since moral discourses themselves contain accounts of practices. Moral discourse is by definition about the question of how ought we behave—what kind of action deserves praise and which admonition? Moral discourse distinguishes between *good* and *evil* acts. One can easily mistake these moral distinctions as a vocabulary of motivation, as C. Wright Mills has pointed out in a pragmatist account of agency. He argues that, from a religious or ethical perspective, "men use vocabularies of moral motives: they call acts and programs 'good' and 'bad,' and impute these qualities to the soul" (Mills 1940, 913). Acts that are evaluated as morally undesirable are attributed to different motivational causes than acts that are evaluated as morally desirable.

From a sociological point of view, such a moral vocabulary is systematically misleading. The moral vocabulary of good and evil is a vocabulary of the ends of action. It is a poor choice when it comes to explaining actual behavior, since action is not exclusively motivated by its final outcome—that is, by the ends of action. As this study has shown, the means of action can have a motivating force as well, in the sense that means and ends can constitute each other reciprocally. Taking means into account and realizing that the means of action, just as much as its ends, can have a motivating force changes how we look at moral discourse. The practice of bird collecting and the early twentieth century discourse that morally condemned shooting birds with the gun and morally praised shooting them with the camera is one of the most striking examples.

Those who turned from the use of guns to cameras continued to draw on the same practical knowledge and skills to approach birds in the field. The very vocabulary of bird collectors is indicative of this continuity. Pioneer bird photographers employed a hunting language to describe their practices in the field and the subsequent use of their exploits. Birds were "stalked" in order to make a "snapshot" that could be "mounted" into one's photo album. This vocabulary was not used metaphorically: it did accurately describe that the practices of collecting birds with the gun and with the camera were in fact identical. The more alert conservationists recognized and harnessed this insight in the identity of practices. Rather than morally condemning those who collected birds with the gun as driven by a desire to destroy and kill, they recognized that bird shooters are driven by the same passion for collecting as bird watchers who aim to protect living species in the field. A moral vocabulary of good and evil—that is, the moral distinction between those who protect birds and those who destroy them—simply misses the point when it comes to accounting for the motivation of action. It is moreover the case that not changes in the moral discourse but improvements in photographic technology brought about the transition from gun collecting to camera collecting.

The fact that bird conservationists were drawn from among those who initially contributed to the destruction and decline of birdlife highlights the importance of studying practices. If one ignores the underlying practices one is left with the puzzle of a miraculous conversion from sinner to saint. If one takes these practices into account, however, one recognizes that very little had changed and that no miracle is needed to account for the transformation.

Classifying practices by the ideals of an accompanying moral discourse deprives us of the possibility to understand the mechanism that produces and sustains the motivation for action. The more perceptive conservationists argued against such moralizing accounts and instead drew on the "true psychology of the collector," as one influential conservationist called it, to set up field-ornithological monitoring schemes that would push collectors to give up the gun and take up binoculars, cameras, and notebooks instead. When we approach the topic with no analytical distinction that separates the moral ends of action from its practical means, we have the best chance to explain what motivates action.

THE RELATION OF PRACTICE TO INSTITUTIONAL CONTEXT

This insight requires a qualification. Means can have a motivating force, yet this force differs by the practice that is involved. In the practice of play means and ends define each other reciprocally, while they are separated from each other in the practice of work. According to John Dewey's pragmatist theory introduced in the first chapter, these differences in the relationship between means and ends are in fact the defining characteristic of the distinction between the practices of work and play. Actors who articulate their concern for birds in the context of the practices of bird collecting as a game do accordingly produce a fundamentally different discourse than those conservationists who encounter birds as a part of the world of work. In the world of play birds appear as toys that are valued for the joys they provide bird watchers, and no distinction between means and ends is made. In play it is the immersion in activity itself that matters. In the world of work, by contrast, a vocabulary of means and ends dominates—a vocabulary that distinguishes between purely instrumental (economic) values on the one hand and purely intrinsic (moral) values on the other.

An explicit moral discourse was most pronounced in the German case. Throughout the late nineteenth and early twentieth centuries actors attributed an intrinsic moral value to birds and condemned the mere instrumental treatment of them for purely economic ends. From a structural-functionalist perspective this would be a privileged site to study morally motivated behavior, since moral values are explicitly evoked by the involved actors and set apart from economic motives. From the point of view of the pragmatist perspective employed in this study, the very opposite is the case. The analysis shows that actors chose a purely moral language to talk about the value of birds because this discourse had no connection to action whatsoever. The pervasiveness of this moral discourse resulted not from the fact that it was a powerful motivating force of behavior; moral discourse loomed so large because it was completely decoupled from action. As such it turns out to be a particularly unsuitable site to study actual behavior.

In order to understand how this discourse came into being we have to understand how the practice of bird conservation was institutionalized. In turn-of-the-century Germany birds were treated as an economic entity; that is, they were addressed as either useful or harmful to agriculture, forestry, and gardening. Every bird species had a peculiar instrumental economic value assigned to it. For some species this instrumental value resulted in practical measures taken for their protection, for other species it resulted in practical measures taken for their destruction. Bird lovers who felt uncomfortable with the latter were among those who objected to instrumental arguments and tried to put moral ones in their place. If every bird species was assigned an intrinsic moral value, even those considered economically harmful had to be protected alongside with the useful ones.

There was, however, very little guidance on what an attribution of such an intrinsic moral value to all bird species would mean in practice, since those who made the argument shared the view that birds are economically significant. Not a single author who made an argument for the existence of an intrinsic moral value of birds was prepared to claim that economic considerations should play no role whatsoever, and that no bird should ever be killed for whatever reason. The discourse on intrinsic values was in consequence completely disconnected from practice. More often than not, the affirmation of an intrinsic moral value of birds could be found in the forewords or conclusions of books on bird conservation, where the authors put their higher moral selves on display, while the main chapters engaged in detailed calculations of which bird species qualified as useful and which as harmful.

The fact that moral discourse was so pronounced was thus not an indicator that it had a particularly strong grip on action but, to the contrary, that it had no grip whatsoever. Moral discourse turned out to be just that, discourse. The real action was somewhere else. Engaging in a discourse on the intrinsic moral value of birds implied a withdrawal from action. The observation resonates with Dewey's account of the separation of means (instrumental values) from ends (intrinsic values) in ethics. In his view such a separation leads to apathy. If actors fail to translate their moral discourse into practical guidelines for action, they will exhibit complete ignorance of these discourses. Such an ignorance is an outcome not of people merely paying lip service to morality but of the specific outlook of a moral discourse or ethic: some ethics are anchored in action and thus have a propensity to guide it, others result from an already existing disconnect and foster further withdrawal. The decoupling of moral discourses on bird conservation from real-life action in turn-of-the-century Germany is an example of an ethic that failed to formulate practical guidelines for action. Conservationists in consequence talked about morality, but they acted economically.

THE DEVELOPMENT OF PRACTICES AND INSTITUTIONS OVER TIME

This account of morality does not imply that conservationists have no deep-seated emotional commitments. That many conservationists employed a moral discourse reflected that the purely economic treatment of birds offended their feelings. Birds they felt attached to were shot, trapped, and poisoned for economic reasons. The argument of this study thus does not rest on a denial of the existence of moral commitments—the point is to advance a different explanation of how these commitments come about and how they are sustained. The argument is that these commitments stand at the end of a temporal process, not at its beginning. Taking this insight seriously implies that searching for the first mention of a moral ideal or concept is not a meaningful exercise. The first articulation of a moral ideal is not at the center of the development but just that, the first articulation. As the study has shown, many of these first articulations of novel concepts were in fact only new labels attached to already existing practices.

The notion of endangered species, for instance, and its accompanying call for action was anything but the outcome of a new moral discourse or a new set of moral ideals. The notion simply resulted from the fact that improvements in the tools for collecting, such as binoculars and cameras, had shifted the passion for rarity collecting from bird bodies to living birds. With this gradual shift to living organisms as the objects of desire rose the notion of threat or endangerment. As long as bird bodies constituted the collectibles, rare birds were not in any way endangered, since they could be preserved in the museum. The shift from the museum to the field made all the difference. It was thus not a moral discourse of endangerment that drove actors into the field to collect data on such species but the already existing practice of field-ornithological data collecting that gave rise to the notion of endangerment. If lines of action transform, the valuations embedded in them transform as well. In the development of the notion of endangerment, such a transformation of practices came first.

This insight also has implications for an assessment of the role of discourse in the rise of environmental movement activism in the time from the 1960s through the 1980s. There are bestselling publications that are associated with this era, one of the most famous among them Rachel Carson's *Silent Spring* (1962b). The book detailed the adverse effects of the misuse of synthetic pesticides and made a moral argument for conservation. It is an episode in which the power of discourse seems to loom large. "Before there was an environmental movement, there was one brave woman and her very brave book," Peter Matthiessen wrote about Carson for the *Time* magazine list of the "100 Most Influential People of the Century" (Matthiessen 1999). The tribute is inspiring but misplaced. There already was a fully institutionalized movement that provided much of the scientific data employed in the book, and there were millions of people like Carson who took joy in watching the natural world around them and who noticed and regretted its destruction.

Carson's interest in the topic began while bird watching in childhood, as did the interest of many of her readers. It is what the author had in common with her audience that made for the success of the book, not what set her apart. *Silent Spring* appealed to the concerns of the natural historian who takes pleasure from observing the outside world, and this practice had been on the rise for decades. Bird watching had become a popular pastime engaged in by a large portion of the population. Action to ban DDT and other pesticides had already been taken before the book was published. The book contributed to the trend, but it did not create it. Regulations grew not from a new moral discourse on the environment but from a set of already established practices that were widespread and well institutionalized by the time the book was written.

This insight is not lost on conservationists. The actors who run major bird conservation organizations do not focus on teaching abstract moral lessons but on drawing people into action and inducing in them the same kind of experiences that proved vital in their own case. The strategies used by these organizations describe the pragmatism of Dewey put into action. Bird watching schemes operated by conservationists assume not that moral valuations and emotional attachments motivate action but that action itself generates valuations and attachments. Yet not everybody who plays the game of bird watching develops a deep emotional attachment to birds. There is a difference between the people at the center of conservation and field-ornithological organizations who initiate these games and devise their rules and those who simply play along.

Organized bird conservation would be much less widespread and less successful, if it relied exclusively on individuals who develop deep moral commitments. The larger impact of the work of deeply committed actors derives from the fact that they can rely on millions of amateur bird watchers who collect field-ornithological data for reasons that require no such commitment. If watching birds was not such a good sport—that is, if birds were not so good to play with—then the entire institutional apparatus built around it would never have come into existence and could not be sustained. The Birds Directive of the European Union, for instance, which protects the majority of endangered bird species in Europe, relies on field-ornithological monitoring data for the selection of protected sites. The enforcement of this environmental policy would be impossible without such data and the thousands of bird watchers who provide it. Many of these bird watchers contribute to bird conservation even in the absence of any deep-seated moral desire to protect birds. The phenomenon of bird conservation can in this sense be described as profane or banal.

Such a notion of the banality of morality that I propose here is an extended version of Hannah Arendt's notion of the banality of evil. Her account of this phenomenon, which she held to be valid in general and not only for the particular case that led her to formulate it (Arendt 1963; Assy 2010), was directed against our tradition of thought—literary, philosophical, and theological—about the phenomenon of evil. In this tradition evil deeds are triggered by weakness, envy, or hate—that is, by motives that are themselves regarded as evil or at least not worthy of praise. Arendt argued that none of these qualities are necessary to produce evil deeds. Actors' deeds might be monstrous and morally horrifying, but the actors can be quite ordinary, commonplace, and neither demonic nor monstrous (Arendt 1978, 4). Banality as used in this context is not assumed to be commonplace or widespread but is considered ordinary in the sense that it has no deeper meaning or deeper motivational forces attached to it.

While denying such depth to the phenomenon of evil, Arendt did assign it to the phenomenon of morality. In her view, evil acts are the outcome of an absence of reflection. Evil can take place because actors lack either the ability or willingness to think and to listen to the voice of their consciousness. Thoughtlessness, to be found in more and less intelligent people alike, makes for evil acts. Arendt argues that one needs philosophy, the exercise of reason as the faculty of thought, to prevent evil. She thus remains enmeshed in the same philosophical tradition that she partially stepped out of. Arendt believed in the possibility of objective insight into the nature of the good as an entity with substance and meaning that exists above and beyond individual judgment and action. These standards are out there, but not everyone listens to their voice and is prepared to act on them. Evil is a banal, shallow, surface phenomenon, while the exercise of thought is deep and goes to the root and true nature of the good. Hers is the age-old philosopher's dream of objective standards for moral judgment that can be discovered by the exercise of reason.

This study takes the initial insight behind Arendt's argument one step further. I argue that the notion of banality can be applied not only to evil but also to morality. Moral concern for birds developed out of a gradual reinterpretation and reappropriation of existing social forms to which no moral quality could previously be assigned. Concern for wild birds in fact developed out of the very practices and institutions of collecting that formerly drove their destruction and decline. This study shows that there is no such thing as a spirit of environmentalism, a Green collective consciousness, or a conservationist creed at the root of the matter. Morality, in this sense, is not deep but a surface phenomenon. Thick descriptions of discourse will accordingly not allow us to understand the significance of practices that sometimes amount to little more than the blink of an eye.

APPENDIX 1: METHOD AND DATA

This study follows the development of organized bird conservation in Great Britain and Germany over an extended period of time. One of the prime challenges in conducting this research was to access data that is consistent across time and space. The life histories of major conservationists proved to be the source of information that was consistently available for both countries stretching over a period of more than two centuries.

The study thus makes systematic use of published biographies and personal interviews. I compare the motives of leading actors in the organizations involved in the conservation of birds and the organizational goals they devised. Throughout two years of fieldwork in Britain and Germany and several subsequent visits, I have accessed the archives and libraries of the *Royal Society for the Protection of Birds* (RSPB) and the *Naturschutzbund Deutschland* (NABU) and interviewed sixty-eight actors involved in various aspects of organized bird watching, field ornithology, and nature conservation in the two countries. Interviews lasted between half an hour and three and a half hours.

Accessing additional sources of data turned out to be more problematic. Metatheoretical discussions on research methods lose much of their practical grip when the real challenge is not to select among different sources of data but to find data in the first place, not a particular kind of data but any data at all. This was not least among the reason why throughout this study the absence of data turned into a source of information itself: looking at the absence of information allowed me to address the question why so much data exists on some species but not on others and why some of this data focuses on species diversity and some focuses on species abundance. Availability of data reflects what matters to people and the absence of data thus indicates what matters less.

While missing data is a valuable source of information, too, some data was nevertheless truly missed. Both organizations, RSPB and NABU, have archives at their headquarters, yet the holdings are limited and largely amount to published material. Those documents that exist are moreover not the documents that would be needed to systematically study the development of these two organizations over time and to compare them with each other. The lack of documentation owes largely to the fact that these organizations were initially rather small enterprises and entirely run by volunteers. Keeping historical records was neither a priority and to some extent not even a possibility. The archive of the NABU in particular comprises mostly printed matter, which owes to the fact that during an organizational downturn in the in mid-twentieth century the then president of the organization disposed all historical documents, including membership files, due to a lack of office space. It was a moment in time when the organization came close to becoming history, rather than having one.

Published annual reports and membership magazines of these organizations are, however, available for the entire period observed and have been systematically analyzed. These publications were the source of the membership numbers of both organizations from their foundation until the present, with only some minor gaps in the record. The membership numbers provide a simple and reliable measure to compare concern for birds between different countries and to assess change over time. Much can be said against the validity of such numbers as measures for the popular resonance of bird conservation at large. What speaks for them is that these data are virtually the only data available for large-scale comparison over time and across space.

Explaining the ups and downs in membership over time is a central part of the qualitative analysis. The main part of this analysis relies on the already mentioned life history data. I made extensive use of autobiographies, travel narratives, and all other available publications in which actors talk about their lives. In the case of those individuals who are still with us, I conducted interviews as oral histories. These interviews document major developments in the field of bird conservation from the 1960s to the present. Collecting oral histories allowed me to look at the entire life course of major actors, following their trajectory in some cases from childhood to retirement. Many crucial organizational changes could moreover often only be studied through interviews, since no other forms of documentation exist.

In the case of some actors it was possible to employ both historical documents and oral histories. While written records have the advantage that they are not subject to distortion through selective memory or forgetfulness, interviews have the advantage that people are generally more willing to freely share sensitive information. There is accordingly a trade-off between the two sources, and I have tried to avoid it by making use of both written documents and oral histories whenever I had the opportunity.

The encounters confirmed the initial assumption that actors are more willing to give away secrets in a personal conversation than in print. Yet this general observation deserves to be qualified. The overall impression that emerged is that actors are the more open to make revealing statements about their actions the more distant in time these actions are. The difference between the information gained from personal interviews and published narratives became the more marginal the further I ventured into the past. Only for the present decade was the published record rather wanting if compared to the oral statement.

The six substantive chapters in this book are split roughly evenly between the sources of data that dominate. Chapters 2 to 4 introduce the central actors based on published narratives, while chapters 5 to 7 rely for the most part on oral histories. The combination of these data allowed for a comparative study of bird watching, field ornithology, and nature conservation in the *longue durée*. It enabled me to produce insights that would have escaped more detailed archival analyses of separate episodes in the past or participant observations of actors at the present.

APPENDIX 2: NAMES AND TRANSLATIONS

The focus of this study is on the development of the primary bird conservation organizations in Great Britain and Germany and their international umbrella organization. The names of some of these organizations changed throughout their history. To prevent confusion, I will document the most crucial changes here.

There was only one major change in name in the British case. The Society for the Protection of Birds was founded in 1889. It gained royal charter in 1904 and was renamed the Royal Society for the Protection of Birds (RSPB). Changes in name are more numerous in the German case, which owes to a large extent to the more radical ruptures in the political history of the country. The German League for Bird Protection (Bund für Vogelschutz) has gone through several name changes since its inauguration, although the organizational history is, as in the case of the RSPB, an unbroken one. The League for Bird Protection became the League for Bird Protection of the Reich (Reichsbund für Vogelschutz) from 1935 to 1945, after which it switched back to its previous name. It was renamed in 1966 the German League for Bird Protection (Deutsche Bund für Vogelschutz) and became the German League for Nature Protection (Naturschutzbund Deutschland) in 1990, when it merged with its East German partner organization. Acronyms changed each time the name changed, from BfV to RbfV and back, and later from DBV to NABU. The main chapters always reference the name that was in use at the time under discussion, while the prologue and chapter 1 use exclusively the contemporary names and acronyms of these two organizations.

The major international umbrella organization of these and many other national bird conservation organizations also changed its name. The International

Council for Bird Preservation (ICBP), founded in 1922, was restructured in 1994 and is now called BirdLife International. The partner organization of the ICBP in Germany was called International Council for Bird Preservation—German Section (Internationale Rat für Vogelschutz—Deutsche Sektion). This organization is still in existence, now called German Council for Bird Protection (Deutsche Rat für Vogelschutz), although it is not part of BirdLife anymore. The national partner organization of this international umbrella organization is now the NABU. For the purpose of recognizability in the international arena, the NABU has created a translation of its name into English. In order to match its German acronym NABU, the German League for Nature Protection uses the title **N**ature **a**nd **B**iodiversity Conservation **U**nion as its official English name. Although almost a literal translation, this name adds the term biodiversity, which is not part of the original German name. The English name is thus to a certain extent misleading, and I therefore use the literal translation rather than the organization's official English name throughout the text.

All translations of interviews originally conducted in German and translations of German publications are my own. Since there is no exception to this rule, I have not specifically indicated the authorship of translations in the text. German publications were always accessed in the original language, including in cases in which English translations are available.

INTERVIEWS

Interviewee	Date	Location
Richard Bashford	07/2006	Sandy, England
Hans-Günther Bauer	07/2007	Radolfzell, Germany
Peter Berthold	07/2007	Radolfzell, Germany
Einhard Bezzel	03/2012	Garmisch-Partenkirchen, Germany
Bernhard Blaszkiewitz	07/2007	Berlin, Germany
Bernd Conrad	03/2012	Recklinghausen, Germany
Adrian Darby	08/2006	Oxford, England
Alasdair Dawes	08/2006	Egleton, England
Rainer Dröschmeister	08/2007	Bonn, Germany
Karl-Heinz Erdmann	07/2007	Bonn, Germany
Birgit Fahrenholz	05/2011	Phone interview
Heike Finke	08/2007	Bad Kreuznach, Germany
Julian Fitter	08/2006	London, England
Martin Flade	07/2007	Brodowin, Germany
Jochen Flasbarth	08/2007	Bonn, Germany
Sylke Frahnert	07/2007	Berlin, Germany
Stephanie Fudge	09/2007	London, England
Alistair Gammell	02/2008	Sandy, England
Rob George	02/2008	Sandy, England
Andrew Gosler	06/2006	Oxford, England
Inge Gotzmann	07/2007	Bonn, Germany
Jeremy Greenwood	08/2006	Thetford, England

Hermann Graf Hatzfeldt	07/2007	Berlin, Germany
Alexander Heyd	08/2007	Bonn, Germany
Peter Holden	02/2008	Sandy, England
Rob Hume	07/2006	Sandy, England
Charles Hussey	08/2006	London, England
Christoph Imboden	03/2012	Phone interview
Sir John Lawton	06/2006	London, England
Robert May, Baron	02/2008	Oxford, England
Claus Mayr	08/2007	Bonn, Germany
Peter Melchett, Baron	09/2007	Phone interview
Bernd-Ulrich Meyburg	06/2007	Berlin, Germany
Sara Morrison	08/2006	London, England
Stephen Moss	06/2006	Hampton, England
Hilmar Freiherr von Münchhausen	07/2007	Hamburg, Germany
Ian Newton	07/2006	Abbots Ripton, England
Manfred Niekisch	07/2007	Greifswald, Germany
Derek Niemann	07/2006	Sandy, England
Sarah Niemann	07/2006	Sandy, England
Harriet Nimmo	06/2006	Bristol, England
Markus Nipkow	05/2011	Berlin, Germany
Stephen Noble	06/2006	Sandy, England
Eugeniusz Nowak	07/2007	Bonn, Germany
Helmut Opitz	03/2012	Berlin, Germany
Myrfyn Owen	08/2006	Slimbridge, England
Bernd Pieper	08/2007	Bonn, Germany
Mike Rands	02/2008	Cambridge, England
Josef Reichholf	07/2007	München, Germany
Adam Rowlands	06/2006	Minsmere, England
Klaus Ruge	08/2012	Phone interview
Norbert Schäffer	02/2008	Sandy, England
Johannes Schwarz	06/2007	Berlin, Germany
John Sheail	07/2006	Abbots Ripton, England
Ken Smith	07/2006	Sandy, England
Oliver Smith	07/2006	Godalming, England
Tony Soper	08/2006	Kingsbridge, England
Axel Ssymank	08/2007	Bonn, Germany
Markus Steigenberger	07/2007	Berlin, Germany
Hans-Jürgen Stork	07/2007	Berlin, Germany
Claus Suffmann	07/2012	Phone interview

Thomas Tennhardt	02/2012	Berlin, Germany
Carl-Albrecht von Treuenfels	07/2007	Neuhorst, Germany
Teresa Wild	02/2008	London, England
Sir Robert Worcester	08/2006	London, England
Graham Wynne	02/2008	Sandy, England
Barbara Young, Baroness	09/2007	London, England
Angelika Zahrnt	07/2007	Berlin, Germany

REFERENCES

Abbott, Andrew. 2001. "On the Concept of Turning Point." In *Time Matters: On Theory and Method*, 240–60. Chicago: University of Chicago Press.

Abend, Gabriel. 2011. "Thick Concepts and the Moral Brain." *European Journal of Sociology* 52 (1): 143–72.

———. 2014. *The Moral Background: An Inquiry into the History of Business Ethics*. Princeton, NJ: Princeton University Press.

Aderhold, R. 1906. "Die kaiserliche Biologischen Anstalt für Land- und Forstwirtschaft in Dahlem bei Steglitz." *Mitteilungen aus der Kaiserlichen Biologischen Anstalt für Land- und Forstwirtschaft* 1:2–20.

AGW-Vorstand und Mitarbeiter. 2005. *40 Jahre Arbeitsgemeinschaft Wanderfalkenschutz—AGW*. Offenburg: Arbeitsgemeinschaft Wanderfalkenschutz im NABU Deutschland.

Alexander, Horace G. 1944. "On Field Identification of Birds." *British Birds* 37:89–93.

———. 1974. *Seventy Years of Birdwatching*. Berkhamsted: T. & A. D. Poyser.

Alexander, Wilfred B., and David Lack. 1944. "Changes in Status among British Breeding Birds." *British Birds* 38:42–45, 62–69, 82–88.

Allen, David Elliston. (1976) 1994. *The Naturalist in Britain: A Social History*. Princeton, NJ: Princeton University Press.

Allen, Glover M. 1942. *Extinct and Vanishing Mammals of the Western Hemisphere, with the Marine Species of All the Oceans*. Cambridge, MA: American Committee for International Wild Life Protection.

Alpin, Oliver V. ca.1900. *The Birds as Labourers*. RSPB Leaflet Series 17.

Altick, Richard D. 1978. *The Shows of London*. Cambridge, MA: Belknap.

Anderson, Nels. 1923. *The Hobo: The Sociology of the Homeless Man*. Chicago: University of Chicago Press.

Anschütz, Ottomar. 1901. *Die Photographie im Hause: Lehrbuch für Amateure*. Berlin: Ottomar Anschütz.

Appeal to British Boys and Girls. n.d. RSPB Leaflet Series 18.

Arendt, Hannah. 1963. *Eichmann in Jerusalem: A Report on the Banality of Evil.* New York: Viking.

———. 1978. *The Life of the Mind: Two Volumes.* New York: Harcourt Brace Jovanovich.

Arnold, Ken. 2006. *Cabinets for the Curious: Looking Back at Early English Museums.* Aldershot: Ashgate.

Asma, Stephen T. 2001. *Stuffed Animals & Pickled Heads: The Culture and Evolution of Natural History Museums.* New York: Oxford University Press.

Assy, Bethania. 2010. "Eichmann, the Banality of Evil, and Thinking in Arendt's Thought." Accessed June 15. https://www.bu.edu/wcp/Papers/Cont/ContAssy.htm.

Axell, Herbert. 1992. *Of Birds and Men.* Sussex: Book Guild.

Axell, Herbert, and Eric Hosking. 1977. *Minsmere: Portrait of a Bird Reserve.* Foreword by HRH Prince Philip, Duke of Edinburgh. London: Hutchinson.

Baker, J. R. 1978. *Julian Huxley: Scientist and World Citizen, 1887–1975.* Paris: UNESCO.

Barber, Lynn. 1980. *The Heyday of Natural History 1820–1870.* London: Jonathan Cape.

Barthel, Peter H. 1993. "Artenliste der Vögel Deutschlands." *Journal für Ornithologie* 134:113–35.

Barthel, Peter H., and Andreas J. Helbig. 2005. "Artenliste der Vögel Deutschlands." *Limicola* 19:89–111.

Barthelmeß, Alfred. 1981. *Vögel—Lebendige Umwelt. Probleme von Vogelschutz und Humanökologie geschichtlich dargestellt und dokumentiert.* Freiburg: Karl Alber.

Batten, L. A., C. J. Bibby, P. Clement, G. D. Elliott, and R. F. Porter, eds. 1990. *Red Data Birds in Britain: Action for Rare, Threatened and Important Species.* Illustrated by Ian Willis. London: T. & A. D. Poyser.

Bau, Alexander. 1905. *Naturgeschichte der Deutschen Vögel einschließlich der sämtlichen Vogelarten Europas / von C. G. Friderich.* Stuttgart: Schweizerbart.

———. 1923. *Naturgeschichte der Vögel Europas / von C. G. Friderich.* Stuttgart: Schweizerbart.

Bauer, Hans-Günther, and Peter Berthold. 1996. *Die Brutvögel Mitteleuropas: Bestand und Gefährdung.* Wiesbaden: Aula-Verlag.

Bechstein, Johann Matthäus. 1789–1795. *Gemeinnützige Naturgeschichte Deutschlands nach allen drey Reichen: Ein Handbuch zur deutlichern und vollständigern Selbstbelehrung besonders für Forstmänner, Jugendlehrer und Oekonomen.* Leipzig: Siegfried Lebrecht Crusius.

———. 1792. *Kurzgefaßte gemeinnützige Naturgeschichte des In- und Auslandes für Schulen und häuslichen Unterricht. Ersten Bandes erste Abtheilung.* Leipzig: Siegfried Lebrecht Crusius.

———. 1792–1812. *Johann Lathams allgemeine Übersicht der Vögel. Aus dem Englischen übersetzt und mit Anmerkungen und Zusätzen versehen von Johann Matthäus Bechstein.* Nürnberg: Weigel und Schneider.

———. 1795. *Naturgeschichte der Stubenvögel oder Anleitung zur Kenntniß und Wartung derjenigen Vögel, welche man in der Stube halten kann.* Gotha: Carl Wilhelm Ettinger.

———. 1798. *Naturgeschichte der schädlichen Waldinsecten.* Nürnberg: Monath & Kußler.

Bechstein, Ludwig. 1855. *Dr. Johann Matthäus Bechstein und die Forstacademie Dreissigacker: Ein Doppel-Denkmal* Meiningen: Brückner & Renner.

Beck, Ulrich. 1992. *Risk Society: Towards a New Modernity.* London: Sage.

———. 1999. *World Risk Society*. Cambridge: Polity Press.

———. 2000. "The Cosmopolitan Perspective: Sociology and the Second Age of Modernity." *British Journal of Sociology* 51:79–105.

Beck, Ulrich, Anthony Giddens, and Scott Lash, eds. 1994. *Reflexive Modernization: Politics, Tradition and Aesthetics in the Modern Social Order*. Cambridge: Polity Press.

Becker, Howard S. 1963. *Outsiders: Studies in the Sociology of Deviance*. London: Free Press of Glencoe.

Bennett, Tony. 1995. *The Birth of the Museum: History, Theory, Politics*. London: Routledge.

Berlepsch, Hans Freiherr von. 1899. *Der gesamte Vogelschutz: Seine Begründung und Ausführung*. Gera-Untermhaus: Köhler.

———. 1929a. *Der gesamte Vogelschutz: Seine Begründung und Ausführung auf wissenschaftlicher, natürlicher Grundlage. Zwölfte Auflage*. Neudamm: J. Neumann.

———. 1929b. "Mein ornithologischer Lebenslauf." In *Der gesamte Vogelschutz: Seine Begründung und Ausführung auf wissenschaftlicher, natürlicher Grundlage. Zwölfte Auflage*, 23–85. Neudamm: J. Neumann.

Berndt, Rudolf, and Fritz Frieling. 1939. "Aus der Muster- and Versuchsstation des Reichsbundes für Vogelschutz: Siedlungs- und brutbiologische Studien an Höhlenbrütern in einem Nordwestsächsischen Park." *Journal für Ornithologie* 87:593–638.

Berthold, Peter. 1974. "Fortschreitende Rückgangerscheinungen bei Vögeln: Vorboten des 'Stummen Frühlings.'" *Mitteilungen der Max Planck Gesellschaft*:41–49.

Berthold, Peter, Einhard Bezzel, and Gerhard Thielcke, eds. 1974. *Praktische Vogelkunde: Ein Leitfaden für Feldornithologen*. Greven: Kilda Verlag.

Berthold, Peter, and Gabriele Mohr. 2006. *Vögel füttern—aber richtig. Anlocken, Schützen, Sicher Bestimmen. Mit einem Vorwort von Heinz Sielmann*. Stuttgart: Franckh-Kosmos.

Beunza, Daniel, and David Stark. 2004. "Tools of the Trade: The Socio-Technology of Arbitrage in a Wall Street Trading Room." *Industrial and Corporate Change* 13:369–401.

Bezzel, Einhard. 1973. *Verstummen die Vögel? Unsere bedrohte Vogelwelt*. München: Ehrenwirth.

———. 1977. *Ornithologie*. Stuttgart: Ulmer.

———. 1982a. *Vögel Beobachten. Wie—Wann—Wo?* München: BLV Verlagsgesellschaft.

———. 1982b. *Vögel in der Kulturlandschaft*. Stuttgart: Ulmer.

———. 1983–1984. *Vögel. 2 Bände*. Zürich: BLV Verlagsgesellschaft.

———. 1984. "125 Bände 'Journal für Ornithologie.'" *Journal für Ornithologie* 125 (4): 381–91.

———. 1985. "Streptopelia decaocto (Frivaldszky 1838)—Türkentaube." In *Kompendium der Vögel Mitteleuropas. Nonpasseriformes—Nichtsingvögel*, edited by Einhard Bezzel, 613–17. Wiesbaden: Aula-Verlag.

———. 1988a. "Die Versammlungen deutscher Ornithologen 1845–1987: Ein Streifzug durch die Geschichte der Deutschen Ornithologen-Gesellschaft." *Journal für Ornithologie* 129:2–21.

———. 1988b. "Die Vorstands- und Ehrenmitglieder der Deutschen Ornitholgen-Gesellschaft." *Journal für Ornithologie* 129:32–42.

———. 1990. *Ornithologie. 2., völlig neubearbeitete und erweiterte Auflage*. Stuttgart: Ulmer.

———. 2000. "Deutsche Ornithologen-Gesellschaft 150 Jahre alt: Das große Jubiläum der Vogelkunde." *Falke* 47:246–49.

Bijker, Wiebe E. 1995. *Of Bicycles, Bakelites, and Bulbs: Toward a Theory of Sociotechnical Change*. Cambridge, MA: MIT Press.

Bijleveld, Maarten. 1974. *Birds of Prey in Europe*. London: Macmillan.

Binot-Hafke, Margret, Horst Gruttke, Gerhard Ludwig, and Uwe Riecken, eds. 1999. *Bundesweite Rote Listen: Bilanzen, Konsequenzen, Perspektiven*. Bonn-Bad Godesberg: Bundesamt für Naturschutz.

BirdLife International 2010. "BirdLife Partners." Accessed October 23. http://www.birdlife.org.

———. 2014. "French Polynesia." Accessed June 15. http://www.birdlife.org/datazone/country/french-polynesia.

"Birds, Insects and Crops." 1917. *Bird Notes and News* 7 (6): 82.

Birkhead, T. R., and S. van Balen. 2008. "Bird-keeping and the Development of Ornithological Science." *Archives of Natural History* 35 (2): 281–305.

Blab, Josef, et al., eds. 2005. *Rote Listen—Barometer der Biodiversität: Entstehungsgeschichte und neuere Entwicklungen in Deutschland, Österreich und der Schweiz*. Bonn-Bad Godesberg: Bundesamt für Naturschutz.

Blab, Josef, and Eugeniusz Nowak, eds. 1989. *Zehn Jahre Rote Liste Gefährdeter Tierarten in der Bundesrepublik Deutschland: Situation, Erhaltungszustand, neuere Entwicklungen*. Greven: Kilda-Verlag.

Blackmore, Michael. 1985. *The Selborne Society: Its Origin and History*. London: Selborne Society.

Blake, Judith , and Kingsley Davis. 1964. "Norms, Values, and Sanctions." In *Handbook of Modern Sociology*, edited by R. E. L. Faris, 456–84. Chicago: Rand McNally.

Blasius, Rudolf. 1890. "Die Vogelwarte in Helgoland." *Ornithologische Monatsschrift* 15:350–65.

Boardman, Robert. 1981. *International Organization and the Conservation of Nature*. Bloomington: Indiana University Press.

———. 2006. *The International Politics of Bird Conservation: Biodiversity, Regionalism and Global Governance*. Cheltenham: Edward Elgar.

Boltanski, Luc, and Laurent Thévenot. 2006. *On Justification: Economies of Worth*. Princeton, NJ: Princeton University Press.

Boraston, John MacLair. 1905. *Birds by Land and Sea. The Record of a Year's Work with Field-Glass and Camera. Illustrated by Photographs taken direct from Nature by the Author*. London: John Lane.

Boschert, Martin. 2003. "Das DDA-Monitoringprogramm seltener Brutvogelarten—Überblick." *Berichte des Landesamtes für Umweltschutz Sachsen-Anhalt. Sonderheft* 1:36–39.

Boye, Peter, M. Vischer-Leopold, C. Paulsch, Axel Ssymank, and F. Beulshausen. 2010. *Drei Jahrzehnte Vogelschutz im Herzen Europas: Rückblick, Bilanz und Herausforderungen*. Bonn-Bad Godesberg: Bundesamt für Naturschutz.

Braverman, Harry. 1974. *Labor and Monopoly Capital: The Degradation of Work in the Twentieth Century*. New York: Monthly Review Press.

"Brean Down: The Hill of Birds." 1912. *Bird Notes and News* 5 (3): 33–34.

Brehm, Alfred Edmund. 1855. *Reiseskizzen aus Nord-Ost-Africa: Oder den unter Egyptischer Herrschaft stehenden Ländern Egypten, Nubien, Sennahr, Rosseeres und Kordofahn gesammelt auf seinen in den Jahren 1847 bis 1852 unternommenen Reisen.* Jena: Mauke.

———. 1861. *Das Leben der Vögel: Dargestellt für Haus und Familie.* Glogau: Flemming.

———. 1864–1869. *Illustriertes Thierleben.* Hildburghausen: Bibliographisches Institut.

———. 1872–1876. *Gefangene Vögel. Ein Hand- und Lehrbuch für Liebhaber und Pfleger einheimischer und fremdländischer Käfigvögel.* Leipzig: Winter.

———. 1874. "Unsere Bodenwirthschaft und die Vögel." *Journal für Ornithologie* 22:26–39.

Brehm, Christian Ludwig. 1823–1824. *Lehrbuch der Naturgeschichte aller europäischen Vögel, 2 Bde.* Jena: Schmid.

———. 1831. *Handbuch der Naturgeschichte alle Vögel Deutschlands.* Ilmenau: Voigt.

———. 1842. *Die Kunst, Vögel als Bälge zu bereiten.* Weimar: Voigt.

———. 1855. *Der vollständige Vogelfang.* Weimar: Voigt.

Brehm Gesellschaft, e.V., ed. 1929. *Das Brehmbuch: Zum 100. Geburtstag von Alfred Brehm.* Berlin: Brehm Verlag.

"British Homes for British Birds." 1915. *Bird Notes and News* 6 (5): 80.

Brightwen, Eliza. 1890. *Wild Nature Won by Kindness. With Illustrations by the Author and F. Carruthers Gould.* London: T. Fisher Unwin.

———. 1892. *More About Wild Nature. With Illustrations by the Author.* London: T. Fisher Unwin.

———. 1895. *Inmates of My House and Garden.* London: T. Fisher Unwin.

———. 1899. *Rambles with Nature Students.* London: Religious Tract Society.

British Ornithologists' Union. 1915. *A List of British Birds, Compiled by a Committee of the British Ornithologists' Union.* Second and revised edition. London: British Ornithologists' Union.

British Ornithologists' Union List Sub-Committee. 1952. *Check-List of the Birds of Great Britain and Ireland.* London: British Ornithologists' Union.

Broom, Leonard, and Philip Selznick. 1963. *Sociology: A Text with Adapted Readings.* 3rd ed. New York: Harper & Row.

Browne, Montagu. 1896. *Artistic and Scientific Taxidermy and Modeling: A Manual of Instruction in the Methods of Preparing and Reproducing the Correct Form of All Natural Objects Including a Chapter on the Modelling of Foliage. With 22 Full-Page Illustrations and 11 Illustrations in Text.* London: Adam and Charles Black.

Bruns, Herbert. 1953. "Behandlung von Vogelnistkästen mit Kontaktinsektiziden und ihre Auswirkung auf Hornissen, Wespen und Vögel." *Anzeiger für Schädlingskunde* 26 (12): 182–85.

———. 1969. "Lebensschutz oder Bioprotektion als Integration von Mensch-, Tier-, Pflanzen-, Natur- und Landschaftsschutz." *Biologische Abhandlungen* 32–33:1–56.

———. 1975. *Ullstein Vogelbuch—Vogelkunde, Vogelbeobachtung, Vogelliebhaberei und Vogelschutz als Wissenschaft und Hobby.* Frankfurt a.M.: Ullstein Verlag.

Bruns, Herbert, and Otto Niebuhr, eds. 1961. *Die Vogelwarten und Vogelschutzwarten Mitteleuropas.* Hamburg-Sasel: Biologische Abhandlungen.

Bruun, Bertel. 1970. *The Hamlyn Guide to Birds of Britain and Europe.* Illustrated by Arthur Singer, consultant editor Bruce Campbell. London: Hamlyn.

———. 1971. *Der Kosmos-Vogelführer: Die Vögel Deutschlands und Europas in Farbe. Illustriert von Arthur Singer. Aus dem Englischen übersetzt und bearbeitet von Claus König*. Stuttgart: Franckh'sche Verlagshandlung.

Buechler, Steven M. 1995. "New Social Movement Theories." *Sociological Quarterly* 36:441–64.

Bullock, William. 1808. *A Companion to the Liverpool Museum*. 6th ed. Hull: Printed for the proprietor by J. Ferraby.

———. 1813. *A Companion to the London Museum and Pantherion*. 15th ed. London: Printed for the proprietor by Whittingham and Rowland.

———. 1819. *Catalogue of the Roman Gallery, of Antiquities and Works of Art, and the London Museum of Natural History*. London.

Bund für Vogelschutz. 1899a. "Aufgaben des Vogelschutzes." In *Kalender für 1900*, edited by Friedrich Arnold, 31–32. Stuttgart: Greiner & Pfeiffer.

———. 1899b. "Aufruf des Bundes für Vogelschutz. Flugblatt I." Stuttgart: Bund für Vogelschutz.

———. 1905. *Jahresbericht*. Stuttgart; Giengen, Brenz (Württemberg): Bund für Vogelschutz.

———. 1906. *Jahresbericht*. Stuttgart; Giengen, Brenz (Württemberg): Bund für Vogelschutz.

———. 1916. *Jahresbericht*. Stuttgart; Giengen, Brenz (Württemberg): Bund für Vogelschutz.

———. 1934. *Jahresbericht*. Stuttgart; Giengen, Brenz (Württemberg): Bund für Vogelschutz.

———. 1960. *Jahresbericht*. Stuttgart; Giengen, Brenz (Württemberg): Bund für Vogelschutz.

———. 1964. *Jahresbericht*. Stuttgart; Giengen, Brenz (Württemberg): Bund für Vogelschutz.

Burawoy, Michael. 1979. *Manufacturing Consent: Changes in the Labor Process under Monopoly Capitalism*. Chicago: University of Chicago Press.

Burfield, Ian, and Franz van Bommel. 2004. *Birds in Europe: Population Estimates, Trends and Conservation Status*. Cambridge: BirdLife International.

Burkhardt, Richard W. 2005. *Patterns of Behavior: Konrad Lorenz, Niko Tinbergen, and the Founding of Ethology*. Chicago: University of Chicago Press.

Butler, Arthur G. 1899. *Foreign Finches in Captivity*. 2nd ed. Hull: Brumby and Clarke.

Caldwell, Lynton Keith. 1996. *International Environmental Policy: From the Twentieth to the Twenty-First Century*. 3rd ed. Durham, NC: Duke University Press.

Callon, Michel. 1986. "Some elements of a sociology of translation: Domestication of the scallops and the fishermen of Saint Brieuc Bay." In *Power, Action and Belief: A New Sociology of Knowledge?* edited by John Law, 196–233. Boston: Routledge.

———. 1987. "Society in the Making: The Study of Technology as a Tool for Sociological Analysis." In *The Social Construction of Technological Systems*, edited by Thomas P. Hughes, Wiebe Bijker, and Trevor Pinch, 83–103. Cambridge, MA: MIT Press.

———. 1998. "Introduction: The Embeddedness of Economic Markets in Economics." In *The Laws of the Markets*, edited by Michel Callon, 1–57. Oxford: Blackwell.

Campbell, Bruce. (1951) 1962. "Birdmanship." In *The Bird Notes Bedside Book*, edited by Gwen Davies, 21–25. Sandy: The Royal Society for the Protection of Birds.

———. 1952. *Bird-watching for Beginners*. With illustrations by R. A. Richardson. Harmondesworth: Penguin Books.

———. 1979. *Birdwatcher at Large: Autobiography of an Ornithologist*. London: J. M. Dent & Sons.

Carrington, Edith. 1898. *The Farmer and the Birds*. London: G. Bell & Sons.

Carson, Rachel. 1962a. *Der stumme Frühling*. München: Biederstein.

———. 1962b. *Silent Spring*. Boston: Houghton Mifflin.

Chernilo, Daniel. 2006. "Social Theory's Methodological Nationalism: Myth and Reality." *European Journal of Social Theory* 9 (1): 5–22.

———. 2007. *A Social Theory of the Nation-State: The Political Forms of Modernity beyond Methodological Nationalism*. London: Routledge.

Chislett, Ralph. 1947. "Outline of Bird Photography." In *Masterpieces of Bird Photography*, edited by Eric Hosking and Harold Lowes, 11–17. London: Collins.

Clare, John. 2007. "Carrot Fields to Cranes in 11 Years." Last modified May 17. https://www.rspb.org.uk/about-the-rspb/about-us/media-centre/releases/details.aspx?id=176159.

Clark, Ronald W. 1960. *Sir Julian Huxley*. London: Phoenix House.

Clark, Terry N., and Ronald Inglehart. 1998. "The New Political Culture: Changing Dynamics of Support for the Welfare State and Other Policies in Postindustrial Societies." In *The New Political Culture*, edited by Terry N. Clark and Vincent Hoffmann-Martinot, 9–73. Boulder, CO: Westview Press.

Clarke, Richard. 2004. *Pioneers of Conservation: The Selborne Society and the Royal SPB*. London: The Selborne Society.

Cocker, Mark. 2001. *Birders: Tales of a Tribe*. London: Jonathan Cape.

Collinge, Walter E. 1913. *The Food of Some British Wild Birds: A Study in Economic Ornithology*. London: Dulau & Co.

———. 1924–1927. *The Food of Some British Wild Birds: A Study in Economic Ornithology*. York: Published by the Author.

Conder, Peter. 1978. *RSPB Guide to Birdwatching*. London: Hamlyn.

Conrad, Bernd. 1977. *Die Giftbelastung der Vogelwelt Deutschlands*. Greven: Kilda-Verlag.

Conrad, Bernd, and Wolfgang Poltz. 1976. *Vogelschutz in Europa: Ein Situationsbericht über den Vogelschutz in den Staaten der Europäischen Gemeinschaft*. Greven: Kilda-Verlag.

Cook, James. (1774) 1961. *The Journals of Captain James Cook on his Voyages of Discovery. Vol. II: The Voyages of the Resolution and Adventure, 1772–1775*. Edited by J. C. Beaglehole. Cambridge: Cambridge University Press.

Cordeaux, John. 1875. "Notes on the Birds of Heligoland in Mr. Gätke's Collection." *Ibis* 17 (2): 172–88.

Corrigan, Philip, and Derek Sayer. 1985. *The Great Arch: English State Formation as Cultural Revolution*. Oxford: Blackwell.

Council of the European Communities. 1973. "Declaration of the Council of the European Communities and of the Representatives of the Governments of the Member States Meeting in the Council of 22 November 1973 on the Programme of Action of the European Communities on the Environment." *Official Journal of the European Communities* 16 (C112): 1–53.

Courtney, Julia. 1989. *Sir Peter Scott: A Founder of the WWF (World Wide Fund for Nature), Who Has Devoted His Life to Conservation*. Watford: Exley.

Cramp, Stanley. 1977. "Bird Conservation in Europe: A Report Prepared for the Environment and Consumer Protection Service, Commission of the European Communities." London: HMSO.

———. 1978. *Schicksal und Zukunft der Vögel Europas*. Greven: Kilda-Verlag.

Cramp, Stanley, and Peter J. Conder. 1961. *The Deaths of Birds and Mammals Connected with Toxic Chemicals in the First Half of 1960*. Report No. 1 of the BTO-RSPB Committee on Toxic Chemicals. London: Royal Society for the Protection of Birds.

Cramp, Stanley, Peter J. Conder, and John S. Ash. 1962. *The Deaths of Birds and Mammals from Toxic Chemicals January–June 1961*. Second Report of the Joint Committee of the British Trust for Ornithology and the Royal Society for the Protection of Birds on Toxic Chemicals, in collaboration with the Game Research Association. Sandy: Royal Society for the Protection of Birds.

———. 1963. *The Deaths of Birds and Mammals from Toxic Chemicals September 1961–August 1962*. Third Report of the Joint Committee of the British Trust for Ornithology and the Royal Society for the Protection of Birds on Toxic Chemicals, in collaboration with the Game Research Association. Sandy: Royal Society for the Protection of Birds.

Cressey, Paul G. 1932. *The Taxi-Dance Hall: A Sociological Study in Commercialized Recreation and City Life*. Chicago: University of Chicago Press.

Csikszentmihalyi, Mihaly. (1975) 2000. *Beyond Boredom and Anxiety: Experiencing Flow in Work and Play*. San Francisco: Jossey-Bass.

Damaschun, Ferdinand, Sabine Hackethal, Hannelore Landsberg, and Reinhold Leinfelder, eds. 2010. *Art, Ordnung, Klasse: 200 Jahre Museum für Naturkunde*. Rangsdorf: Basilisken-Presse.

Dean, Alan R. 2007. "The British Birds Rarities Committee: A Review of Its History, Publications and Procedures." *British Birds* 100:149–76.

Deutscher Bund für Vogelschutz. 1967. *Jahresheft*. Stuttgart: Deutscher Bund für Vogelschutz.

Dewey, John. 1916. *Democracy and Education: An Introduction to the Philosophy of Education*. New York: Macmillan.

———. 1922. *Human Nature and Conduct: An Introduction to Social Psychology*. New York: Henry Holt and Co.

———. 1929. *The Quest for Certainty: A Study of the Relation of Knowledge and Action*. New York: Balch.

———. 1939. *Theory of Valuation*. Chicago: University of Chicago Press.

———. 1949. "The Field of Value." In *Value: A Cooperative Inquiry*, edited by Ray Lepley, 64–77. New York: Columbia University Press.

———. 1966. "Means and Ends." In *Their Morals and Ours: Marxist versus Liberal Views on Morality; Four Essays by Leon Trotsky, John Dewey, George Novack*, edited by George Novack, 55–60. New York: Merit Publishers.

Dewey, John, and Arthur F. Bentley. 1949. *Knowing and the Known*. Boston: Beacon Press.

Diaz, Carolina Lasén. 2001. "The EC Habitats Directive Approaches its Tenth Anniversary: An Overview." *Review of European Community & International Environmental Law* 10 (3): 287–95.

DiMaggio, Paul J., and Walter W. Powell. 1983. "The Iron Cage Revisited: Institutional Isomorphism and Collective Rationality in Organizational Fields." *American Sociological Review* 48 (2): 147–60.

———. 1991. "Introduction." In *The New Institutionalism in Organizational Analysis*, edited by Walter W. Powell and Paul J. DiMaggio, 1–38. Chicago: University of Chicago Press.

Dixon, Charles. 1888. *Our Rarer Birds: Being Studies in Ornithology and Oology. With Twenty Illustrations by Charles Whymper and a Frontispiece by J. G. Keulemans*. London: Richard Bentley & Son.

———. 1898. *Lost and Vanishing Birds: Being a Record of some Remarkable Extinct Species and a Plea for some Threatened Forms. With Ten Plates by Charles Whymper*. London: John Macqueen.

Dobinson, Humphrey M. 1976. *Bird Count: A Practical Guide to Bird Surveys. Illustrated by Roy Wiltshire and Robert Micklewright*. Harmondsworth: Kestrel Books.

Donald, Paul F., Fiona J. Sanderson, Ian J. Burfield, Stijn M. Bierman, Richard D. Gregory, and Zoltan Waliczky. 2007. "International Conservation Policy Delivers Benefits for Birds in Europe." *Science* 317 (August 10): 810–13.

Dougall, Robert, and Herbert Axell. 1982. *Birdwarch Round Britain: A Personal Selection of Britain's Bird Reserves*. London: Collins and Harvill.

Doughty, Robin W. 1975. *Feather Fashions and Bird Preservation: A Study in Nature Protection*. Berkeley: University of California Press.

Dröschmeister, Rainer. 2000. "Conceptual Requirements on Monitoring for Nature Conservation at a European Level." In *European Monitoring for Nature Conservation: Proceedings of the International Symposium "Monitoring for Nature Conservation at a European Level,"* edited by Corinna Bischoff and Rainer Dröschmeister, 7–19. Bonn-Bad Godesberg: Bundesamt für Naturschutz.

Droste-Hülshoff, Ferdinand. 1872. *Die Vogelschutzfrage: Ein Referat*. Münster: Brunn.

Dudley, Steve P., Mike Gee, Chris Kehoe, Tim M. Melling, and The British Ornithologists' Union Records Committee. 2006. "The British List: a Checklist of Birds of Britain (7th edition)." *Ibis* 148 (3): 526–63.

Dugmore, A. Radclyffe. 1902. *Nature and the Camera. How to Photograph Live Birds and their Nests; Animals, Wild and Tame; Reptiles; Insects; Fish and other Aquatic Forms; Flowers, Trees, and Fungi. Illustrated from Photographes by the Author*. London: Doubleday, Page & Company.

Eaton, Mark A., Andy F. Brown, David G. Noble, Andy J. Musgrove, Richard D. Hearn, Nicholas J. Aebischer, David W. Gibbons, Andy Evans, and Richard D. Gregory. 2009. "Birds of Conservation Concern 3: The Population Status of Birds in the United Kingdom, Channel Islands and Isle of Man." *British Birds* 102:296–341.

Ecke, Herbert. 1964. "Hermann Hähnle 85 Jahre: Festsprache am 5. Juni 1964 in Stuttgart." *Internationaler Rat für Vogelschutz Deutsche Sektion, Bericht* 4:24–27.

Edwards, Richard. 1979. *Contested Terrain: The Transformation of the Workplace in the Twentieth Century*. New York: Basic Books.

Effenberger, Walter. 1926. "Die Photographie im Dienste des Naturschutzes." In *Wege Zum Naturschutz*, edited by Walther Schoenichen, 158–78. Breslau: Ferdinand Hirt.

Elsner, John, and Roger Cardinal, eds. 1994. *The Cultures of Collecting*. Cambridge, MA: Harvard University Press.

Engels, Jens Ivo. 2006. "Von der Heimat-Connection zur Fraktion der Ökopolemiker." In *Nützliche Netzwerke und korrupte Seilschaften*, edited by Arne Karsten and Hillard von Thiessen, 18–45. Göttingen: Vandenhoeck & Ruprecht.

Engling, Silke. 2003. "Der Kranich." In *Das große Buch des WWF. 40 Jahre Naturschutz für und mit den Menschen*, edited by Klaus-Henning Groth, 72 . Steinfurt: Tecklenborg Verlag

Ennion, Eric A. R. 1963. *Birdwatching*. London: Michael Joseph.

Ertel, Rainer. 1976. "Füttert die hungernden Vögel." *Wir und die Vögel* 8:14–15.

———. 1981. "Jahreshauptversammlung: Erweiterter Aufgabenbereich nun in der Satzung. DBV hat sich vom 'Vogelschutzclub' zum Naturschutzverband gemausert." *Wir und die Vögel* 13 (5): 34–35.

Etzioni, Amitai. 1988. *The Moral Dimension: Toward a New Economics*. New York: Free Press.

European Commission. 2004. *LIFE for Birds. 25 Years from the Birds Directive: The Contribution of LIFE Nature Projects*. Luxembourg: Office for Official Publications of the European Communities.

———. 2017. "The LIFE Programme." Accessed March 20. http://ec.europa.eu/environment/life/about/index.htm#evaluation.

European Environment Agency. 2014. "Data and Maps." Accessed June 15. www.eea.europa.eu/data-and-maps.

FACE. 2012. *The FACE Annual Report 2012*. Brussels: FACE.

Farber, Paul Lawrence. 1976. "The Type-Concept in Zoology during the First Half of the Nineteenth Century." *Journal of the History of Biology* 9:93–119.

———. 1977. "The Development of Taxidermy and the History of Ornithology." *Isis* 68: 550–66.

———. 1982. *The Emergence of Ornithology as a Scientific Discipline, 1760–1850*. Boston: Kluwer.

"The Farmer and the Rook." 1916. *Bird Notes and News* 7 (2): 13–14.

Ferguson-Lees, James, Quentin Hockliffe, and Ko Zweeres, eds. 1975. *A Guide to Bird-Watching in Europe*. London: Bodley Head.

Ferguson-Lees, James, and Ian Willis. 1987. *Vögel Mitteleuropas. Übersetzung und Bearbeitung der deutschen Ausgabe: Einhard Bezzel. Konzeption der deutschen Ausgabe: Josef Reichholf*. München: BLV.

Findlen, Paula. 1994. *Possessing Nature: Museums, Collecting, and Scientific Culture in Early Modern Italy*. Berkeley: University of California Press.

Fine, Gary Alan, ed. 1995. *A Second Chicago School? The Development of a Postwar American Sociology*. Chicago: University of Chicago Press.

"The First Fifty Years." 1957. *British Birds* 50 (6): 213–23.

Fischer, Wolfgang. 1967. *Der Wanderfalk*. Wittenberg: Ziemsen Verlag.

Fisher, James. (1939) 1954. *Birds as Animals*. London: Hutchinson's University Library.

———. (1940) 1953. *Watching Birds*. Harmondsworth: Penguin Press.

———. 1942. *The Birds of Britain*. William Collins: London.

———. 1947–1955. *Bird Recognition*. 3 vols. Harmondsworth: Penguin Books.

———. 1953. "The Collared Turtle Dove in Europe." *British Birds* 46:153–81.

———. 1966. *The Shell Bird Book*. London: Ebury Press and Michael Joseph.

Fisher, James, Noel Simon, and Jack Vincent. 1969. *The Red Book: Wildlife in Danger*. Foreword by Harold J. Coolidge and Peter Scott. London: Collins.

Fitter, Richard S. R. (1954) 1968. *Collins Pocket Guide to Nests and Eggs.* Illustrated by Richard A. Richardson. London: Collins.

———. 1963. *Collins Guide to Bird-watching. With 49 Line Drawings by R. A. Richardson and 112 Photographs by Eric Hosking, John Markham and others.* London: Collins.

———. 1968. *Vanishing Wild Animals of the World.* London: Midland Bank.

———. 1974. "25 Years On: A Look at Endangered Species." *Oryx* 12 (3): 341–46.

Fitter, Richard S. R., and Maisie Fitter, eds. 1987. *The Road to Extinction: Problems of Categorizing the Status of Taxa Threatened with Extinction.* Gland: International Union for the Conservation of Nature and Natural Resources.

Fitter, Richard S. R., and Richard A. Richardson. 1952. *Collins Pocket Guide to British Birds.* Foreword by Peter Scott. London: Collins.

Flade, Martin. 1994. *Die Brutvogelgemeinschaften Mittel- und Norddeutschlands: Grundlagen für den Gebrauch vogelkundlicher Daten in der Landschaftsplanung.* Eching: IHW-Verlag.

———. 1998. "Neue Prioritäten für den Vogelschutz: Kleiber oder Wiedehopf?" *Falke* 45:348–55.

———. 2006. "Literaturbesprechung: Peter Berthold und Gabriele Mohr: Vögel füttern - aber richtig." *Die Vogelwelt* 127:264.

Flade, Martin, Harald Plachter, Eberhard Henne, and Kenneth Anders, eds. 2003. *Naturschutz in der Agrarlandschaft: Ergebnisse des Schorfheide-Chorin-Projektes.* Wiebelsheim: Quelle & Meyer Verlag.

Flade, Martin, Harald Plachter, Rolf Schmidt, and Armin Werner, eds. 2006. *Nature Conservation in Agricultural Ecosystems: Results of the Schorfheide-Chorin Research Project.* Wiebelsheim: Quelle & Meyer Verlag.

Flasbarth, Jochen. 1986. "Winterfütterung." *Naturschutz Heute* 18:10.

———. 2001. "Die Bedeutung der Ornithologie in der Naturschutzarbeit." *Journal für Ornithologie* 142 Sonderheft 1:172–81.

Floericke, Kurt. 1892. *Versuch einer Avifauna von Preussisch-Schlesien.* Marburg: Univiversitäts-Buchdruckerei.

———. 1907. *Deutsches Vogelbuch für Forst- und Landwirte, Jäger, Naturfreunde und Vogelliebhaber, Lehrer und die reifere Jugend und für alle Gebildeten des deutschen Volkes gemeinverständlich geschildert.* Stuttgart: Franckh.

———. 1912. *Taschenbuch zum Vogelbestimmen: Praktische Anleitung zur Bestimmung unserer Vögel in freier Natur nach Stimme, Flug, Bewegungen usw. nebst Tabellen zur Bestimmung toter Vögel, der Nester und Eier.* Stuttgart: Franckh.

Föhr, Gerhard. 2005. *Nistkästen und Vogelschutz: Im Wandel der Zeit.* Hohenwarsleben: Westarp Wissenschaften.

Forster, Johann Reinhold. 1771. *A Catalogue of the Animals of North America, to which added short Directions for Collecting, Preserving and Transporting all kinds of Natural History Curiosities.* London: B. White.

Foucault, Michel. (1975) 1979. *Discipline and Punish: The Birth of the Prison.* New York: Vintage Books.

Fourcade, Marion. 2009a. "Cents and Sensibility: Economic Valuation and the Nature of 'Nature.'" *American Journal of Sociology* 116 (6): 1721–77.

———. 2009b. *Economists and Societies: Discipline and Profession in the United States, Britain, and France, 1890s to 1990s*. Princeton, NJ: Princeton University Press.

Fourcade, Marion, and Kieran Healy. 2007. "Moral Views of Market Society." *Annual Review of Sociology* 33:285–311.

Franke, Ulrich. 2009. *Dr. Kurt Floericke: Naturforscher, Ornithologe, Schriftsteller. Mit einer ersten umfassenden Bibliographie seiner Schriften*. Norderstedt: Books on Demand.

Friderich, Carl G. 1849. *Naturgeschichte aller deutschen Zimmer-, Haus- und Jagdvögel, nebst einem Anhange über die ausländischen Vögel, welche in Deutschland im Handel vorkommen*. Stuttgart: Hoffmann.

———. 1863. *Vollständige Naturgeschichte der deutschen Zimmer-, Haus- und Jagdvögel sammt allen übrigen in Deutschland vorkommenden Vögeln. 2. sehr vermehrte und verbesserte Auflage*. Stuttgart: Thienemann.

———. 1876. *Vollständige Naturgeschichte der deutschen Zimmer-, Haus- und Jagdvögel, nebst einem Anhange über die ausländischen Vögel, welche in Deutschland im Handel vorkommen. 3. sehr vermehrte und verbesserte Auflage*. Stuttgart: Hoffmann.

———. 1891. *Naturgeschichte der Deutschen Vögel einschliesslich der sämtlichen Vogelarten Mittel-Europas: Bearbeitet nach vielen eigenen Erfahrungen und den besten Quellen für Liebhaber der Zimmer- und Hausvögel, für Ornithologen und Eiersammler, für Jäger und Jagdfreunde, sowie zur Belehrung der reifern Jugend. 4. sehr vermehrte und verbesserte Auflage*. Stuttgart: Hoffmann.

Frieling, Heinrich. 1950. *Was fliegt denn da? Tabellen zum Bestimmen von 396 Vogelarten Mitteleuropas und der angrenzenden Länder*. Stuttgart: Franckh'sche Verlagshandlung.

Friends of the Earth International. 2011. *Annual Report 2010*. Amsterdam: Friends of the Earth International.

Frost, Christopher. 1987. *A History of British Taxidermy*. Sudbury: C. Frost.

Fuller, Errol. 2001. *Extinct Birds*. Ithaca, NY: Comstock.

"The Future of the 'British Birds' Ringing Scheme: Transfer to the British Trust for Ornithology." 1937. *British Birds* 31 (1): 5–6.

Gadamer, Hans-Georg. (1960) 1989. *Truth and Method*. 2nd rev. ed. London: Sheed & Ward.

Gardiner, Linder. 1900. *Birds and Boys*. RSPB Leaflet Series 34.

———. 1905. *Our Ally the Bird*. RSPB Leaflet Series 54.

Garling, Max. 1940. "Beiträge zur Vogelfauna der Berliner Rieselfelder." *Märkische Tierwelt* 4:141–62.

Gates, Barbara T. 1998. *Kindred Nature: Victorian and Edwardian Women Embrace the Living World*. Chicago: University of Chicago Press.

Gätke, Heinrich. 1891. *Die Vogelwarte Helgoland. Herausgegeben von Rudolf Blasius*. Braunschweig: J. H. Meyer.

———. 1895. *Heligoland as an Ornithological Observatory: The Result of Fifty Years' Experience*. Translated by Rudolph Rosenstock. Edinburgh: Douglas.

Gedeon, Kai, Stefan Fischer, and Christoph Sudfeldt. 2003. "Steckbyer Grundsätze und Ziele zum Vogelmonitoring in Deutschland." In *Vogelmonitoring in Deutschland*, edited by Landesamt für Umweltschutz Sachsen-Anhalt, 147–49. Halle: Landesamt für Umweltschutz Sachsen-Anhalt

Geertz, Clifford. 1973. "Deep Play: Notes on the Balinese Cockfight." In *The Interpretation of Cultures: Selected Essays*, 412–53. New York: Basic Books.

Gernsheim, Helmut, and Alison Gernsheim. 1969. *The History of Photography: From the Camera Obscura to the Beginning of the Modern Era*. London: Thames and Hudson.

Gibbons, David W. 2000. "Pan-European Breeding Bird Monitoring." In *European Monitoring for Nature Conservation: Proceedings of the International Symposium "Monitoring for Nature Conservation at a European Level,"* edited by Corinna Bischoff and Rainer Dröschmeister, 173–78. Bonn-Bad Godesberg: Bundesamt für Naturschutz.

Gibbons, D. W., M. I. Avery, S. R. Baillie, R. D. Gregory, J. Kirby, R. F. Porter, G. M. Tucker, and G. Williams. 1996. "Bird Species of Conservation Concern in the United Kingdom, Channel Islands and Isle of Man: revising the Red Data List." *RSPB Conservation Review* 10:7–18.

Giddens, Anthony. 1990. *The Consequences of Modernity*. Stanford, CA: Stanford University Press.

Gizncki, Georg von. 1876. "Über die moralische Berechtigung der Vogelliebhaberei." *Monatsschrift des Sächisch-Thüringischer Verein für Vogelkunde und Vogelschutz* 1:190–94.

Glasewald, Konrad. 1933. "Die Vogelwelt eines fiskalischen Kiefernforstes der Mark Brandenburg." *Beiträge zur Naturdenkmalpflege* 11:135–57.

———. 1937. *Vogelschutz und Vogelhege*. Neudamm: Neumann.

Glaubrecht, Matthias, and Jürgen Haffer. 2010. "Classifying Nature: Constantin W. L. Gloger's (1803–1863) Tapestry of a 'Natural System of the Animal Kingdom.'" *Zoosystematics and Evolution* 86:81–115.

Gloger, Constantin W. L. 1833. *Schlesiens Wirbelthier-Fauna: Ein Systematischer Überblick der in dieser Provinz vorkommenden Säugethiere, Vögel, Amphibien und Fische; mit Rücksicht auf den allgemeinen Charakter des Landes, sowie auf das locale und quantitative Vorkommen seiner Thiere*. Breslau: Grass.

———. 1834. *Vollständiges Handbuch der Naturgeschichte der Vögel Europa's, mit besonderer Rücksicht auf Deutschland. Erster Theil: Die deutschen Landvögel enthaltend*. Breslau: August Schulz.

———. 1842. *Gemeinnütziges Hand- und Hilfsbuch der Naturgeschichte: Für gebildete Leser aller Stände, besonders für die reifere Jugend und ihre Lehrer. Erster Band: Enthaltend die erste Hälfte der Naturgeschichte der Thiere*. Breslau: August Schulz.

———. 1847. *Der Walfisch Fang und seine Beförderung in Deutschland*. Berlin: Hirschwald.

———. 1850. *Eine freundliche Mahnung zur Vermehrung der Bienenzucht in den meisten Gegenden Deutschlands überhaupt und Preußens ins Besondere*. Berlin: Verlag der Deckerschen Geheimen Ober-Hofbuchdruckerei.

———. 1853. "Zu der Frage über den Land-, Forst- und Volkswirthschaftlich so Wünschenswerthen Schutz der Höhlenbrüter." *Journal für Ornithologie* 1:110–29.

———. 1855a. "Nachträgliches über Nist-Vorrichtungen für Höhlenbrüter." *Journal für Ornithologie* 3:80–87.

———. 1855b. *Wohlfeileres Fleisch für die Armen, mehr und besseres für Alle, und als Mittel hierzu die Einführung der englischen Fleisch-Verkaufs-Weise auch in Deutschland*. Berlin: Gebauer.

———. 1858a. *Die nützlichsten Freunde der Land- und Forstwirthschaft unter den Thieren, als die von der Natur bestellten Verhüter und Bekämpfer von Ungezieferschäden und Mäusefraß: zur Belehrung für Landleute und Land-Schullehrer*. Berlin: Allgemeine Deutsche Verlags-Anstalt.

———. 1858b. *Kleine Ermahnung zum Schutze nützlicher Tiere, als naturgemässer Abwehr von Ungeziefer und Mäusefrass*. Berlin: Allgemeine Deutsche Verlags-Anstalt.

———. 1865. *Die Hegung der Höhlenbrüter mit besonderer Rücksicht auf die Nachteile des Vogelfanges für die Land- und Forstwirtschaft*. Berlin: Allgemeine Verlags-Anstalt.

Gmelin, Johann Friedrich. 1788–1789. *Caroli a Linné [. . .] Systema Naturae per Regna Tria Naturae, Secondum Classes, Ordines, Genera, Species, Cum Characteribus, Differentiis, Synonymis, Locis. Editio Decima Tertia, Aucta Reformata. Vol. 1(1 and 2)*. Leipzig: Georg Emanuel Beer.

Gooders, John. 1975. *How to Watch Birds*. London: André Deutsch.

Götz, Wilhelm H. J., and Alois Kosch. 1936. *Was fliegt denn da?: Tabelle zur Bestimmung der Vögel Mitteleuropas*. Stuttgart: Franckh'sche Verlagshandlung.

Great Crane Project. 2017. "The Great Crane Project." Accessed March 20. http://www.thegreatcraneproject.org.uk/project.

Greene, W. T. 1891. *Favourite Foreign Birds for Cages and Aviaries*. London: L. Upcott Gill.

Greenway, James C. 1958. *Extinct and Vanishing Birds of the World*. New York: American Committee for International Wild Life Protection.

———. 1967. *Extinct and Vanishing Birds of the World*. 2nd rev. ed. New York: Dover Publication.

Greenwood, Jeremy. 1999. "Why Conduct Bird Census and Atlas Work in Europe?" *Vogelwelt* 120 (Supplement): 11–23.

———. 2000. "How BTO's Monitoring of Birds Contributes to Conservation." In *European Monitoring for Nature Conservation: Proceedings of the International Symposium "Monitoring for Nature Conservation at a European Level,"* edited by Corinna Bischoff and Rainer Dröschmeister, 105–17. Bonn-Bad Godesberg: Bundesamt für Naturschutz.

———. 2005. "Science with a Team of Thousands: The British Trust for Ornithology." In *Participating in the Knowledge Society: Researchers Beyond the University Walls*, edited by Ruth Finnegan, 152–65. Houndmills: Palgrave Macmillan.

———. 2007. "Citizens, Science and Bird Conservation." *Journal of Ornithology* 148 (Supplement 1): 77–124.

Greenwood, Jeremy, and Nick Carter. 2003. "Organizing a National Bird Monitoring by the BTO-Experiences from Britain." *Berichte des Landesamtes für Umweltschutz Sachsen-Anhalt. Sonderheft* 1:14–26.

Greenpeace International. 2011. *Annual Report 2010*. Amsterdam: Greenpeace International.

Gregory, Richard D., Jeremy Greenwood, and E. J. M. Hagemeijer. 1998. "The EBCC Atlas of European Breeding Birds: A Contribution to Science and Conservation." *Biologia e Conservazione della Fauna* 102:38–49.

Gregory, Richard D., Nicholas I. Wilkinson, David G. Noble, James A. Robinson, Andrew F. Brown, Julian Hughes, Deborah Procter, David W. Gibbons, and Colin A. Galbraith. 2002. "The Population Status of Birds in the United Kingdom, Channel Islands and Isle of Man: An Analysis of Conservation Concern 2002–2007." *British Birds* 95:410–48.

Gregory, Richard D., and Petr Vorisek. 2003. "Report on the Pan-European Common Bird Monitoring Workshop." *Bird Census News: Newsletter of the European Bird Census Council* 16 (1): 4–40.

Grimmett, R. F. A., and T. A. Jones. 1989. *Important Bird Areas in Europe.* ICBP Technical Publication No. 9. Cambridge: International Council for Bird Preservation.

Groth, Klaus-Henning, ed. 2003. *Das große Buch des WWF. 40 Jahre Naturschutz für und mit den Menschen.* Steinfurt: Tecklenborg Verlag.

Gründungs-Ausschuss. 1867. "Statut der 'deutschen ornithologischen Gesellschaft zu Berlin.'" *Journal für Ornithologie* 15 (6): 423–26.

Guenther, Konrad. 1907. *Rückkehr zur Natur? Eine Betrachtung über das Verhältnis des Menschen zur Natur und die Entwicklung desselben.* Leipzig: J. A. Barth.

———. 1910. *Der Naturschutz.* Freiburg i.Br.: Fehsenfeld.

Guggisberg, C. A. W. 1977. *Early Wildlife Photographers.* Foreword by Eric Hosking. Newton Abbot: David & Charles.

HRH The Prince Philip, Duke of Edinburgh, and James Fisher. 1970. *Wildlife Crisis.* Foreword by HRH The Prince of the Netherlands and Peter Scott. Published with the co-operation of the World Wildlife Fund. London: Hamish Hamilton.

Habermas, Jürgen. 1981. "New Social Movements." *Telos* 49:33–37.

———. 1984–1987. *The Theory of Communicative Action.* 2 vols. Boston: Beacon Press.

Haemmerlein, Hans-Dieter. 1985. *Der Sohn des Vogelpastors: Szenen, Bilder, Dokumente aus dem Leben von Alfred Edmund Brehm.* Berlin: Evangelische Verlagsanstalt.

Haffer, Jürgen. 1994. "Die Seebohm-Hartert-'Schule' der europäischen Ornithologie." *Journal für Ornithologie* 135:37–54.

———. 1996. "Vogelarten im Werk von Christian Ludwig Brehm (1787–1864)." *Anzeiger des Vereins Thüringer Ornithologen* 3:1–31.

———. 2001a. "Die 'Stresemann Revolution' in der Ornithologie des frühen 20. Jahrhunderts." *Journal für Ornithologie* 142:381–89.

———. 2001b. "Ornithological Research Traditions in Central Europe during the 19th and 20th Centuries." *Journal für Ornithologie* 142 (Sonderheft 1): 27–93.

———. 2006. "Altmeister der Feld-Ornithologie in Deutschland." *Blätter aus dem Naumann-Museum* 25:1–55.

———. 2007. "The Development of Ornithology in Central Europe." *Journal of Ornithology* 148 (Supplement 1): 125–53.

Haffer, Jürgen, and H. Hude. 2007. "Naturphilosophie im Leben des Ornithologen Constantin Gloger (1803–1863)." *Blätter aus dem Naumann-Museum* 26:11–29.

Haffer, Jürgen, Erich Rutschke, and Klaus Wunderlich. 2004. *Erwin Stresemann (1889–1972): Leben und Werk eines Pioniers der wissenschaftlichen Ornithologie. 2. korrigierte Auflage.* Stuttgart: Wissenschaftliche Verlagsgesellschaft.

Hagemeijer, Ward J. M., and Michael J. Blair. 1997. *The EBCC Atlas of European Breeding Birds: Their Distribution and Abundance.* London: T. & A. D. Poyser.

Hähnle, Hermann. 1936. "Das Schutzgebiet Behr-Steckby (Anhalt) des Reichsbundes für Vogelschutz." *Veröffentlichungen der Württembergischen Landesstelle für Naturschutz,* 167–83.

———. 1955. “Das Raumbild in der Tierphotographie.” In *Die Photographie freilebender Tiere. 2. verbesserte und erweiterte Auflage*, edited by Fritz Steiniger, 172–82. Baden-Baden: Verlag für angewandte Wissenschaften.

Hammond, Nicholas. 1978. *The Young Birdwatcher*. London: Hamlyn.

Hanemann, Horst, and Jürgen M. Simon. 1987. *Bund für Vogelschutz: Die Chronik eines Naturschutzverbandes von 1899–1984*. Wiesbaden: Wirtschaftsverlag.

Harper, Francis. 1945. *Extinct and Vanishing Mammals of the Old World*. New York: American Committee for International Wild Life Protection.

Hartert, Ernst, F. C. R. Jourdain, N. F. Ticehurst, and H. F. Witherby. 1912. *A Hand-List of British Birds, with an Account of the Distribution of Each Species in the British Isles and Abroad*. London: Witherby.

Hartlaub, Carl Johann Gustav. 1894. “Über die Gätke’sche Vogelsammlung.” *Ornithologische Monatsberichte* 2:143–45.

Hayden, Sherman Strong. 1942. *The International Protection of Wild Life: An Examination of the Treaties and Other Agreements for the Preservation of Birds and Mammals*. New York: Columbia University Press.

Hays, Samuel P. 1959. *Conservation and the Gospel of Efficiency: The Progressive Conservation Movement, 1890–1920*. Cambridge, MA: Harvard University Press.

Healy, Kieran. 2000. “Embedded Altruism: Blood Collection Regimes and the European Union’s Donor Population.” *American Journal of Sociology* 105 (6): 1633–57.

———. 2004. “Altruism as an Organizational Problem: The Case of Organ Procurement.” *American Sociological Review* 69:387–404.

———. 2006. *Last Best Gifts: Altruism and the Market for Human Blood and Organs*. Chicago: University of Chicago Press.

Heath, Melanie F., and Michael I. Evans. 2000. *Important Bird Areas in Europe: Priority Sites for Conservation*. 2 vols. Cambridge: BirdLife International.

Heger, Wolfgang. 1993. “Lina Hähnle (1851–1941)—Nicht nur Vogelmutter.” In *Frauen im deutschen Südwesten*, edited by Birgit Knorr and Rosemarie Wehling, 170–75. Stuttgart: Kohlhammer.

Heinzel, Hermann, Richard Fitter, and John Parslow. 1972a. *Pareys Vogelbuch: Alle Vögel Europas, Nordafrikas und des Mittleren Ostens. Übersetzt und bearbeitet von Günther Niethammer und H. E. Wolters*. Hamburg: Parey.

———. 1972b. *The Birds of Britain and Europe with North Africa and the Middle East*. London: Collins.

Helfer, Hermann. 1941. “Lina Hähnle: Der Gründerin des Reichsbundes für Vogelschutz zu ihrem 90. Geburtstage.” *Deutsche Vogelwelt* 66:2–6.

———. 1954. “Hermann Hähnle: Zum 75. Geburtstage am 5. Juni 1954.” *Ornithologische Mitteilungen: Hermann-Hähnle-Festschrift. Herausgegeben von Dr. Herbert Bruns* 6:101–03.

Hennicke, Carl R., ed. 1893. *Hofrat Professor Dr. K. Th. Liebes Ornithologische Schriften. Hauptband*. Leipzig: Malende.

———, ed. 1895. *Hofrat Professor Dr. K. Th. Liebes Ornithologische Schriften. Ergänzungsband*. Gera: Köhler.

———. 1912. *Handbuch des Vogelschutzes*. Magdeburg: Creutzsche Verlagsbuchhandlung.

———. 1917a. "Etwas über die hygienische und wirtschaftliche Bedeutung der Vögel." *Journal für Ornithologie* 65:96–111.

———. 1917b. *Schwindende Vogelarten in Deutschland*. Berlin: Bornträger.

Henze, Otto. 1936. *Kontrollbuch für Vogelnistkästen in der Land- und Forstwirtschaft*. Berlin: Reichsnährstand.

———. 1943. *Vogelschutz gegen Insektenschaden in der Forstwirtschaft*. München: Bruckmann.

———. 1949. *Kontrollbuch für Vogelnistkästen in der Forstwirtschaft. 2. Auflage*. Garmisch-Partenkirchen: Staatliche Vogelschutzwarte.

———. 1958. *Kontrollbuch für Vogelnistkästen und Nisthöhlen in der Forstwirtschaft. 3. Auflage*. Garmisch-Partenkirchen: Selbstverlag.

———. 1983. *Kontrollbuch für Vogelnistkästen in Wald und Garten. 4. Auflage*. Überlingen am Bodensee.

———. 1991. *Die richtigen Vogelnistkästen in Wald und Garten: Über die Lebensweise und Bedeutung aller ihrer Bewohner. Ein Sach- und Kontrollbuch. 5. völlig überarbeitete Auflage*. Konstanz: Südkurier.

Heredia, Borja, Laurence Rose, and Mary Painter, eds. 1996. *Globally Threatened Birds in Europe: Action Plans*. Strasbourg: Council of Europe Publishing.

Hey, Christian, and Uwe Brendle. 1994. *Umweltverbände und EG. Strategien, politische Kulturen und Organisationsformen*. Opladen: Westdeutscher Verlag.

Heyd, Alexander. 2006. "Das Jahr der Netze." *Artenschutzbrief* 10:6–9.

Hickling, Ronald. 1983. *Enjoying Ornithology: A Celebration of Fifty Years of the British Trust for Ornithology, 1933–1983*. Calton: T. & A. D. Poyser.

Hiesemann, Martin. 1907. *Lösung der Vogelschutzfrage nach Freiherrn von Berlepsch: Im Auftrage der "Kommission zur Förderung des Vogelschutzes."* Leipzig: Wagner.

———. 1911a. *How to Attract and Protect Wild Birds. A Full Description of Successful Methods*. Translated by Emma S. Buchheim, with an introduction by Her Grace The Duchess of Bedford. Second Edition with Many Revisions with Many Illustrations. London: Witherby & Co.

———. 1911b. *Lösung der Vogelschutzfrage nach Freiherrn von Berlepsch: Im Auftrage der "Kommission zur Förderung des Vogelschutzes." 4., ergänzte und verbesserte Auflage, bearbeitet durch Major z.D. Henrici*. Leipzig: Wagner.

"Hiesemann on the Protection of Wild Birds." 1911. *Ibis* 53:386–88.

Hildebrandt, Hugo. 1933. "Johann Matthaeus Bechstein: The German Ornithologist (1757–1822)." *Ibis* 75 (2): 219–28.

Hinze, Gustav. 1937. *Biber in Deutschland*. Berlin-Lichterfelde: Bermühler.

Hitlin, Steven, and Stephen Vaisey. 2013. "The New Sociology Morality." *Annual Review of Sociology* 39:51–68.

Holden, Peter, and Geoffrey Abbott. 2008. *RSPB Handbook of Garden Wildlife*. London: Christopher Helm.

Holden, Peter, and Tim Cleeves. 2002. *RSPB Handbook of British Birds*. London: Christopher Helm.

———. 2006. *RSPB Handbook of British Birds*. 2nd ed. London: Christopher Helm.

Holden, Peter, and John T. R. Sharrock. 2002. *The RSPB Guide to British Birds*. London: Macmillan.

Holdgate, Martin. 1999. *The Green Web: A Union for World Conservation*. London: Earthscan Publications.

"Homes for Birds." 1912. *Bird Notes and News* 5 (1): 14.

Hornaday, William T. 1913. *Our Vanishing Wild Life: Its Extermination and Preservation*. New York: New York Zoological Society.

Hosking, Eric, and Harold Lowes, eds. 1947. *Masterpieces of Bird Photography*. London: Collins.

Hudson, William H. 1891. *Osprey, or Egrets or Aigrettes*. RSPB Leaflet Series 3.

———. 1893. *Feathered Women*. RSPB Leaflet Series 10.

Huxley, Elspeth. 1993. *Peter Scott: Painter and Naturalist*. Introduction by Sir David Attenborough. London: Faber and Faber.

Huxley, Julian S. 1916. "Bird-Watching and Biological Science." *Auk* 33:142–61, 256–70.

———. 1930. *Bird-Watching and Bird Behaviour*. London: Chatto & Windus.

———. 1946a. "The New Bird-Watching: Evolution at Work." *The Countryman* 33 (1): 68–75.

———. 1946b. "What We Are Learning from Birds." *The Countryman* 33 (2): 241–44.

———. 1950. "Birds and Science." *Country Life* 108 (August 4): 364–67.

———. 1970–1973. *Memories*. 2 vols. New York: Harper & Row.

Inglehart, Ronald. 1977. *The Silent Revolution: Changing Values and Political Styles*. Princeton, NJ: Princeton University Press.

———. 1990. "Values, Ideology, and Cognitive Mobilization in New Social Movements." In *Challenging the Political Order*, edited by Russell J. Dalton and Manfred Kuechler, 43–66. Cambridge: Polity Press.

———. 1997. *Modernization and Postmodernization: Cultural, Economic, and Political Change in 43 Societies*. Princeton, NJ: Princeton University Press.

IUCN 2010. "IUCN Red List of Threatened Species. Version 2010.4." Accessed October 23. http://www. iucnredlist.org.

Jahn, Ilse. 1985. "Zur Vertretung der Zoologie und zur Entwicklung ihrer institutionellen Grundlage an der Berliner Universität von ihrer Gründung bis 1920." *Wissenschaftliche Zeitschrift der Humboldt-Universität zu Berlin, Mathematisch-Naturwissenschaftliche Reihe* 34:260–80.

Jarvis, Charlie, Christine Leon, and Sara Oldfield. 1981. "Bibliography of Red Data Books and Threatened Plant Lists." In *The Biological Aspects of Rare Plant Conservation*, edited by Hugh Synge, 513–40. Chichester: John Wiley & Sons.

Jasanoff, Sheila, ed. 2004. *States of Knowledge: The Co-Production of Science and the Social Order*. London: Routledge.

Jaskolla, Dieter. 2005. "Gründung der 'Kaiserlich Biologischen Anstalt für Land- und Fortswirtschaft' als selbstständige Reichbehörde vor 100 Jahren am Wissenschaftsstandort Berlin-Dahlem." *Nachrichtenblatt des Deutschen Pflanzenschutzdienstes* 57 (5): 110–14.

Joas, Hans. 1993. "Pragmatism in American Sociology." *Pragmatism and Social Theory*, 14–51. Chicago: University of Chicago Press.

———. 1996. *The Creativity of Action*. Chicago: University of Chicago Press.

———. 2000. *The Genesis of Values*. Chicago: University of Chicago Press.

Johler, Reinhard. 1997. "Vogelmord und Vogelliebe: Zur Ethnographie konträrer Leidenschaften." *Historische Anthropologie* 5:1–35.

Kear, Janet. 1990. *Man and Wildfowl*. London: T. & A. D. Poyser.

Kearton, Richard. 1895. *British Birds' Nests: How, Where, and When to Find and Identify Them. Illustrated from photographs by Cherry Kearton*. London: Cassell.

———. 1898. *Wild Life at Home: How to Study and Photograph It; Fully Illustrated by Photographs Taken Direct from Nature by C. Kearton*. London: Cassell.

———. 1926. *A Naturalist's Pilgrimage*. With eight plates. London: Cassell.

Kennedy, Alexander W. M. Clark. 1868. *Birds of Berkshire and Buckinghamshire: A Contribution to the Natural History of the Two Counties*. London: Simpkin, Marshall, and Company.

King, Warren B. 1981. *Endangered Birds of the World: The ICBP Bird Red Data Book*. Washington, DC: Smithsonian Institution Press.

Kinkela, David. 2011. *DDT and the American Century: Global Health, Environmental Politics, and the Pesticide that Changed the World*. Chapel Hill: University of North Carolina Press.

Knorr-Cetina, Karin, and Urs Bruegger. 2002. "Global Microstructures: The Virtual Societies of Financial Markets." *American Journal of Sociology* 107:905–50.

Knorr-Cetina, Karin, and Barbara Grimpe. 2008. "Global Financial Technologies: Scoping Systems that Raise the World." In *Living in a Material World*, edited by Trevor Pinch and Richard Swedberg, 161–89. Cambridge, MA: MIT Press.

König, Claus. 1966. *Europäische Vögel. Band 1: Ziegenmelker, Segler, Racken, Spechte, Sperlingsvögel*. Stuttgart: Belser.

———. 1967. *Europäische Vögel. Band 2: Sumpf- und Wasservögel, Greifvögel, Hühnervögel, Kraniche, Tauben, Kuckucksvögel, Eulen*. Stuttgart: Belser.

König, Klaus, and Friedrich Schilling. 1970. "Beeinflussen Pestizide die Populationsentwicklung des Wanderfalken (*Falco peregrinus*) in Baden-Württemberg?" *Die Vogelwelt* 91:170–76.

Köstering, Susanne. 2003. *Natur zum Anfassen: Das Naturkundemuseum des deutschen Kaiserreichs 1871–1914*. Köln: Böhlau Verlag.

Krause, Monika. 2014. *The Good Project: Humanitarian Relief NGOs and the Fragmentation of Reason*. Chicago: University of Chicago Press.

Lack, David. 1937. "A Review of Bird Census Work and Bird Population Problems." *Ibis* 14 (1): 369–95.

Lainer-Vos, Dan. 2013. *Sinews of the Nation: Constructing Irish and Zionist Bonds in the United States*. Cambridge: Polity Press.

Lamont, Michèle. 1992. *Money, Morals, and Manners: The Culture of the French and American Upper-Middle Class*. Chicago: University of Chicago Press.

Lamont, Michèle, and Virág Molnár. 2002. "The Study of Boundaries in the Social Sciences." *Annual Review of Sociology* 28:167–95.

Lamont, Michèle, and Laurent Thévenot, eds. 2000. *Rethinking Comparative Cultural Sociology: Repertoires of Evaluation in France and the United States*. Cambridge: Cambridge University Press.

Lang, John Temple. 1982. "The European Community Directive on Bird Conservation." *Biological Conservation* 22:11–25.

Latham, John. 1781–1785. *A General Synopsis of Birds*. 3 vols. London: Leigh & Sotheby.

———. 1787. *Supplement to the General Synopsis of Birds*. London: Leigh & Sotheby.

———. 1790. *Index Ornithologicus, Sive Systema Ornithologicae; Complectens Avium Divisionem in Classes, Ordines, Genera, Species, Ipsarumque Varietates: Adjectis Synonymis, Locis, Descriptionibus, &c. 2 Volumes*. London: Leigh & Sotheby.

———. 1801. *Supplement II to the General Synopsis of Birds*. London: Leigh, Sotheby & Son.

———. 1821–1828. *A General History of Birds*. Winchester: Printed by Jacob and Johnson.

Latour, Bruno. 1987. *Science in Action: How to Follow Scientists and Engineers through Society*. Cambridge, MA: Harvard University Press.

———. 1992. "Where Are the Missing Masses? The Sociology of a Few Mundane Artifacts." In *Shaping Technology/Building Society: Studies in Sociotechnical Change*, edited by Wiebe E. Bijker and John Law, 225–58. Cambridge, MA: MIT Press.

———. 1996. "On Actor-Network Theory: A Few Clarifications." *Soziale Welt* 47 (4): 369–81.

———. 1999. "Circulating Reference: Sampling the Soil in the Amazon Forest." In *Pandora's Hope: Essays on the Reality of Science*, 24–79. Cambridge, MA: Harvard University Press.

———. 2002. "Morality and Technology: The End of the Means." *Theory, Culture & Society* 19:247–60.

———. 2005. *Reassembling the Social: An Introduction to Actor-Network-Theory*. Oxford: Oxford University Press.

Law, John. 1987. "Technology and Heterogeneous Engineering: The Case of the Portuguese Expansion." In *The Social Construction of Technological Systems*, edited by Thomas P. Hughes Wiebe Bijker, and Trevor Pinch, 111–34. Cambridge, MA: MIT Press.

———. 1992. "Notes on the Theory of Actor-Network: Ordering, Strategy, and Heterogeneity." *Systems Practice* 5:379–93.

Lawton, John H., and Robert M. May. 1995. *Extinction Rates*. Oxford: Oxford University Press.

Lear, Linda. 1997. *Rachel Carson: Witness for Nature*. New York: Henry Holt.

Lemon, Margarette L. 1943. "The Story of the R.S.P.B." *Bird Notes & News* 20:67–68, 84–87, 100–102, 16–18.

Leverian Museum. 1790. *A Companion to the Museum, (late Sir Ashton Lever's) removed to Albion Street, the Surry End of Black Friars Bridge*. London: Leverian Museum.

Lévi-Strauss, Claude. (1962) 1966. *The Savage Mind*. Chicago: University of Chicago Press.

———. (1962) 1963. *Totemism*. Boston: Beacon.

Liebe, Karl Theodor. 1876. "Über Züchtungserfolge mit Steinröteln (Monticola saxatilis L.)." *Monatsschrift des Sächisch-Thüringischer Verein für Vogelkunde und Vogelschutz* 1:7–8.

———. 1878. "Die Brutvögel Ostthüringens und ihr Bestand." *Journal für Ornithologie* 26:1–88.

———. 1894. "Lernt erst das Leben der Vögel genau kennen, wenn Ihr sie mit rechtem Erfolge schützen wollt." *Ornithologische Monatsschrift* 19:250–58.

Lockley, R. M. 1968. *The Book of Bird-watching*. London: Arthur Barker.

Luther, Dieter. 1986. *Die Ausgestorbenen Vögel der Welt. 3. überarbeitete Auflage*. Wittenberg Luthertstadt: A. Ziemsen Verlag.

Lytle, Mark Hamilton. 2007. *The Gentle Subversive: Rachel Carson, Silent Spring, and the Rise of the Environmental Movement*. New York: Oxford University Press.

MacGregor, Arthur. 2007. *Curiosity and Enlightenment: Collectors and Collections from the Sixteenth to the Nineteenth Century*. New Haven, CT: Yale University Press.

Maddox, John. 1972. *The Doomsday Syndrome*. London: Macmillan.

Mansfeld, Karl. 1934. *Vogelschutz im Garten*. Seebach: Vogelschutzwarte.

———. 1935. *Der Friedhof als Vogelschutzstätte*. Seebach: Vogelschutzwarte.

———. 1937a. *Bedeutung und Durchführung des Vogelschutzes*. Seebach: Vogelschutzwarte.

———. 1937b. *Selbstbau von Nistkästen und Futterstellen*. Seebach: Vogelschutzwarte.

———. 1938. "Die Deutschen Vogelschutzwarten und ihre Aufgaben." *Deutsche Vogelwelt* 63 (1): 18–22.

———. 1944. *Vogelschutz in Wald, Feld und Garten. 6. Auflage*. Seebach: Vogelschutzwarte.

MANU 2010. "Important Bird Areas." Accessed April 4. http://www.manu.pf.

Manza, Jeff, Michael Sauder, and Nathan Wright. 2010. "Producing Textbook Sociology." *European Journal of Sociology* 51 (2): 271–304.

Martins, Herminio. 1974. "Time and Theory in Sociology." In *Approaches to Sociology*, edited by J. Rex, 246–94. London: Routledge & Kegan Paul.

Marwinski, Felicitas. 2004. *Karl Theodor Liebe: Gymnasialprofessor, Geologe und Beobachter der heimischen Vogelwert*. Weimar: Hain Verlag.

Marx, Karl, and Friedrich Engels. (1845) 1998. *The German Ideology*. Amherst: Prometheus Books.

———. (1848) 1988. *The Communist Manifesto: Annotated Text*. New York: WW Norton.

Masefield, John R. B. 1897. *Wild Bird Protection and Nesting Boxes. With Illustrations of Various Designs of Boxes, Brackets, etc., That Have Actually Been Used by Wild Birds for Nidification, and a Full List of the Orders Made under the "Wild Birds Protection Acts" on the Application of County Councils, with the Names of the Species Protected*. Leeds: Taylor Brothers.

"Masefield on Wild-Bird Protection and Nesting-Boxes." 1897. *Ibis* 3:619–20.

Matthiessen, Peter. 1999. "Environmentalist Rachel Carson," *Time*, March 29. http://content.time.com/time/magazine/article/0,9171,990622,00.html.

May, Helge. 2003. "100 Jahre NABU—Ein historischer Abriss: 1899–1999." Unpublished manuscript. NABU, Bonn.

May, Robert M. 1988. "How Many Species Are There on Earth?" *Science* 241 (4872): 1441–49.

———. 1990. "How Many Species?" *Philosophical Transactions: Biological Sciences* 330 (1257): 293–303.

McCormick, John. 1989. *The Global Environmental Movement: Reclaiming Paradise*. London: Belhaven Press.

———. 2001. *Environmental Policy in the European Union*. Houndmills: Palgrave.

McTaggart, David Fraser, and Robert Hunter. 1978. *Greenpeace III: Journey into the Bomb*. London: Collins.

Mearns, Barbara, and Richard Mearns. 1998. *The Bird Collectors*. San Diego, CA: Academic Press.

Mebs, Theodor. 1966. "Die Bestandsverhältnisse des Wanderfalken (*Falco peregrinus*) in Westdeutschland." *Natur und Landschaft* 41:162–66.

Medway, David G. 1979. "Some Ornithological Results of Cook's Third Voyage." *Journal of the Society for the Bibliography of Natural History* 9:315–51.

Meerwarth, Hermann. 1905. *Photographische Naturstudien: Eine Anleitung für Amateure und Naturfreunde*. Eßlingen: Schreiber.

Meerwarth, Hermann, and Karl Soffel, eds. 1908–1912. *Lebensbilder aus der Tierwelt. 6 Bände*. Leipzig: Voigtländer.

Meise, Wilhelm, ed. 1957. *Fünfzig Jahre Seevogelschutz. Festschrift des Vereins Jordsand zur Begründung von Vogelfreistätten an den deutschen Küsten*. Hamburg: Verein Jordsand.

Mellanby, Kenneth. 1967. *Pesticides and Pollution*. London: Collins.

Melucci, Alberto. 1985. "The Symbolic Challenge of Contemporary Movements." *Social Research* 52:789–816.

———. 1989. *Nomads of the Present: Social Movements and Individual Needs in Contemporary Society*. Philadelphia: Temple University Press.

Mey, Eberhard. 2003. "Johann Matthäus Bechstein (1757–1822): Vater der deutschen Vogelkunde." *Rudolstädter Naturhistorische Schriften* 11:63–100.

Meyer, Jan-Henrik. 2010a. "Greening Europe? Environmental Interest Groups and the Europeanization of a New Policy Field." *Comparativ: Zeitschrift für Globalgeschichte und vergleichende Gesellschaftsforschung* 20 (3): 83–104.

———. 2010b. "Saving Migrants: A Transnational Network Supporting Supranational Bird Protection Policy." In *Transnational Networks in Regional Integration: Governing Europe 1945–83*, edited by Wolfram Kaiser, Brigitte Leucht, and Michael Gehler, 176–98. Basingstoke: Palgrave Macmillan.

———. 2011. "Green Activism. The European Parliament's Environmental Committee Promoting a European Environment Policy in the 1970s." *Journal of European Integration History* 17:73–85.

Meyer, John W., and Brian Rowan. 1977. "Institutionalized Organizations: Formal Structure as Myth and Ceremony." *American Journal of Sociology* 83 (2): 340–63.

Miller, Edward. 1973. *That Noble Cabinet: A History of the British Museum*. London: Deutsch.

Miller, Stephen. 1973. "Ends, Means, and Galumphing: Some Leitmotifs of Play." *American Anthropologist* 75:87–98.

Mills, C. Wright. 1940. "Situated Actions and Vocabularies of Motive." *American Sociological Review* 5 (6): 904–13.

Mitchell, W. R. 2001. *Watch That Birdie: The Life and Times of Richard and Cherry Kearton, Pioneers of Nature Photography*. Giggleswick/Settle: Castleberg.

Mitschke, Alexander, Christopher Sudfeldt, Holger Heidrich-Riske, and Rainer Dröschmeister. 2005. "Das neue Brutvogelmonitoring in der Normallandschaft

Deutschlands—Untersuchungsgebiete, Erfassungsmethode und erste Ergebnisse." *Vogelwelt* 126:127–40.

Moore, Norman W. 1965. "Pesticides and Birds—A Review of the Situation in Great Britain in 1965." *Bird Study* 12 (3): 222–52.

———. 1966. "A Pesticide Monitoring System with Special Reference to the Selection of Indicator Species." *Journal of Applied Ecology* 3:261–69.

———. 1973. "Indicator Species." *Nature in Focus* 14:3–6.

———. 1975. "Pesticides Monitoring from the National and International Points of View." *Agriculture and Environment* 2:75–83.

———. 1987. *The Bird of Time: The Science and Politics of Nature Conservation: A Personal Account*. Cambridge: Cambridge University Press.

Moss, Stephen. 2004. *A Bird in the Bush: A Social History of Birdwatching*. London: Aurum Press.

———. 2006a. *The Garden Bird Handbook*. London: New Holland.

———. 2006b. *How to Birdwatch*. London: New Holland.

———. 2006c. *This Birding Life: The Best of the Guardian's Birdwatch*. London: Aurum Press.

———. 2007. *Remarkable Birds*. London: Collins.

Mountfort, Guy. 1959. "One Hundred Years of the British Ornithologists' Union." *Ibis* 101:8–18.

———. 1991. *Memories of Three Lives*. Braunton: Merlin Books.

Murphy, Priscilla Coit. 2005. *What a Book Can Do: The Publication and Reception of Silent Spring*. Amherst: University of Massachusetts Press.

Museums: Their History and Their Use: With a Bibliography and List of Museums in the United Kingdom. 1904. 3 vols. Glasgow: James MacLehose and Sons.

NABU. 2010. *Vogelschutz in Deutschland: Das NABU-Grundsatzprogramm Vogelschutz*. Berlin: NABU.

Natusch, Sheila, and Geoffrey M. Swainson. 1987. *William Swainson of Fern Grove: The Anatomy of a Nineteenth-Century Naturalist*. Wellington: Published by the Authors with the Aid of New Zealand Founders Society.

Naumann, Johann Andreas. 1789. *Der Vogelsteller oder die Kunst allerley Arten von Vögeln sowohl ohne als auch auf dem Vogelheerd bequem und in Menge zu fangen nebst den dahin gehörigen Kupfern und einer Naturgeschichte der bekannten und neu entdeckten Vögel*. Leipzig: Schwickert.

———. 1791. *Der Philosophische Bauer, oder Anleitung die Natur durch Beobachtung und Versuche zu erforschen*. Leipzig: Schwickertsche Verlag.

———. 1797–1805. *Naturgeschichte der Land- und Wasser-Vögel des nördlichen Deutschlands und angränzender Länder, nach eigenen Erfahrungen entworfen, und nach dem Leben gezeichnet*. Köthen: J. A. Aue.

Naumann, Johann Andreas, and Johann Friedrich Naumann. 1820–1860. *Naturgeschichte der Vögel Deutschlands. Nach eigenen Erfahrungen entworfen*. Leipzig: Fleischer.

Naumann, Johann Friedrich. 1815. *Taxidermie oder die Lehre, Thiere aller Klassen am einfachsten und zweckmäßigsten für Kabinette auszustopfen und aufzubewahren*. Halle: Hemmerde & Schwetschke.

———. 1846. "Über den Vogelzug mit besonderer Hinsicht auf Helgoland." *Rhea* 1:18–27.

———. 1848. *Taxidermie oder die Lehre, Thiere aller Klassen am einfachsten und zweckmäßigsten für Kabinette auszustopfen und aufzubewahren. 2. Auflage*. Halle: Schwetschke.

———. 1849. "Beleuchtung der Klage: Über Verminderung der Vögel in der Mitte von Deutschland." *Rhea* 2:131–44.

———. 1854. "Reminiscenzen über stufenweise Entwicklung der vaterländischen Ornithologie in der ersten Hälfte unseres Jahrhunderts." *Naumannia* 4:149–60.

"Nesting-Boxes." 1913. *Bird Notes and News* 5 (7): 106.

"Nesting Boxes for Birds." 1904. *Irish Naturalist* 13:140.

"Nesting-Box Making as a British Industry." 1914. *Bird Notes and News* 6 (4): 57.

Newton, Ian. 1979. *Population Ecology of Raptors*. Berkhamsted: T. & A. D. Poyser.

———. 1986. *The Sparrowhawk*. Calton: T. & A. D. Poyser.

———. 1989. "Organochlorines and Mercury in the Eggs of British Peregrines *Falco peregrinus*." *Ibis* 131:355–76.

———. 1998. *Population Limitation in Birds*. San Diego, CA: Academic Press.

Nicholson, Edward Max. 1926. *Birds in England: An Account of the State of Our Bird-Life and a Criticism of Bird Protection*. London: Chapman and Hall.

———. 1927. *How Birds Live: A Brief Account of Bird-Life in the Light of Modern Observation*. London: Williams & Norgate.

———. 1929. "Report on the 'British Birds' Census of Heronries, 1928: Part I and II." *British Birds* 22 (11): 270–323, 34–72.

———. 1931. *The Art of Bird-Watching: A Practical Guide to Field Observation*. London: H. F. & G. Witherby.

———. 1983. "The Trust—Origins and Early Days." In *Enjoying Ornithology: A Celebration of Fifty Years of The British Trust for Ornithology, 1933–1983*, edited by Ronald Hickling, 15–28. Calton: T. & A. D. Poyser.

"Nicholson on Bird-watching." 1932. *Ibis* 13 (2): 186–88.

Niethammer, Günther, ed. 1937–1942. *Handbuch der Deutschen Vogelkunde. Im Auftrag der Deutschen Ornithologischen Gesellschaft. Drei Bände*. Leipzig: Akademische Verlagsgesellschaft.

———. 1941. "Beobachtungen über die Vogelwelt von Auschwitz (Ost-Oberschlesien)." *Annalen des Naturhistorischen Museums in Wien* 52:164–99.

Nipkow, Markus. 2006. "Prioritäre Arten für den Vogelschutz in Deutschland." *Berichte zum Vogelschutz* 42:123–35.

Norris, Cuthbert Antony. 1960. "The Breeding Distribution of Thirty Bird Species in 1952." *Bird Study* 7:129–84.

Nowak, Eugeniusz. 1965. *Die Türkentaube*. Wittenberg: Ziemsen.

———. 2003. "Professor Erwin Stresemann (1889–1972)." *Mitteilungen des Vereins Sächsischer Ornithologen* 9, Sonderheft 2:1–95.

Nowak, Eugeniusz. 2010. *Wissenschaftler in turbulenten Zeiten. Erinnerungen an Ornithologen, Naturschützer und andere Naturkundler. Überarbeitete und erweiterte Auflage*. Hohenwarsleben: Westarp Wissenschaften.

Offe, Claus. 1985. "New Social Movements: Challenging the Boundaries of Institutional Politics." *Social Research* 52:817–68.

Ogilvie, Malcom, James Ferguson-Lees, and Richard Chandler. 2007. "A History of British Birds." *British Birds* 100 (1): 3–15.

Opitz, Helmut. 1997. "Die Liste der Erwählten: Vogel des Jahres—ein Vierteljahrhundert Artenschutzpolitik?" In *Ornithologen-Kalender*, 223–40. Wiebelsheim: Aula-Verlag.

———. 2006. "Buchbesprechung: Peter Berthold und Gabriele Mohr: Vögel füttern—aber richtig." *Berichte zum Vogelschutz* 43:150–52.

Osieck, E. R., and M. F. Mörzer Bruyns. 1981. "Important Bird Areas in the European Community." Cambridge: International Council for Bird Preservation.

Papastergiadis, Nikos. 2000. *The Turbulence of Migration: Globalization, Deterritorialization and Hybridity*. Cambridge: Polity Press.

Papazoglou, Clairie, Konstantin Kreiser, Zoltan Waliczky, and Ian Burfield. 2004. "Birds in the European Union: A Status Assessment." Wageningen, The Netherlands: BirdLife International.

Parslow, John. 1973. *Breeding Birds of Britain and Ireland: Historical Survey*. Berkhamsted: T. & A. D. Poyser.

Parslow, John, and Michael Everett. 1981. *Birds in Need of Special Protection in Europe*. Strasbourg: Council of Europe.

Parsons, Talcott. 1935. "The Place of Ultimate Values in Sociological Theory." *International Journal of Ethics* 45:282–316.

———. (1937) 1968. *The Structure of Social Action: A Study in Social Theory with Special Reference to a Group of Recent European Writers*. 2 vols. New York: Free Press.

———. 1951. *The Social System*. Glencoe: Free Press.

———. 1969. "On the Concept of Value-Commitments." In *Politics and Social Structure*, edited by Talcott Parsons, 439–72. New York: Free Press.

Parsons, Talcott, and Edward A. Shils, eds. 1951. *Toward a General Theory of Action*. Cambridge, MA: Harvard University Press.

PECBMS. 2006. "The State of Europe's Common Birds 2005." Prague: CSO/RSPB.

———. 2007. "The State of Europe's Common Birds 2007." Prague: CSO/RSPB.

Peterson, Roger Tory. 1934. *A Field Guide to the Birds: Giving Field Marks of All Species Found in Eastern North America*. Boston: Houghton Mifflin Co.

Peterson, Roger Tory, and James Fisher. 1955. *Wild America: The Record of a 30,000 Mile Journey around the Continent by a Distinguished Naturalist and His British Colleague*. Boston: Houghton Mifflin.

Peterson, Roger Tory, Guy Mountfort, and P. A. D. Hollom. 1954a. *A Field Guide to the Birds of Britain and Europe*. With an introduction by Julian Huxley. London: Collins.

———. 1954b. *Die Vögel Europas: Ein Taschenbuch für Ornithologen und Naturfreunde über alle in Europa lebenden Vögel. Übersetzt und bearbeitet von Günther Niethammer*. Hamburg Parey.

———. 1966. *A Field Guide to the Birds of Britain and Europe*. Revised and enlarged edition in collaboration with I. J. Ferguson-Lees and D. I. M. Wallace. Introduction by Sir Julian Huxley. London: Collins.

Pfauch, Wolfgang. 1966. "Der Naturforscher und Forstmann Johann Matthäus Bechstein." *Abhandlungen und Berichte des Naturkundemuseums Gotha* 3:27–54.

Pfauch, Wolfgang, and Reinhard Röder. 1991. "Der internationale Aspekt der Bechstein'schen 'Stubenvögel.'" *Beiträge zur Vogelkunde* 37:161–85.

Pfeiffer, Sebastian. 1936. *Die Vögel unserer Heimat*. Frankfurt a.M.: Kramer.

———. 1952. *Taschenbuch der deutschen Vogelwelt*. Frankfurt a.M.: Kramer.

———. 1955a. "Ergebnisse zweier Versuche zur Steigerung der Siedlungsdichte der Vögel auf forstlicher Kleinfläche und benachbarter Großfläche." *Biologische Abhandlungen*.

———. 1955b. *Experimentelle Untersuchungen und Freilandbeobachtungen zur Feststellung der Vertilgung des Kartoffelkäfers (leptinotarsa decemlineata Say) durch mitteleuropäische Vogelarten*. Berlin: Parey.

———, ed. 1957. *Taschenbuch für Vogelschutz*. Frankfurt a.M.: Limpert.

———, ed. 1980. *Taschenbuch für Vogelschutz. 5. Auflage*. Frankfurt a.M.: Strobach.

Pfeiffer, Sebastian, and Werner Keil. 1958. "Versuche zur Steigerung der Siedlungsdichte höhlen- und freibrütender Vogelarten und ernährungsbiologische Untersuchungen an Nestlingen einiger Singvogelarten in einem Schadgebiet des Eichenwicklers (*Tortrix viridana L.*) im Osten von Frankfurt am Main." *Biologische Abhandlungen* 15–16:1–52.

Pfeiffer, Sebastian, and Kurt Ruppert. 1953. "Versuche zur Steigerung der Siedlungsdichte höhlen- und buschbrütender Vogelarten." *Biologische Abhandlungen* 6:1–28.

Phillips, Mrs. Edward. 1890. Destruction of Ornamental-Plumaged Birds. RSPB Leaflet Series 1.

Pike, Oliver G. 1900. *In Bird-Land with Field-Glass and Camera. Illustrated with 83 Photographs Taken Direct from Nature by the Author*. London: T. Fisher Unwin.

———. 1943. *Nature and Camera: Summing up Fifty Years' Experience of Photographing Flora and Fauna*. London: Focal Press.

Pomian, Krzysztof. 1990. *Collectors and Curiosities: Paris and Venice, 1500–1800*. Cambridge: Polity Press.

Powell, Jillian. 2004. *WWF: The Global Environment Network*. London: Franklin Watts.

"Preservation of Birds." 1917. *Wild Life* 9:269–70.

Prinzinger, Roland. 2001. "150 Jahre 'Deutsche Ornithologen-Gesellschaft.'" *Journal für Ornithologie* 142 Sonderheft 1:2–26.

Przygodda, Wilfried. 1955. *Pflanzenschutzmittel und Vogelwelt mit Berücksichtigung der übrigen freilebenden Tierwelt*. Würzburg-Versbach: Verlag Biologische Abhandlungen.

Rainwater, Clarence Elmer. 1921. *The Play Movement in the United States: A Study of Community Recreation*. Chicago: University of Chicago Press.

Ratcliffe, Derek A. 1958. "Broken Eggs in Peregrine Eyries." *British Birds* 51:23–26.

———. 1962. "Breeding Density in the Peregrine *Falco peregrinus* and Raven *Corvus corax*." *Ibis* 104:13–39.

———. 1963. "The Status of the Peregrine in Great Britain." *Bird Study* 10 (2): 56–90.

———. 1967. "Decrease in Eggshell Weight in Certain Birds of Prey." *Nature* 215:208–10.

———. 1969. "Population Trends of the Peregrine Falcon in Great Britain." In *Peregrine Falcon Populations: Their Biology and Decline*, edited by Joseph J. Hickey, 239–69. Madison: University of Wisconsin Press.

———. 1970. "Changes Attributable to Pesticides in Egg Breakage Frequency and Eggshell Thickness in some British Birds." *Journal of Applied Ecology* 1:67–115.

———. 1980. *The Peregrine Falcon*. Calton: T. & A. D. Poyser.

———. 2000. *In Search of Nature*. Leeds: Peregrine Books.

Reichenow, Anton. 1889. *Systematisches Verzeichnis der Vögel Deutschlands und des angrenzenden Mittel-Europas*. Berlin: Linnaea.

———. 1900–1905. *Die Vögel Afrikas. Drei Bände*. Neumann: Neudamm.

———. 1902. *Die Kennzeichen der Vögel Deutschlands: Schlüssel zum Bestimmen, deutsche und wissenschaftle Benennungen, geographische Verbreitung, Brut- und Zugzeiten der deutschen Vögel*. Neumann: Neudamm.

Reichsbund für Vogelschutz. 1943. *Jahresbericht*. Stuttgart; Giengen, Brenz (Württemberg): Reichsbund für Vogelschutz.

Risely, K., D. G. Noble, and S. R. Baillie. 2008. "The Breeding Bird Survey 2007. BTO Research Report 508." Thetford: British Trust for Ornithology.

Rockenbauch, Dieter. 1970. "Vom Rückgang und Schutz der Wanderfalken (*Falco peregrinus*)." *Ornithologische Mitteilungen* 22:35–38.

———, ed. 1998. *Der Wanderfalke in Deutschland und umliegenden Gebieten. Band 1: Verbreitung, Bestand, Gefährdung und Schutz*. Ludwigsburg: Hölzinger.

———, ed. 2002. *Der Wanderfalke in Deutschland und umliegenden Gebieten. Band 2: Jahresablauf und Brutbiologie, Beringungsergebnisse, Jagdverhalten und Ernährung, Verschiedenes*. Ludwigsburg: Hölzinger.

Rohweder, J. 1905. "Zur Vorgeschichte der Vogelwarte Helgoland." *Ornithologische Monatsschrift* 30:333–49.

Rojek, Chris. 1992. " 'The Eye of Power': Moral Regulation and the Professionalization of Leisure Management from the 1830s to the 1950s." *Society and Leisure* 15 (1): 355–73.

Rook, Dorothy. 1966. "Protecting Britain's Birds." *Birds: The R.S.P.B. Magazine* 1 (4): 65–69.

Roose, Jochen. 2003. *Die Europäisierung von Umweltorganisationen. Die Umweltbewegung auf dem langen Weg nach Brüssel*. Wiesbaden: Westdeutscher Verlag.

Rörig, Georg. 1903a. "Studien über die wirtschaftliche Bedeutung der insektenfressenden Vögel." *Arbeiten aus der Biologischen Abteilung für Land- und Forstwirtschaft am Kaiserlichen Gesundheitsmat* 4:1–50.

———. 1903b. "Untersuchungen über die Nahrung unserer heimischen Vögel, mit besonderer Berücksichtigung der Tag- und Nachtraubvögel." *Arbeiten aus der Biologischen Abteilung für Land- und Forstwirtschaft am Kaiserlichen Gesundheitsmat* 4:51–117.

———. 1905. "Magenntersuchungen heimischer Raubvögel." *Arbeiten aus der Biologischen Abteilung für Land- und Forstwirtschaft am Kaiserlichen Gesundheitsmat* 5:237–55.

———. 1909. "Magen- und Gewölluntersuchungen heimischer Raubvögel." *Mitteilungen aus der Kaiserlichen Biologischen Anstalt für Land- und Forstwirtschaft* 7:473–520.

———. 1910. "Die Wirtschaftliche Bedeutung der Vogelwelt als Grundlage des Vogelschutzes." *Mitteilungen aus der Kaiserlichen Biologischen Anstalt für Land- und Forstwirtschaft* 9:3–48.

Roselaar, C. S. 1985. "Streptopelia decaocto Collared Dove." In *Handbook of the Birds of Europe the Middle East and North Africa: The Birds of the Western Palearctic. Volume IV: Terns to Woodpeckers*, edited by Stanley Cramp et al., 340–53. Oxford: Oxford University Press.

Rothschild, Walter. 1907. *Extinct Birds. An Attempt to unite in one Volume a short Account of those Birds which have become extinct in historical Times—that is, within the last six or seven hundred Years. To which are added a few which still exist, but are on the verge of Extinction. With 45 Coloured Plates, embracing 63 Subjects, and other Illustrations*. London: Hutchinson & Co.

Royal Society for the Protection of Birds. 1936. *Annual Report*. London: Royal Society for the Protection of Birds.

———. 1938. *Annual Report*. London: Royal Society for the Protection of Birds.

———. 1945. *Annual Report*. London: Royal Society for the Protection of Birds.

———. 1951. *Annual Report*. London: Royal Society for the Protection of Birds.

———. 1970/71. *Report & Accounts*. Sandy: Royal Society for the Protection of Birds.

"The R.S.P.B. in War-time." 1915. *Bird Notes and News* 6 (7): 97–100.

Rucht, Dieter. 1994. *Modernisierung und neue soziale Bewegungen. Deutschland, Frankreich und USA im Vergleich*. Frankfurt: Campus.

Ruge, Klaus. 1988. "Vom Nistkasten zur Politik." *Naturschutz Heute* 20:4–9.

———. 1989. *Vogelschutz: Ein praktisches Handbuch*. Ravensburg: Otto Maier.

Ruß, Karl. 1869. "Vorläufige Mittheilungen über die Zucht fremdländischer Vögel." *Journal für Ornithologie* 17:73–82.

———. 1870. "Jahresbericht aus meine Vogelstube." *Journal für Ornithologie* 18:25–31.

———. 1871–1873. *Handbuch für Vogelliebhaber, -Züchter und -Händler*. Hannover: Rümpler.

———. 1877. *Die Brieftaube: Ein Hand- und Lehrbuch für ihre Verpflegung, Züchtung und Abrichtung*. Hannover: Rümpler.

———. 1882. *Zum Vogelschutz. Eine Darstellung der Vogelschutzfrage in ihrer geschichtlichen Entwicklung bis zur Gegenwart nebst Besprechung aller bisherigen Maßnahmen sowie der Gesetz-Vorschläge*. Leipzig: Hugo Voigt.

———. 1884. *Das Huhn als Nutzgeflügel für die Stadt- und Landwirthschaft*. Magdeburg: Creutz.

Ruß, Karl, and Bruno Düring. 1876–1881. *Gloger's Vogelschutzschriften I-IV*. Berlin: Verlag von Hugo Voigt.

Russell, Sarah. 1996. *A History of WWF*. Gland: World Wide Fund for Nature.

Sahlins, Marshall. 1981. *Historical Metaphors and Mythical Realities: Early History of the Sandwich Islands Kingdom*. Ann Arbor: University of Michigan Press.

———. 1985. *Islands of History*. Chicago: University of Chicago Press.

Samstag, Tony. 1988. *For Love of Birds: The Story of the Royal Society for the Protection of Birds, 1889–1988*. Bedfordshire: Royal Society for the Protection of Birds.

Sassen, Saskia. 1991. *The Global City: New York, London, Tokyo* Princeton, NJ: Princeton University Press.

———. 2001. "Cracked Casings: Notes Towards an Analytics for Studying Transnational Processes." In *New Transnational Social Spaces*, edited by L. Pries, 187–207. London: Routledge.

Saunders, Howard. 1899. *An Illustrated Manual of British Birds*. Second ed., revised and enlarged. London: Gurney and Jackson.

———. 1907. "Additions to the List of British Birds since 1899." *British Birds* 1 (1): 1–16.

Schäffer, Anita, and Norbert Schäffer. 2006. *Gartenvögel: Naturbeobachtungen vor der eigenen Haustür*. Wiebelsheim: Aula-Verlag.

Schäffer, Norbert. 2007. "Von Rosenheim bis Buenos Aires: BirdLife International entwickelt sich." In *Der Falke Taschenkalender für Vogelbeobachter 2008*, edited by Hans H. Bergmann, Siegfried Klaus, and Daniel Doer, 189–96. Wiebelsheim: Aula-Verlag.

Schalow, Herman. 1901. "Ein Rückblick auf die Geschichte der Deutschen Ornithologischen Gesellschaft." *Journal für Ornithologie* 49: 6–25.

Schiermann, Gottfried. 1930. "Studien über die Siedlungsdichte im Brutgebiet." *Journal für Ornithologie* 78 (2): 137–80.

———. 1934. "Studien über die Siedlungsdichte im Brutgebiet. II." *Journal für Ornithologie* 82 (4): 455–86.

Schilling, Friedrich, and Klaus König. 1980. "Die Biozidbelastung des Wanderfalken (*Falco peregrinus*) in Baden-Württemberg und ihre Auswirkung auf die Populationsentwicklung." *Journal für Ornithologie* 121: 1–35.

Schilling, Friedrich, and Peter Wegner. 2001. *Der Wanderfalke in der DDT-Ära*. Stuttgart: Ulmer.

Schillings, Carl Georg. 1905a. *Mit Blitzlicht und Büchse. Neue Beobachtungen und Erlebnisse in der Wildnis inmitten der Tierwelt von Äquatorial-Ost-Afrika. Mit 302 urkundtreu in Autotypie wiedergegebenen photographischem Original-Tag- und Nacht-Aufnahmen des Verfassers. Zweiter Abdruck*. Leipzig: R. Voigtländer.

———. 1905b. *With Flash-light and Rifle: Photographing by Flash-light at Night the Wild Animal World of equatorial Africa*. Translated and abridged by Henry Zick. New York: Harper & Brothers.

———. 1906. *Der Zauber des Elelescho. Mit 318 Abbildungen*. Leipzig: R. Voigtländer.

———. 1911. "Die Tragödie des Paradiesvogels und des Edelreihers. Wesentlich erweiterter Sonderdruck aus dem Juniheft 1911 der Süddeutschen Monatshefte." n.p.

Schmitz, Siegfried. 1984. *Tiervater Brehm: Seine Reisen, Sein Leben, Sein Werk*. München: Harnack.

Schmoll, Friedemann. 2004. *Erinnerung an die Natur. Die Geschichte des Naturschutzes im deutschen Kaiserreich*. Frankfurt a.M.: Campus.

Schulz, Georg E. F. 1926. "Film und Naturschutz." In *Wege Zum Naturschutz*, edited by Walther Schoenichen, 179–91. Breslau: Ferdinand Hirt.

———. 1941. "Über den Reichsbund für Vogelschutz und die deutsche Naturphotographie." *Deutsche Vogelwelt* 66: 9–13.

Schulz, Heinrich. 1947. *Die Welt der Seevögel: Ein Führer durch die Vogelbrutstätten der deutschen Küsten*. Hamburg: Anton Lettenbauer.

Schulze-Hagen, K., F. D. Steinheimer, R. Kinzelbach, and C. Gasser. 2003. "Avian Taxidermy in Europe from the Middle Ages to the Renaissance." *Journal für Ornithologie* 144 (4): 459–78.

Sclater, Philip L. 1909. "A Short History of the British Ornithologists' Union." *Ibis* (Jubilee Supplement): 19–64.

Scott, Peter. 1957a. "50th Anniversary Messages: From the Director of the Wildfowl Trust." *British Birds* 50 (6): 229.

———. 1957b. *A Coloured Key to the Wildfowl of the World*. Slimbridge: Wildfowl Trust.
———. 1966. *The Eye of the Wind: An Autobiography*. London: Hodder and Stoughton.
———. 1980. *Observations of Wildlife*. Oxford: Phaidon.
Scott, Peter, John A. Burton, and Richard S. R. Fitter. 1987. "Red Data Books: The Historical Background." In *The Road to Extinction. Problems of Categorizing the Status of Taxa Threatened with Extinction*, edited by Richard S. R. Fitter and Maisie Fitter, 1–5. Gland: International Union for the Conservation of Nature and Natural Resources.
Scott, W. Richard. 2013. *Institutions and Organizations: Ideas, Interests, and Identities*. 4th ed. Thousand Oaks: Sage.
Seebohm, Henry. 1877. "Supplementary Notes on the Ornithology of Heligoland." *Ibis* 19 (2): 156–65.
———. 1892. "List of the Birds of Heligoland as recorded by Herr Gätke." *Ibis* 34 (1): 1–32.
Selous, Edmund. 1901. *Bird Watching*. London: J. M. Dent.
———. 1927. *Realities of Bird Life: Being Extracts from the Diaries of a Life-loving Naturalist*. With an introduction by Julian S. Huxley. London: Constable.
Sharp, Granville. 1902. *Birds in the Garden: Studies with a Camera. With over 100 Illustrations including eight Photogravures*. London: J. M. Dent & Co.
Sharrock, J. T. R., and E. M. Sharrock. 1976. *Rare Birds in Britain and Ireland*. Berkhamsted: T. & A. D. Poyser.
Sheail, John. 1985. *Pesticides and Nature Conservation. The British Experience 1950–1975*. Oxford: Clarendon Press.
Shelton, Anthony Alan. 1994. "Cabinets of Transgression: Renaissance Collections and the Incorporation of the New World." In *The Cultures of Collecting*, edited by John Elsner and Roger Cardinal, 177–203. Cambridge, MA: Harvard University Press.
Shun Aigrettes and Save the Herons. n.d. RSPB Leaflet Series 5.
Simon, Jürgen M., and Horst Hanemann. 1989. *Es begann mit einer Insel Deutscher Bund für Vogelschutz e.V. Die Bilder-Chronik eines Naturschutzverbandes*. Wiesbaden: Wirtschaftsverlag.
Society for the Protection of Birds. 1891. *Report*. London: J. Davy & Sons, Dryden Press.
Spates, James L. 1983. "The Sociology of Values." *Annual Review of Sociology* 9:27–49.
Spencer, Robert, et. al. 1992. "The Rare Breeding Birds Panel." *British Birds* 85:117–22.
Stanford, J. K. 1950. *Bledgrave Hall*. London: Faber and Faber.
———. 1954. *A Bewilderment of Birds*. London: Rupert Hart-Davis.
Stark, David. 2009. *The Sense of Dissonance: Accounts of Worth in Economic Life*. Princeton, NJ: Princeton University Press.
Stearn, William T. 1981. *The Natural History Museum at South Kensington: A History of the British Museum (Natural History), 1753–1980*. London: Heinemann.
Steinbacher, Georg. 1942. "Die Siedlungsdichte in der Parklandschaft." *Journal für Ornithologie* 90:342–60.
Steinheimer, Frank D. 2005. "The Whereabouts of Pre-nineteenth Century Bird Collections." *Zoologische Mededelingen Leiden* 79 (3): 45–67.
———. 2006. "Sammlungsgeschichte und Präparationsstil von fünf Exemplaren der Kleidervogelart Ou Psittirostra psittacea (Gmelin, 1789) der dritten Weltumsegelung von James

Cook." *Sitzungsberichte der Gesellschaft Naturforschender Freunde zu Berlin NF.* 44:119–31.
Steiniger, Fritz, ed. 1955. *Die Photographie freilebender Tiere. 2. verbesserte und erweiterte Auflage.* Baden-Baden: Verlag für angewandte Wissenschaften.
Stresemann, Erwin. 1923. "Die Anfänge ornithologischer Sammlungen." *Journal für Ornithologie* 71 (1): 112–27.
———. 1943. "Anton Reichenow." *Journal für Ornithologie* 91:111–20.
———. 1949. "Birds Collected in the North Pacific Area during Capt. James Cook's last Voyage (1778 and 1779)." *Ibis* 91:244–55.
———. 1950. "Birds Collected during Capt. James Cook's Last Expedition (1776–1780)." *Auk* 67:66–88.
———. (1951) 1975. *Ornithology from Aristotle to the Present.* Cambridge, MA: Harvard University Press.
———. 1953. "Additions and Corrections to 'Birds Collected in the North Pacific Area during Capt. James Cook's last Voyage' ('Ibis' 1949: 244–255)."' *Ibis* 95:371.
———. 1967. "Vor- und Frühgeschichte der Vogelforschung auf Helgoland." *Journal für Ornithologie* 108:377–429.
Stresemann, Erwin, and Eugeniusz Nowak. 1958. "Die Ausbreitung der Türkentaube in Asien und Europa." *Journal für Ornithologie* 99:243–96.
Sucker, Ulrich. 1998. *Anfänge der modernen Phytomedizin: Die Gründungsgeschichte der Biologischen Anstalt für Land- und Forstwirtschaft (1898 bis 1919)—zugleich ein Beitrag zur Disziplingenese der Phytomedizin.* Berlin: Parey.
Südbeck, Peter, Hans-Günther Bauer, Martin Boschert, Peter Boye, and Wilfried Knief. 2007. "Rote Liste der Brutvögel Deutschlands. 4. Fassung, 30. November 2007." *Berichte zum Vogelschutz* 44:23–81.
Sudfeldt, Christoph, et al. 2012. *Vogelmonitoring in Deutschland: Programme und Anwendungen.* Bonn-Bad Godesberg: Bundesamt für Naturschutz.
Swainson, William. 1820–1833. *Zoological Illustrations or Original Figures and Descriptions of New Rare or Interesting Animals.* London: Printed by R. and A. Taylor for Baldwin, Cradock, and Joy; and W. Wood.
———. 1840. *Taxidermy with the Biography of Zoologists and other Notices of their Works.* London: Printed for Longman, Orme, Brown, Green & Longmans.
Swidler, Ann. 1986. "Culture in Action: Symbols and Strategies." *American Sociological Review* 51:273–86.
———. 2001a. *Talk of Love: How Culture Matters.* Chicago: University of Chicago Press.
———. 2001b. "What Anchors Cultural Practices." In *The Practice Turn in Contemporary Theory*, edited by Theodore R. Schatzki, Eike von Savigny, and Karin Knorr-Cetina, 83–101. New York: Routledge.
———. 2008. "Comment on Stephen Vaisey's 'Socrates, Skinner, and Aristotle: Three Ways of Thinking about Culture in Action.'" *Sociological Forum* 23 (3): 614–18.
Taylor, Charles. 1985. *Human Agency and Language.* Cambridge: Cambridge University Press.
Taylor, S. M. 1965. "The Common Birds Census—Some Statistical Aspects." *Bird Study* 12 (4): 268–86.

Thielcke, Gerhard. 1969. "Der Bestand der Greifvogelarten in der Bundesrepublik Deutschland und die Ursachen ihres Rückgangs." *Internationaler Rat für Vogelschutz Deutsche Sektion* 9:22–30.

Thiele. 1884. "Generalversammlung zu Zeitz am 17. Januar 1883." *Ornithologische Monatsschrift* 8:2–5.

Thomas, Nicholas. 1991. *Entangled Objects: Exchange Material Culture and Colonialism in the Pacific*. Cambridge, MA: Harvard University Press.

———. 1994. "Licensed Curiosity: Cook's Pacific Voyages." In *The Cultures of Collecting*, edited by John Elsner and Roger Cardinal, 116–36. Cambridge, MA: Harvard University Press.

Thomas, William I. 1901. "The Gaming Instinct." *American Journal of Sociology* 6 (6): 750–63.

———. 1923. *The Unadjusted Girl: With Cases and Standpoint for Behavior Analysis*. Boston: Little, Brown.

Thomas, William I., and Florian Znaniecki. 1918–1920. *The Polish Peasant in Europe and America: Monograph of an Immigrant Group*. Chicago: University of Chicago Press.

Thomsen, Peter, and Erwin Stresemann. 1957. *Johann Friedrich Naumann der Altmeister der deutschen Vogelkunde. Sein Leben und sein Werk*. Leipzig: Johann Ambrosius Barth Verlag.

Thrasher, Frederic M. . 1927. *The Gang: A Study of 1,313 Gangs in Chicago*. Chicago: University of Chicago Press.

Touraine, Alain. 1981. *The Voice and the Eye: An Analysis of Social Movements*. Cambridge: Cambridge University Press.

———. 1985. "An Introduction to the Study of Social Movements." *Social Research* 52:749–87.

Touraine, Alain, Zsuzsa Hegedus, François Dubet, and Michel Wieviorka. 1983. *Anti-Nuclear Protest: The Opposition to Nuclear Energy in France*. Cambridge: Cambridge University Press.

The Trade in Bird Feathers. n.d. RSPB Leaflet Series 28.

Treuenfels, Carl-Albrecht von. 1968. *Photographieren und Filmen von Wild und Vögeln*. Hamburg: Verlag Paul Parey.

———. 1979. *Photographieren und Filmen von Wild und Vögeln. 3. neubearbeitete und erweiterte Auflage*. Hamburg: Verlag Paul Parey.

———. 1984. *Abenteuer Naturschutz in Deutschland*. Hamburg: Hoffmann und Campe.

———. 1986. *Für unsere Natur. Der WWF-Ratgeber*. Hamburg: Rasch und Röhring.

———. 1989. *Kraniche Vögel des Glücks*. Hamburg: Rasch und Röhrig.

———. 2005. *Zauber der Kraniche*. München: Knesebeck.

Tucker, Graham M., and Melanie F. Heath. 1994. *Birds in Europe: Their Conservation Status*. Cambridge: BirdLife International.

Urry, John. 2000. *Sociology Beyond Societies*. London: Routledge.

Vaan, Mathijs de, David Stark, and Balazs Vedres. 2015. "Game Changer: The Topology of Creativity." *American Journal of Sociology* 120:1–51.

Vaisey, Stephen. 2008a. "Socrates, Skinner, and Aristotle: Three Ways of Thinking about Culture in Action." *Sociological Forum* 23 (3): 603–61.

———. 2008b. "Reply to Ann Swidler." *Sociological Forum* 23 (3): 619–22.

———. 2009. "Motivation and Justification: A Dual-Process Model of Culture in Action." *American Journal of Sociology* 114 (6): 1675–715.

Vauk, Gottfried. 1957. "Das Vorkommen der Türkentaube auf Helgoland." *Ornithologische Mitteilungen* 9:218.

Vauk, Gottfried, unter Mitarbeit von Wilhelm Bindig, Friedrich Goethe und Dietmar Gruner. 1977. *Geschichte der Vogelwarte und der Vogelforschung auf der Insel Helgoland*. Otterndorf: Verlagsdruckerei H. Huster.

Veblen, Thorstein. (1899) 2001. *The Theory of the Leisure Class*. New York: Modern Library.

Vedres, Balazs, and David Stark. 2010. "Structural Folds: Generative Disruption in Overlapping Groups." *American Journal of Sociology* 115:1150–90.

Verein Jordsand. 2007. "100 Jahre Seevogelschutz an deutschen Küsten." *Seevögel* Band 28, Sonderband.

Vogelwarten und Vogelschutzwarten. 1955. Dresden: Sachsenverlag.

Vorstand des Deutschen Vereins zum Schutze der Vogelwelt. 1884. "Auszug aus der Generalversammlung zu Weißenfels am 17. Januar 1884 von Vereinsvorsitzenden W. Thienemann gehaltenen Rede." *Ornithologische Monatsschrift* 9:2–4.

———. 1899. "An die geehrten Vereinsmitglieder." *Ornithologische Monatsschrift* 24:2–3.

Wahl, Johannes, and Christoph Sudfeldt. 2010. "Ehrenamtliches Engagement im Vogelmonitoring in Deutschland." In *Drei Jahrzehnte Vogelschutz im Herzen Europas: Rückblick, Bilanz und Herausforderungen*, edited by Peter Boye, M. Vischer-Leopold, C. Paulsch, Axel Ssymank, and F. Beulshausen, 199–230. Bonn-Bad Godesberg: Bundesamt für Naturschutz.

Wallace, Ian. 2004. *Beguiled by Birds: Ian Wallace on British Birdwatching*. London: Christopher Helm.

Warren, Rachel L. M. 1966–1973. *Type-Specimens of Birds in the British Museum (Natural History)*. 3 vols. London: British Museum (Natural History).

Wegner, Julius. 1919. *Festschrift Dr. Kurt Floericke zum 50. Geburtstag und zum 25. Doktorjubiläum*. Stuttgart: E. H. Moritz.

Wegner, Peter, Gert Kleinstäuber, Frank Baum, and Friedrich Schilling. 2005. "Long-Term Investigation of the Degree of Exposure of German Peregrine Falcons (*Falco peregrinus*) to Damaging Chemicals from the Environment." *Journal of Ornithology* 146:34–54.

Whitehead, P. J. P. 1969. "Zoological Specimens from Captain Cook's Voyages." *Journal of the Society for the Bibliography of Natural History* 5:161–201.

Whitford, Josh. 2002. "Pragmatism and the Untenable Dualism of Means and Ends: Why Rational Choice Theory Does Not Deserve Paradigmatic Privilege." *Theory and Society* 31:325–63.

Wigglesworth, V. B. 1945. "DDT and the Balance of Nature." *Atlantic Monthly* 176:107–13.

"Wild Bird Protection and Nesting-Boxes." 1897. *Nature* 56 (1439): 78.

The Wildfowl Trust Slimbridge, Gloucestershire, Peakirk, Northamptonshire: An Introduction to the Collection and Work of the Wildfowl Trust. 1968. Slimbridge: Wildfowl Trust.

Wilkening, Bettina. 2004. "Kranich (Grus grus)." In *Brutvögel in Deutschland*, edited by K. Gedeon, A. Mitschke, and C. Sudfeldt, 12–13. Hohenstein-Ernstthal: Stiftung Vogelmonitoring Deutschland.

Williamson, Kenneth, and R. C. Homes. 1964. "Methods and Preliminary Results of the Common Birds Census, 1962–63." *Bird Study* 11 (4): 240–56.

Wils, Wouter P. J. 1994. "The Birds Directive 15 years Later: A Survey of the Case law and a Comparison with the Habitats Directive." *Journal of Environmental Law* 6:219–42.

Wimmer, Andreas, and Nina Glick Schiller. 2002. "Methodological Nationalism and Beyond: Nation-State Building, Migration and the Social Sciences." *Global Networks* 2 (4): 1470–2266.

———. 2003. "Methodological Nationalism, the Social Sciences, and the Study of Migration: An Essay in Historical Epistemology." *International Migration Review* 37 (3): 576–610.

Wirzberger, Karl-Heinz, ed. 1970. *200 Jahre Museum für Naturkunde an der Humboldt-Universität zu Berlin. Bände 2–3 der Wissenschaftlichen Zeitschrift der Humboldt-Universität zu Berlin. Mathematisch- Naturwissenschaftliche Reihe.* Berlin: Humboldt-Universität zu Berlin.

Witherby, Harry F. 1894. *Forest Birds Their Haunts and Habits: Short Studies from Nature.* London: Kegan Paul & Co.

———. 1902. *Bird Hunting on the White Nile: A Naturalist's Experiences in the Soudan.* London: The Office of "Knowledge."

———. 1920–24. *A Practical Handbook of British Birds.* 2 vols. London: Witherby.

Witherby, Harry F., F. C. R. Jourdain, and Norman F. Ticehurst. 1930. "Sight Records." *British Birds* 23 (12): 342–44.

Witherby, Harry F., F. C. R. Jourdain, Norman F. Ticehurst, and Bernard W. Tucker. 1938–41. *The Handbook of British Birds.* 5 vols. London: H. F. & G. Witherby.

Witherby, Harry F., and W. P. Pycraft. 1907. "Editorial." *British Birds* 1 (1): 1–3.

———. 1908. "Preface." *British Birds* 2 (1): iii-iv.

Witherby, Harry F., and Norman F. Ticehurst. 1907–1908. "On the More Important Additions to Our Knowledge of British Birds since 1899." *British Birds* 1:52–57, 81–85, 109–14, 47–52, 78–84, 246–56.

Witt, Klaus. 1983. "Bestanderfassung einiger ausgewählter gefährdeter Vogelarten (1977–1980)." *Vogelwelt* 104:233–40.

———. 2003. "Eine Geschichte der Feldornithologen und Vogelbeobachter in Deutschland." *Falke* 50:264–68.

Wittgenstein, Ludwig. (1953) 2001. *Philosophical Investigations.* The German text, with a revised English translation. 3rd ed. Translated by G. E. M. Anscombe. Oxford: Blackwell.

Wöbse, Anna-Katharina. 2003a. "Lina Hähnle—eine Galionsfigur der frühen Naturschutzbewegung." In *Naturschutz hat Geschichte*, edited by Stiftung Naturschutzgeschichte, 113–30. Essen: Klartext Verlag.

———. 2003b. "Soziale Bewegung im Gleichschritt: Lina Hähnle und der Reichsbund für Vogelschutz." In *Naturschutz und Nationalsozialismus*, edited by Joachim Radkau and Frank Uekötter, 309–28. Frankfurt a.M.: Campus.

———. 2004. "Als eine Mode untragbar wurde—Die Kampagne gegen den Federschmuck im Deutschen Kaiserreich." In *Federn Kitzeln die Sinne*, edited by Dorothea Deterts, Michael Gautier, and Andrea Müller, 43–50. Bremen: Überseemuseum.

Wolff, Jonathan. 2011. "Karl Marx." In *The Stanford Encyclopedia of Philosophy*, edited by Edward N. Zalta. Stanford, CA: Center for the Study of Language and Information. http://plato.stanford.edu/archives/sum2011/entries/marx/.

Wood, J. Duncan. 2003. *Horace Alexander: 1889 to 1989: Birds and Binoculars*. York: Ebor Press.

WWF. 2011. *WWF Annual Review 2010*. Gland: WWF.

Yarrell, William. 1843. *A History of British Birds*. Illustrated by 520 wood-engravings. Three volumes. London: John Van Voorst.

———. 1871–1885. *A History of British Birds. Four Volumes*. 4th ed., revised and enlarged, by Alfred Newton and Howard Saunders. London: John Van Voorst.

Zelizer, Viviana A. 1979. *Morals and Markets: The Development of Life Insurance in the United States*. New York: Columbia University Press.

———. 1985. *Pricing the Priceless Child: The Changing Social Value of Children*. New York: Basic Books.

———. 1994. *The Social Meaning of Money*. New York: Basic Books.

———. 2005. *The Purchase of Intimacy*. Princeton, NJ: Princeton University Press.

INDEX